W9-AVK-325

Ad Hoc Mobile Wireless Networks:

Protocols and Systems

C.-K. Toh, Ph.D.

Prentice Hall PTR
Upper Saddle River, New Jersey 07548
www.phptr.com

Library of Congress Cataloging-in-Publication Data

Toh, C.-K. (Chai-Keong)
Ad hoc mobile wireless networks: protocols and systems / C.-K. Toh.
p. cm.
Includes index.
ISBN 0-13-007817-4
1. Wireless communications systems. I. Title.
TK5103.2 T64 2002
384.3--dc21 2001036953
CIP

Editorial/Production Supervision: *Nick Radhuber*
Acquisitions Editor: *Bernard Goodwin*
Editorial Assistant: *Michelle Vincente*
Marketing Manager: *Dan DePasquale*
Manufacturing Buyer: *Alexis Heydt-Long*
Cover Design: *Talar Boorujy*
Cover Design Direction: *Jerry Votta*

Prentice-Hall, Inc.
Upper Saddle River, NJ 07458

Prentice Hall books are widely used by corporations and government agencies for training, marketing, and resale.

The publisher offers discounts on this book when ordered in bulk quantities. For more information, contact Corporate Sales Department, phone: 800-382-3419; fax: 201-236-7141; email: corpsales@prenhall.cor
Or write: Corporate Sales Department, Prentice Hall PTR, One Lake Street, Upper Saddle River, NJ 07458.

Printed in the United States of America

10 9 8 7 6 5 4 3 2 1

ISBN 0-13-007817-4

Pearson Education LTD.
Pearson Education Australia PTY, Limited
Pearson Education Singapore, Pte. Ltd
Pearson Education North Asia Ltd
Pearson Education Canada, Ltd.
Pearson Educación de Mexico, S.A. de C.V.
Pearson Education—Japan
Pearson Education Malaysia, Pte. Ltd
Pearson Education, Upper Saddle River, New Jersey

DEDICATION

So many changes have occurred in my life over the past few years. I moved from Asia to Europe and subsequently to America. I also realized just how much one could do if one is determined and motivated. Besides having a passion for technology, I have a great appreciation of life. Only by striking a balance can one truly enjoy the meaning of our existence. I would like to dedicate this book to:

My Parents

My Sister

My parents not only gave me life, but also the opportunity of a superb education. Without their support and love, this book would not have been written.

CONTENTS

ABOUT THE AUTHOR

C.-K. Toh was born in Singapore in 1965. He received his BEng (First Class Honors) and PhD degrees in Electrical Engineering and Computer Science from University of Manchester Institute of Science and Technology (UMIST) and the University of Cambridge, England, respectively. While at Cambridge, he was an Honorary Cambridge Commonwealth Trust Scholar and a Fellow of the Cambridge Commonwealth Society and Cambridge Philosophical Society. He is an editor for the *IEEE Journal on Selected Areas in Communications*, *IEEE Network*, *Journal on Communication Networks*, and *Personal Technologies Journal*. He is a senior member of IEEE and is listed in MARQUIS Who's Who in the World and Who's Who in Science and Engineering. He has been a consultant for Hughes, DARPA, the U.S. Navy, and TRW. He is Chairman of the IEEE TCPC Subcommittee on Ad Hoc Mobile Wireless Networks and Director of the Ad Hoc Wireless Networking Consortium. He had held visiting professorships at the Royal Institute of Technology (Sweden), Mid-Sweden University, and Swiss Federal Institute of Technology (Switzerland).

C.-K. Toh revisiting the "backs," Cambridge University, England, summer 2000.

PREFACE

So many advances have been made in the field of infrastructureless wireless networks that it is time to consolidate the insights and technical know-how of this exciting and state-of-the-art technology into a book for educational and research purposes.

This book presents an introduction to wireless networks and packet radio networks in Chapters 1 and 2. These chapters allow the reader to gain a broader understanding of existing wireless networks before embarking into the details of ad hoc wireless networks. Packet radio networks are considered an earlier form of ad hoc networks since they address *mobility* and allow repeaters to be mobile. Chapter 3 provides an introduction to ad hoc wireless networks, highlighting the presence of device heterogeneity, different traffic profiles, mobility, and technical challenges. Chapter 4 exposes the problems associated with media access in ad hoc networks and presents some suitable solutions. Chapter 5 then gives an overview of existing ad hoc routing protocols, highlighting their features and differences. A new routing protocol based on the concept of long-life routing, or associativity, is presented in Chapter 6.

Insights into the implementation of a practical ad hoc wireless network are revealed in Chapter 7. Many may wonder about the communication performance of a practical ad hoc wireless network. Chapter 8 discusses this and presents results obtained from practical field trials. The importance of power conservation and the impact of periodic beaconing on battery life are discussed in Chapter 9. Support for multicast communications in an ad hoc wireless network and the various existing multicast routing protocols are discussed in Chapter 10. Support for reliable TCP communications over an ad hoc mobile environment is presented in Chapter 11.

Users of a network must be able to discover the presence of services in the network. Hence, Chapter 12 explains some existing service discovery methods. Fundamentals of the evolving Bluetooth technology are given in Chapter 13. The principles of Wireless Application Protocol (WAP) are explained in Chapter 14. Possible ad hoc mobile applications are revealed and discussed in Chapter 15. Finally, a conclusion is presented in Chapter 16.

ACKNOWLEDGMENTS

Many parts of this book were written while I was traveling outside the United States, in particular, in England (London, Cambridge, and Manchester), Sweden, France, Finland, Canada, Cyprus, Singapore, and Switzerland.

I would like to thank Victor Li, Joseph Hui, Jon W. Mark, and Victor Leung for their encouragement over the years. I truly respect and admire their accomplishments, professional practice, and preserverance. Kevin and Melody provided a great source of motivation when it came to facing the difficult times in life. Their strong faith in God also helped me to overcome my many difficulties and doubts. Vincent and Araceli are such caring friends with strong faith and great companions that I will not forget them. My special thanks to Hillary for her encouragement and appreciation.

With a mixture of Asian, British, and American cultures in me, it is a gift from God that I am unique. Best of all, I am still able to maintain strong traditional family and professional values despite the very sophisticated working world.

I also want to thank Kings College, Cambridge for their support during my PhD at Cambridge University, England, from 1993–1996. In particular, Rob Wallach deserves special mention. From him, I learned far more about life than research and grades. While time causes us to age, knowledge and virtues will prevail forever.

QUOTES & WORDS OF WISDOM

"I have not powered down. I have just rebooted."

"Do not bow to conspiracy and dishonesty."

"No to technical achievements made through politics."

"No to promoting technology for a degradation in humanity."

"Dictatorship is never workable in our society. It is not even respectable."

"A man removes a mountain by first carrying away small stones."

"The greatest reward for a man's turmoil is not what he gets for it but what he becomes by it."

"Knowledge is Power."

Chapter 1

INTRODUCTION TO WIRELESS NETWORKS

Wireless communications have become very pervasive. The number of mobile phones and wireless Internet users has increased significantly in recent years. Traditionally, first-generation wireless networks were targeted primarily at voice and data communications occurring at low data rates.

Recently, we have seen the evolution of second- and third-generation wireless systems that incorporate the features provided by broadband. In addition to supporting *mobility*, broadband also aims to support *multimedia traffic*, with quality of service (QoS) assurance. We have also seen the presence of different air interface technologies, and the need for interoperability has increasingly been recognized by the research community.

Wireless networks include local, metropolitian, wide, and global areas. In this chapter, we will cover the evolution of such networks, their basic principles of operation, and their architectures.

1.1 Evolution of Mobile Cellular Networks

1.1.1 First-Generation Mobile Systems

The first generation of analog cellular systems included the Advanced Mobile Telephone System (AMPS) [1] which was made available in 1983. A total of 40MHz of spectrum was allocated from the 800MHz band by the Federal Communications Commission (FCC) for AMPS. It was first deployed in Chicago, with a service area of 2100 square miles [2]. AMPS offered 832 channels, with a data rate of 10 kbps. Although omnidirectional antennas were used in the earlier AMPS implementation, it was realized that using directional antennas would yield better cell reuse. In fact, the smallest reuse factor that would fulfill the 18db signal-to-interference ratio (SIR) using 120-degree directional antennas was found to be 7. Hence, a 7-cell reuse pattern was adopted for AMPS. Transmissions from the base stations to mobiles occur over the *forward channel* using frequencies between 869-894 MHz. The *reverse channel* is used for transmissions from mobiles to base station, using frequencies between 824-849 MHz.

In Europe, TACS (Total Access Communications System) was introduced with 1000 channels and a data rate of 8 kbps. AMPS and TACS use the frequency modulation (FM) technique for radio transmission. Traffic is multiplexed onto an FDMA (frequency division multiple access) system. In Scandinavian countries, the Nordic Mobile Telephone is used.

1.1.2 Second-Generation Mobile Systems

Compared to first-generation systems, second-generation (2G) systems use digital multiple access technology, such as TDMA (time division multiple access) and CDMA (code division multiple access). Global System for Mobile Communications, or GSM [3], uses TDMA technology to support multiple users.

Examples of second-generation systems are GSM, Cordless Telephone (CT2), Personal Access Communications Systems (PACS), and Digital European Cordless Telephone (DECT [4]). A new design was introduced into the mobile switching center of second-generation systems. In particular, the use of base station controllers (BSCs) lightens the load placed on the MSC (mobile switching center) found in first-generation systems. This design allows the interface between the MSC and BSC to be standardized. Hence, considerable attention was devoted to *interoperability* and *standardization* in second-generation systems so that carriers could employ different manufacturers for the MSC and BSCs.

In addition to enhancements in MSC design, the mobile-assisted handoff mechanism was introduced. By sensing signals received from adjacent base stations, a mobile unit can trigger a handoff by performing explicit signalling with the network.

Second generation protocols use digital encoding and include GSM, D-AMPS (TDMA) and CDMA (IS-95). 2G networks are in current use around the world. The protocols behind 2G networks support voice and some limited data communications, such as Fax and short messaging service (SMS), and most 2G protocols offer different levels of encryption, and security. While first-generation systems support primarily voice traffic, second-generation systems support voice, paging, data, and fax services.

1.1.3 2.5G Mobile Systems

The move into the 2.5G world will begin with General Packet Radio Service (GPRS). GPRS is a radio technology for GSM networks that adds packet-switching protocols, shorter setup time for ISP connections, and the possibility to charge by the amount of data sent, rather than connection time. Packet switching is a technique whereby the information (voice or data) to be sent is broken up into packets, of at most a few Kbytes each, which are then routed by the network between different destinations based on addressing data within each packet. Use of network resources is optimized as the resources are needed only during the handling of each packet.

The next generation of data heading towards third generation and personal multimedia environments builds on GPRS and is known as Enhanced Data rate for GSM Evolution (EDGE). EDGE will also be a significant contributor in 2.5G. It will allow GSM operators to use existing GSM radio bands to offer wireless multimedia IP-based services and applications at theoretical maximum speeds of 384 kbps with a bit-rate of 48 kbps per timeslot and up to 69.2 kbps per timeslot in good radio conditions. EDGE will let operators function without a 3G license and compete with 3G networks offering similar data services. Implementing EDGE will be relatively painless and will require relatively small changes to network hardware and software as it uses the same TDMA (Time Division Multiple Access) frame structure, logic channel and 200 kHz carrier bandwidth as today's GSM networks. As EDGE progresses to coexistence with 3G WCDMA, data rates of up to ATM-like speeds of 2 Mbps could be available.

GPRS will support flexible data transmission rates as well as continuous connection to the network. GPRS is the most significant step towards 3G.

1.1.4 Third-Generation Mobile Systems

Third-generation mobile systems are faced with several challenging technical issues, such as the provision of seamless services across both wired and wireless networks and universal mobility. In Europe, there are three evolving networks under investigation: (a) UMTS (Universal Mobile Telecommunications Systems), (b) MBS (Mobile Broadband Systems), and (c) WLAN (Wireless Local Area Networks).

The use of hierarchical cell structures is proposed for IMT2000. The overlaying of cell structures allows different rates of mobility to be serviced and handled by different cells. Advanced multiple access techniques are also being investigated, and two promising proposals have evolved, one based on wideband CDMA and another that uses a hybrid TDMA/CDMA/FDMA approach.

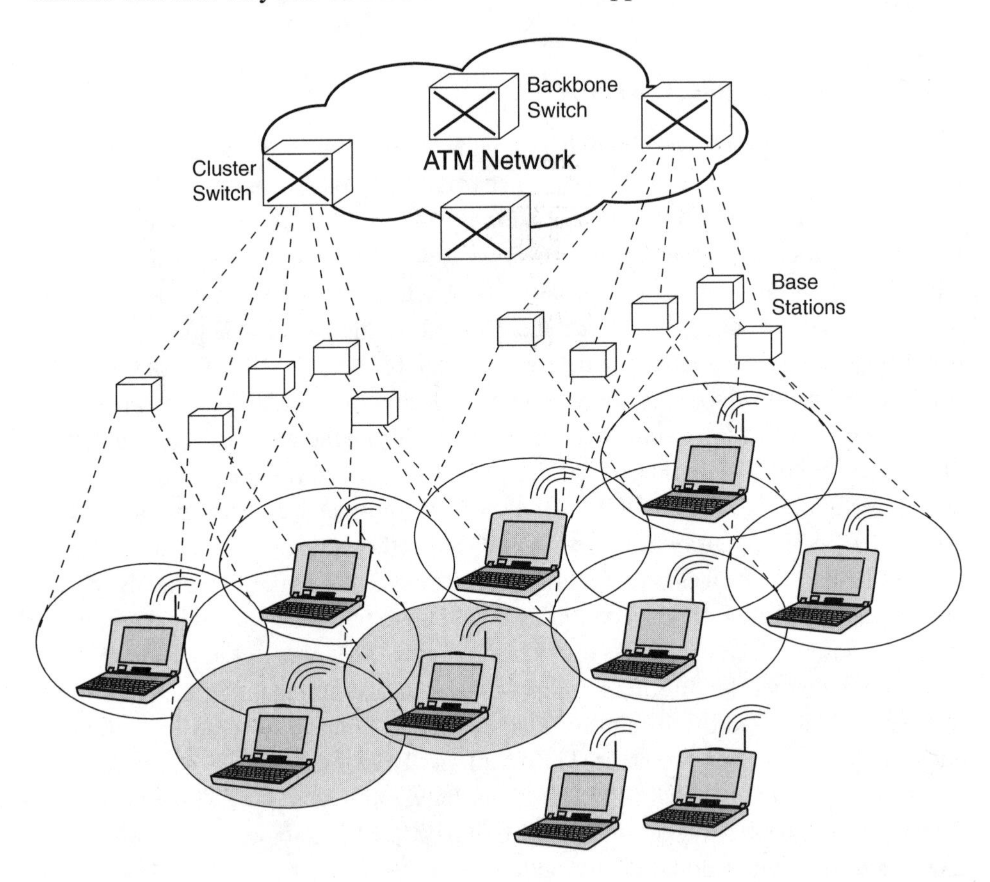

Figure 1.1. The architecture of a cellular wireless network based on ATM.

1.2 Global System for Mobile Communications (GSM)

GSM is commonly referred to as the second-generation mobile cellular system. GSM has its own set of communication protocols, interfaces, and functional entities. It is capable of supporting roaming, and carrying speech and data traffic.

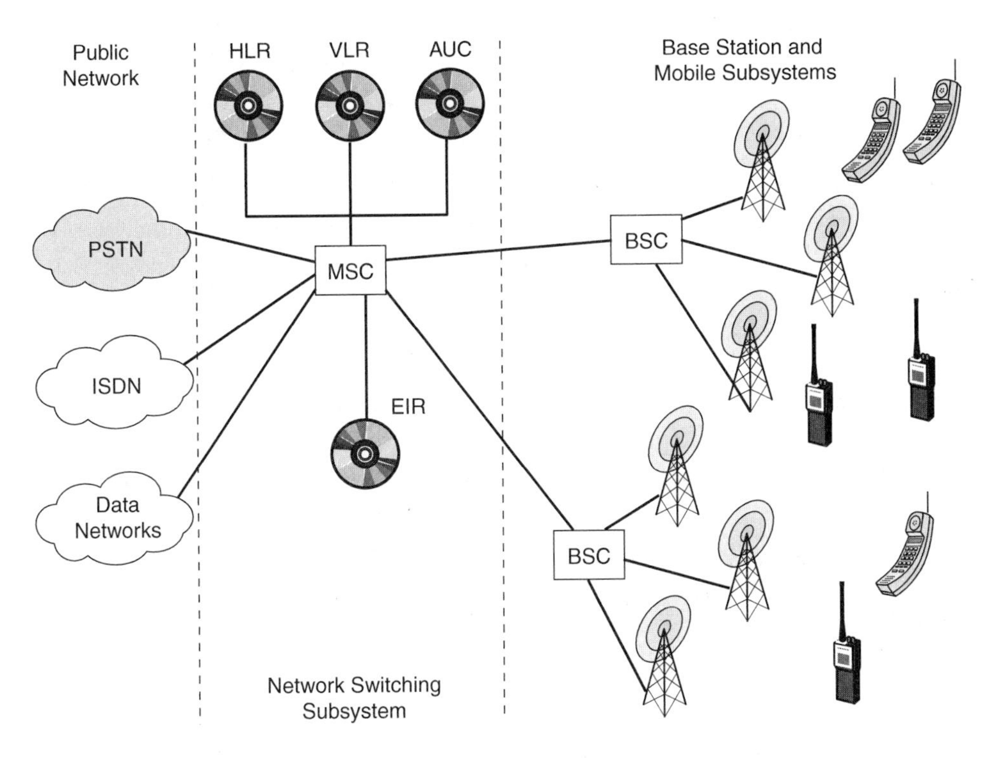

Figure 1.2. The network architecture of GSM.

The GSM network architecture (see Figure 1.2) comprises several base transceiver stations (BTS), which are clustered and connected to a base station controller (BSC). Several BSCs are then connected to an MSC. The MSC has access to several databases, including the visiting location register (VLR), home location register (HLR), and equipment identity register (EIR). It is responsible for establishing, managing, and clearing connections, as well as routing calls to the proper radio cell. It supports call rerouting at times of mobility. A gateway MSC provides an interface to the public telephone network.

The HLR provides identity information about a GSM user, its home subscription base, and service profiles. It also keeps track of mobile users registered within its home area that may have roamed to other areas. The VLR stores information

Table 1.1. The IMSI in GSM

Mobile Country Code
Mobile Network Code
Mobile Subscriber Identification Code

about subscribers visiting a particular area within the control of a specific MSC.

The authentication center (AC) is used to protect subscribers from unauthorized access. It checks and authenticates when a user powers up and registers with the network. The EIR is used for equipment registration so that the hardware in use can be identified. Hence if a device is stolen, service access can be denied by the network. Also, if a device has not been previously approved by the network vendor (perhaps subject to the payment of fees by the user), EIR checks can prevent the device from accessing the network.

In GSM, each mobile device is uniquely identified by an IMSI (international mobile subscriber identity). It identifies the country in which the mobile system resides, the mobile network, and the mobile subscriber. The IMSI is stored on a subscriber identity module (SIM), which can exist in the form of a plug-in module or an insertable card. With a SIM, a user can practically use any mobile phone to access network services.

1.3 General Packet Radio Service (GPRS)

The GSM general packet radio service (GPRS) is a data overlay over the voice-based GSM cellular network. It consists of a packet wireless access network and an IP-based backbone. GPRS is designed to transmit small amounts of frequently sent data or large amounts of infrequently sent data. GPRS has been seen as an evolution toward UMTS (Universal Mobile Telecommunications Systems). Users can access IP services via GPRS/GSM networks.

GPRS services include both point-to-point and point-to-multipoint communications. The network architecture of GPRS is shown in Figure 1.3. Gateway GSN (GGSN) nodes provide interworking functions with external packet-switched networks. A serving GPRS support node (SGSN), on the other hand, keeps track of an individual mobile station's location and provides security and access control. As shown in Figure 1.3, base stations (BSSs) are connected to SGSNs, which

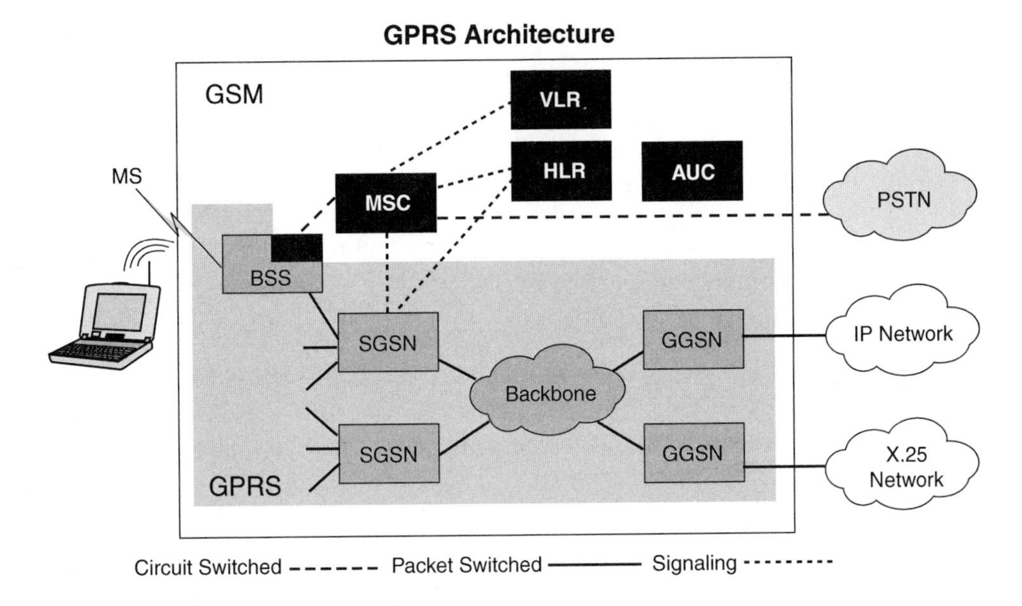

Figure 1.3. Architecture of GSM general packet radio service.

are subsequently connected to the backbone network. SGSNs interact with MSCs and various databases to support mobility management functions. The BSSs provide wireless access through a TDMA MAC protocol. Both the mobile station (MS) and SGSNs execute the SNDCP (Subnetwork-Dependent Convergence Protocol), which is responsible for compression/decompression and segmentation and reassembly of traffic. The SGSNs and GGSNs execute the GTP (GPRS Tunnelling Protocol), which allows the forwarding of packets between an external public data networks (PDN) and mobile unit (MU). It also allows multiprotocol packets to be tunneled through the GPRS backbone.

1.4 Personal Communications Services (PCSs)

The FCC defines PCS [5] as "Radio communications that encompass mobile and ancillary fixed communication that provides services to individuals and business and can be integrated with a variety of competing networks." However, the Telecommunications Industry Association (TIA) has a different definition for PCS:

> *A mobile radio voice and data service for the provision of unit-to-unit communicatoins, which can have the capability of public switched telephone network access, and which is based on microcellular or other technologies*

that enhance spectrum capacity to the point where it will offer the potential of essentially ubiquitous and unlimited, untethered communications.

PCS can also be defined in a broader sense [6] as a set of capabilities that allows some combination of *personal* mobility and *service* management. In short, PCS [7] is a commonly used term that defines the next generation of advanced wireless networks providing *personalized* communication services. In Europe, the term "personal communication networks (PCNs)" is used instead of PCS.

The basic requirements for a PCS are:

- Users must be able to make calls wherever they are
- Offered services must be reliable and of good quality
- Provision of multiple services such as voice, fax, video, paging, etc., must be available.

Unlike AMPS, PCS is aimed at the personal consumer industry for mass consumption. The FCC's view of PCS is one where the public switched telephone network (PSTN) is connected to a variety of other networks, such as CATV (cable television), AMPS cellular systems, etc.

1.5 Wireless LANs (WLANS)

Wireless LAN technology has evolved to extend to existing wired networks. Local area networks (LANs) are mostly based on Ethernet media access technology that consists of an interconnection of hosts and routers. LANs are restricted by distance. They are commonly found in offices and inside buildings. Interconnection using wires can be expensive when it comes to relocating servers, printers, and hosts.

Now, more wireless LANs (WLANs) are being deployed in offices. Most WLANs are compatible with Ethernet, and hence, there is no need for protocol conversion. The IEEE has standardized 802.11 protocols to support WLANs media access. A radio base station can be installed in a network to serve multiple wireless hosts over 100-200 m. A host (for example, a laptop) can be wirelessly enabled by installing a wireless adapter and the appropriate communication driver. A user can perform all network-related functions as long as he or she is within the coverage area of the radio base station. This gives the user the capability to perform work beyond his or her office space.

As shown in Figure 1.4, several overlapping radio cells can be used to provide wireless connectivity over a desired region. If a wireless host migrates from one

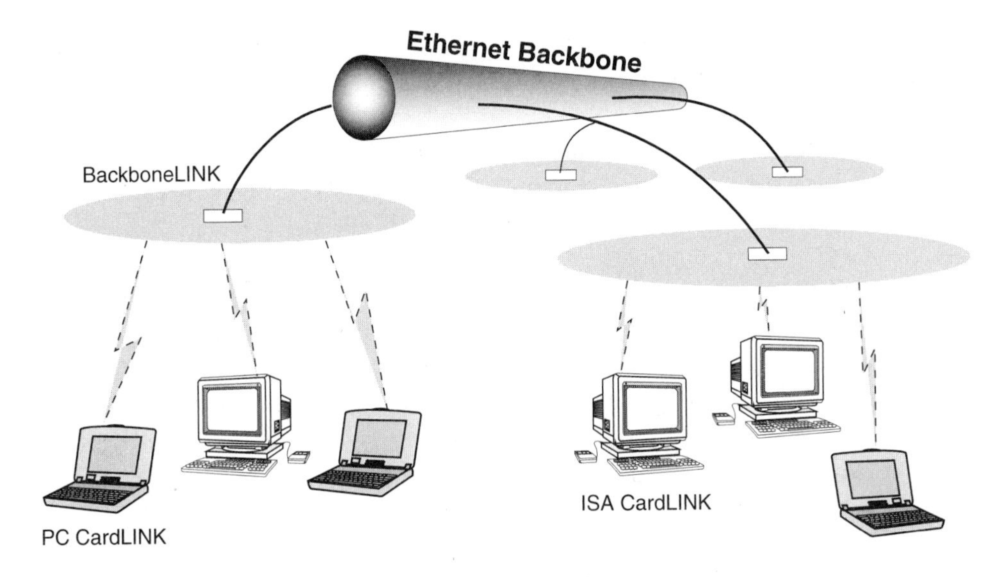

Figure 1.4. A WLAN with an Ethernet wired backbone.

radio cell to another within the same subnet, then there is no handoff. It is basically bridging, since the host's packet will eventually be broadcast onto the same Ethernet backbone.

WLANs support existing TCP/IP-based applications. There has been considerable debate in the past as to the low throughput WLANs provide compared to high-speed wired networks. It was not long ago that switched Ethernet technology [8] evolved, bringing the communication throughput of Ethernet into the gigabit range.

The desire to support higher throughput and ad hoc mobile communications has prompted the ETSI (European Communications Standard Institute) to produce a standard for high-performance Radio LAN (HIPERLAN), at 20Mbps throughput with a self-organizing and distributed control network architecture. HIPERLAN II is a wireless ATM system operating at the 17GHz band.

1.6 Universal Mobile T elecommunicationsSystem (UMTS)

The Universal Mobile Telecommunications System (UMTS) is commonly referred to as a third-generation system. It is targeted to be deployed in 2002. UMTS employs an ATM-based switching network architecture and aims to provide services for both *mobile* and *fixed* subscribers by common call-processing procedures.

The UMTS architecture is split into *core* (switching) networks, *control* (service) networks, and *access* networks. The core network is responsible for performing switching and transmission functions. The control network supports roaming through the presence of mobility management functions. Finally, the radio access network provides channel access to mobile users and performs radio resource management and signalling. UMTS will include both terrestrial and global satellite components.

The UMTS network comprises: (a) the mobile terminal, (b) the base transceiver station (BTS), (c) the cell site switch (CSS), (d) mobile service control points (MSCP), and (e) the UMTS mobility service (UMS). UMTS employs a hierarchical cell structure, with macrocells overlaying microcells and picocells. Highly mobile traffic is operated on the macrocells to reduce the number of handoffs required. UMTS aims to support roaming across different networks.

The UMTS Radio Access System (UTRA) will provide at least 144 kbps for full-mobility applications, 384 kbps for limited-mobility applications, and 2.048 Mbps for low-mobility applications. UMTS terminals will be multiband and multimode so that they can work with different standards.

UMTS is also designed to offer data rate on-demand. The network will react to a user's needs, based on his/her profile and current resource availability in the network. UMTS supports the virtual home environment (VHE) concept, where a personal mobile user will continue to experience a consistent set of services even if he/she roams from his/her home network to other UMTS operators. VHE supports a consistent working environment regardless of a user's location or mode of access. UMTS will also support adaptation of requirements due to different data rate availability under different environments, so that users can continue to use their communication services.

To support universal roaming and global coverage, UMTS will include both terrestrial and satellite systems. It will enable roaming with other networks, such as GSM. UMTS will provide a flexible broadband access technology that supports both IP and non-IP traffic in a variety of modes, such as packet, circuit-switched, and virtual circuit.

1.7 IMT2000

The ITU (International Telecommunications Union) has introduced a new framework of standards by the name IMT2000, which is a federation of systems for third-generation mobile telecommunications. IMT2000 aims to provide: (a) high-speed access, (b) support for broadband multimedia services, and (c) universal mobility. Frequency spectrum has been allocated for IMT2000 by the ITU. Several multiple-access protocols based on code division have been proposed by many different countries. The ITU has approved the CDMA2000 radio access system as the CDMA multicarrier member of the IMT2000 family of standards. CDMA2000 is capable of supporting IS-41 and GSM-MAP to ensure backward compatibility. IS-41 is a network protocol standard that supports interoperator roaming [2]. It allows MSCs of different service providers to exchange information about their subscribers to other MSCs on-demand.

1.8 IS-95, cdmaOne and cdma2000 Evolution

The IS-95 [9] air interface was standardized by TIA in July 1993. Networks that utilize IS-95 CDMA [10] air interface and the ANSI-41 network protocol are known as cdmaOne networks. IS-95 networks use one or more 1.25 MHz carriers and operate within the 800 and 1900 MHz frequency bands.

Following the launch of the first cdmaOne network in Hong Kong in 1995, the number of cdmaOne subscribers has grown into millions. cdmaOne networks provide soft handoffs and higher capacity than traditional AMPS networks, with data rates up to 14.4 kbps. CdmaOne is based on IS-95A technology. IS-95B improves this technology further by providing higher data rates for packet- and circuit-switched CDMA data, with data rates up to 115 kbps.

This evolution continues with cdma2000, which is the third generation verion of IS-95. This new standard is developed to support third generation services as defined by ITU. cdma2000 is divided into two parts, namely: (a) IS-2000/cdma200 1X, and (b) IS-2000A/cdma2000 3X. cdma2000 1X standard delivers twice the voice capacity of cdmaOne with a data rate of 144 kbps. The term 1X, as derived from 1XRTT (radio transmission technology), is used to signify that the standard carrier on the air interface is 1.25 MHz, which is similar to IS-95A and IS-95B. In cdma2000 3x, the term 3X, derived from 3XRTT, is used to signify three times 1.25 MHz, i.e., 3.75 MHz. cdma2000 3X offers greater capacity than 1X with data rates up to 2 Mbps while retaining backward compatibility with earlier 1X and cdmaOne deployments.

Lately, 3GPP (Third Generation Partnership Project) [11] is formed to defined standards for third generation all-IP networks. It is also responsible for the production of globally applicable technical specifications and reports for a 3G mobile system based on evolved GSM core networks and the radio access technologies that they support (i.e., Universal Terrestrial Radio Access (UTRA) both Frequency Division Duplex (FDD) and Time Division Duplex (TDD) modes).

1.9 Organization of this Book

This book is organized in a manner that allows a gradual progression of the subject toward more advanced topics. Chapter 2 discusses the origin of ad hoc networks, in paticular, the DARPA packet radio networks. Chapter 3 presents a current version of ad hoc networks and related challenges. Since there are no static base stations present in ad hoc wireless networks, centralized access control becomes a problem. Chapter 4 presents current channel access protocols and discusses some emerging protocols. Chapter 5 provides an overview of current ad hoc mobile routing protocols, discussing their principles, features, and operation. Chapter 6 presents a new era of routing known as longevity, or associativity-based routing. This protocol is a major deviation from traditional routing protocols, which use shortest path as the main routing metric. Chapter 7 provides a narration of the implementation of an ad hoc wireless network using a new routing protocol and current-off-the-shelf (COS) hardware. It also provides a discussion of the experimental results obtained via campus field trials. Chapter 8 continues with a discussion of the communication performance of ad hoc wireless networks so that readers can understand the capabilities of such networks and what potential applications can be supported.

The advancement in CPU technology has way surpassed that of battery technology. Hence, Chapter 9 discusses how device power life can affect communication performance and protocol design for ad hoc wireless networks. Multicasting has been widely used to support multiparty communications and conferencing. Chapter 10 provides insight on how ad hoc mobile multicasting can be achieved and presents a survey of current multicasting protocols. It also reveals how associativity or longevity can be applied to ad hoc multicasting.

Since the Internet Protocol (IP) provides unreliable datagram delivery, transmission control protocol has been introduced to provide reliable delivery of information over the internet. Chapter 11 discusses the problems associated with TCP in an ad hoc wireless network environment. Ad hoc networks should provide services to users. Chapter 12 presents existing service discovery protocols that will allow an ad hoc mobile host to discover services present in the network and to access such services.

Commercial realization of ad hoc networks has taken the form of Bluetooth. Chapter 13, therefore, presents a case study of this technology. Prior to the arrival of Bluetooth, the Wireless Access Protocol (WAP) was a popular technology since it enabled a cellular network to support data in addition to voice traffic. Chapter 14, therefore, provides a discussion of WAP. Many people have been wondering about the potential applications of ad hoc networking; Chapter 15 addresses this issue. A conclusion is finally presented in Chapter 16.

Chapter 2

ORIGINS OF AD HOC: PACKET RADIO NETWORKS

2.1 Introduction

The merits of having an infrastructureless network were discovered in the 1970s. At that time, computers were bulky and so were radio transceivers. DARPA had a project known as packet radio, where several wireless terminals could communicate with one another on a battlefield. Packet radio was a technology that extended the concept of packet switching (which evolved from point-to-point communication metworks) to the domain of broadcast radio networks. During the 1970s, the ALOHA [12] project at the University of Hawaii demonstrated the feasibility of using the broadcasting property of radios to send/receive data packets in a single radio hop system. The ALOHA project later led to the development of a *multi-hop* multiple-access packet radio network (PRNET) under the sponsorship of the

Advanced Research Project Agency (ARPA) [13]. Unlike ALOHA, PRNET permits multi-hop communications over a wide geographical area. ARPA itself has a history in terms of its name and roles. This is summarized below.

Source DARPA -

DoD directive 5105.15 establishing the Advanced Research Projects Agency (ARPA) was signed on February 7, 1958. The directive gave ARPA the responsibility "for the direction or performance of such advanced projects in the field of research and development as the Secretary of Defense shall, from time to time, designate by individual project or by category."

DARPA - On March 23, 1972, by DoD Directive, the name was changed to the Defense Advanced Research Projects Agency (DARPA). DARPA was established as a separate defense agency under the Office of the Secretary of Defense.

ARPA - On February 22, 1993, DARPA was redesignated the Advanced Research Projects Agency (ARPA) – as the agency was known before 1972. The change was outlined in President Bill Clinton's strategy paper, "Technology for America's Economic Growth, A New Direction to Build Economic Strength."

DARPA - On February 10, 1996, Public Law 104-106, under Title IX of the Fiscal Year 1996 Defense Authorization Act, directed an organizational name change to the Defense Advanced Research Projects Agency (DARPA).

Several pioneers contributed to the field of PRNET, including Robert Khan, Barry Leiner, Leonard Kleinrock, and John Jubin. Technical issues addressed for PRNET included media access, error and flow control, addressing, routing, network initialization and control, etc. One of the most attractive features of PRNET is rapid deployment. Once installed, the system is self-initializing and self-organizing. This implies that network nodes should be able to discover radio connectivity among neighboring nodes and organize routing strategies based on this connectivity. PRNETs are expected to require no system administration and can be left unattended.

In this chapter, we will discuss challenges, network architectures, and routing principles of packet radio, followed by a description of its communication performance.

2.2 T echnicalChallenges

PRNETs are different from wired networks in many aspects. They have an infrastructureless backbone and network nodes that act as routers or packet switches to forward packets from one node to another. Routers are connected without wires and routers themselves can be mobile. The introduction of *wireless connectivity* and the presence of *mobility* result in great technical challenges in the field of computer communications.

In fact, PRNET is the network that attempts to merge computer communications with telecommunciations. It allows networks to be formed and de-formed on-the-fly, through a set of innovative and adaptive communication protocols. The technical challenges for PRNET can be summarized as:

- Flow control over a wireless multi-hop communication route
- Error control over wireless links
- Deriving and maintaining network topology information
- Deriving accurate routing information
- Mechanisms to handle router mobility
- Shared channel access by multiple users
- Processing capability of terminals
- Size and power requirements

2.3 Architecture of PRNETs

As depicted in Figure 2.1, a PRNET consists of several mobile radio repeaters, wireless terminals, and dedicated mobile stations. The role of a repeater is to relay packets from one repeater to another, until the packets eventually reach the destination host. The mobile station is present to derive routes from one host to another. As network conditions change (terminal movement, repeater failures or recovery, changes in hop reliability, and network congestion state), routes are dynamically reassigned by the station to satisfy minimum delay criteria. Hosts and terminals attached to the PRNET are unaware of the station's assignment and reassignment of communication routes.

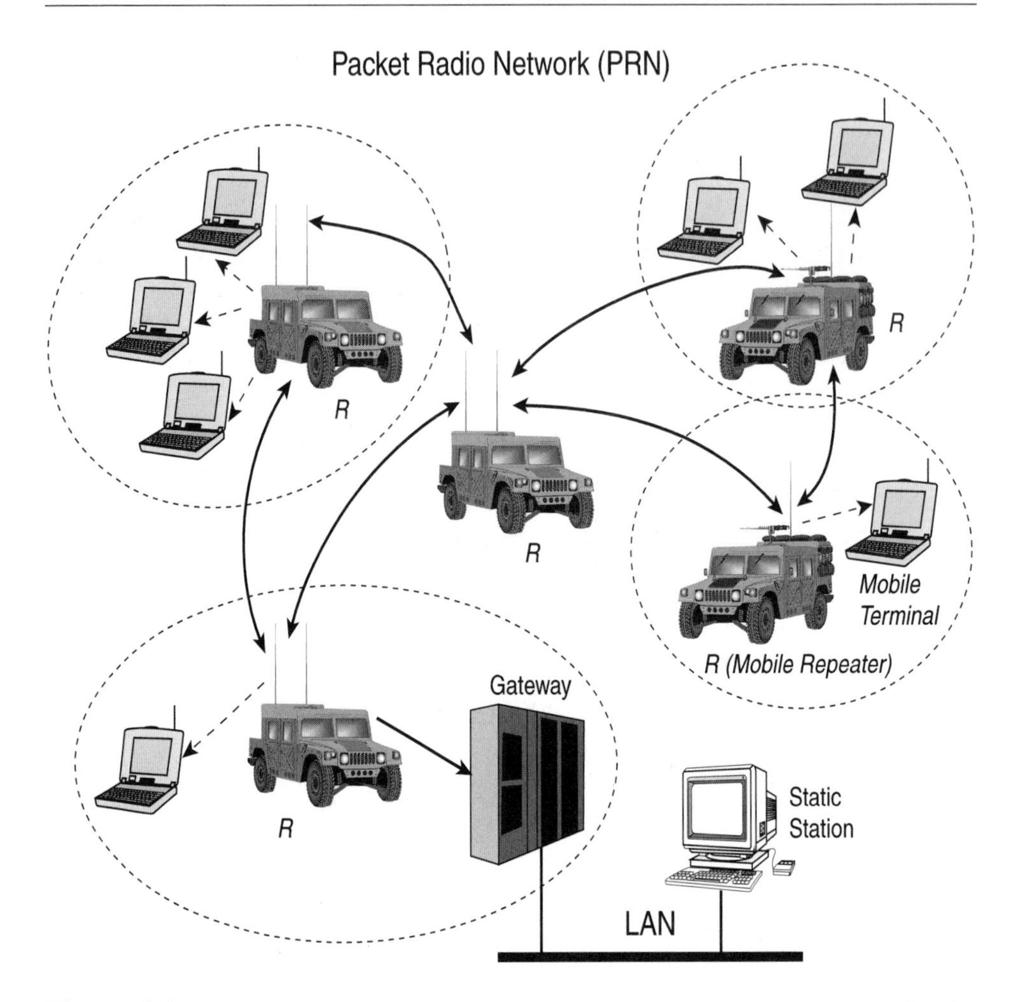

Figure 2.1. The network architecture of PRNETs, which comprises mobile devices/terminals, packet radios, and repeaters. The static station is optional.

2.4 Components of Packet Radios

As shown in Figure 2.2, the user computer is interfaced to a radio via the terminal-network controller (TNC). The user computer is commonly referred to as the mobile device/terminal, while the radio and TNC logic are commonly referred to as the packet radio. The packet radio, therefore, implements functions related to protocol layers 1, 2, and 3. It is an intermediate system (IS) in the ISO context. A packet radio network (PRN) is a collection of packet radios, with some packet radios connected to user devices while others are not.

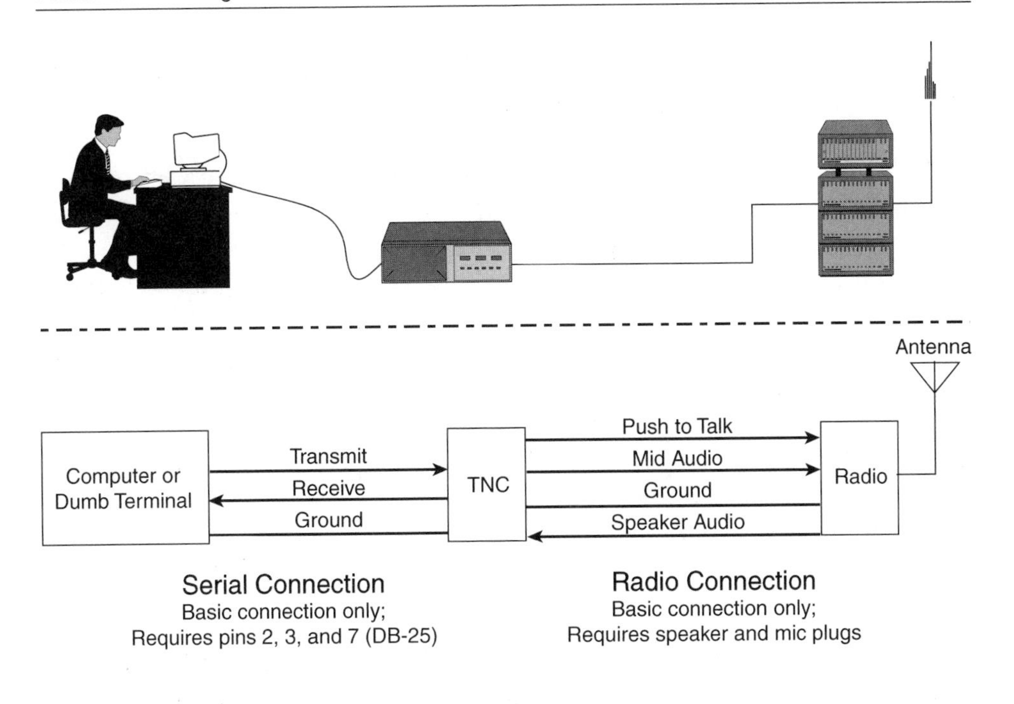

Figure 2.2. The interface of a data terminal to a packet radio.

2.5 Routing in PRNETs

2.5.1 Point-to-Point Routing

PRNETs support point-to-point communications through point-to-point routing. Here, a packet originating at one part of the network moves through a series of one or more repeaters until it eventually reaches the final destination. This point-to-point route is an ordered set of repeater addresses that is determined by the mobile station. This station is the only element in the network that has knowledge of the overall network connectivity, that is, the network topology. With network topology information, the mobile station computes the best point-to-point route and distributes this information to all repeaters in the route or directly to the source packet radio. This scheme was found to be suitable for slow moving user terminals.

2.5.2 Broadcast Routing

Radio technology provides very good broadcasting properties. Broadcasting information to all radios in a network is equivalent to flooding. To ensure that each packet radio only forwards a packet once, each repeater has to maintain a list of packet identifiers for previously broadcast packets that it recently had received and forwarded.

In *broadcast routing*, a packet radiates away from the source packet radio in a wave-like fashion, that is, the packet ripples away from the source. Although broadcasting is very robust (since a packet will be received by every node in the non-partitioned network), it is not efficient for two-party communications since all other nodes in the network must participate in the transmission and reception of packets that are not intended for them. Hence, when *broadcast routing* is used for point-to-point communication, the destination host address is included in each data packet. No specific routes are derived prior to data transmission; hence, routing decisions are not centralized. Packets will eventually reach the destination host if the network is not partitioned. For fast moving user terminals, broadcast routing was found to be useful as it avoids the need to process rapidly changing routes.

2.5.3 Packet Forwarding

The *connectionless* approach to packet forwarding requires some background operation to maintain up-to-date network topology and link information in each node. This means that as network topology changes, the background routing traffic can be substantial. This is commonly associated with broadcast routing, where each packet carries sufficient routing information for it to arrive at the destination.

In the *connection-oriented* packet forwarding approach, however, an explicit route establishment phase is required before data traffic can be transported. This approach is commonly associated with point-to-point routing, where each node in a route has a lookup table for forwarding incoming packets to the respective outgoing links. Hence, if a topology changes, a route re-establishment phase is needed.

2.5.4 Impact of Mobility

In a PRNET, all elements of the network can be mobile. Some move relatively slowly (for example, the repeaters), and hence, a topological change in the backbone network is not frequent. The assumption made in a PRNET is that user terminals normally move slowly enough such that the assigned point-to-point routes are valid for at least a few seconds before another route must be chosen.

When the user rate of mobility is increased, point-to-point routing may not be practical since most of the time will be spent in computing alternate point-to-point routes instead of forwarding the packets to their intended destinations. Under such circumstances, very poor communication performance will be observed. Broadcast routing is less affected by user mobility since the packets do not follow a specific point-to-point route. Instead, every node is supposed to relay the packets, and hence, the destination host will receive the packet eventually. There is, therefore, no need to cope with rapidly changing routes in broadcast routing under conditions of rapid host mobility. However, broadcasting is power inefficient.

2.6 Route Calculation

Each packet radio operates in a fully distributed manner. It gathers and maintains information about current network topology so that it can make independent decisions about how to route packets toward their destinations. Each node maintains the following tables:

- Neighbor table
- Tier table
- Device table

• Neighbor Table

As a packet radio is powered up, it broadcasts a PROP (packet radio organization packet) every 7.5 seconds, announcing its existence and information about the network topology from its own perspective at that time. Given the broadcast nature of the PROP, neighboring nodes will receive this packet and will update their neighbor tables, which will now contain information about their neighbors. The table tracks the bidirectional quality of the link to and from those neighbors (see Table 2.1).

Table 2.1. The Structure of a Neighbor Table in a Packet Radio

Neighboring PR	Link Quality
Node 1	5/8
Node 7	1/3
Node 3	3/4
Node 5	7/8

In PRNETs, link quality is measured by taking the ratio of the number of packets correctly received from the transmitting packet radio during a PROP interval to the number of packets that the transmitting packet radio actually transmitted at that same interval. This information will be subsequently used by the routing algorithm.

• Tier Table

Routing in PRNETs relies on each packet radio maintaining adequate knowledge of the best packet radio to forward packets to for every prospective destination. The tier information ripples outward from each packet radio at an average rate of 3.75 seconds per hop and eventually reaches all packet radios. Hence, eventually, every packet radio knows its distance in tiers (or radio hops) from itself to every prospective destination and the next-hop packet radio (see Table 2.2). This is, in principle, similar to the early ARPANET routing algorithm [12], which is based on the classical Bellman-Ford routing.

Table 2.2. The Structure of Tier Table in a Packet Radio

Destination PR	**Next-Hop PR**	**Tier Count**
Node 1	Node 15	2
Node 7	Node 7	0
Node 3	Node 4	1
Node 5	Node 2	1

In PRNETs, the **best route** is defined as the **shortest route with good connectivity on each hop**. However, a best route could change over time. When a link (for example, from node $\mathcal{A}$ to $\mathcal{B}$) to a neighboring packet radio turns bad, all routes in node $\mathcal{A}$'s tier table for which node $\mathcal{B}$ is the next-hop packet radio will also be marked bad. In addition, the news of bad tier data is also propagated via PROPs.

• Device Table

Given the network configuration of packet radios where there are mobile terminals and mobile repeaters, there is a need to maintain device-to-packet radio mapping. With the presence of mobility, this logical mapping will have to be updated. Each mobile device/terminal periodically sends a control packet across the wired interface to its attached packet radio. A packet radio keeps track of affiliated devices and propagates this mapping information via a PROP to other packet radios in the network at an average rate of 3.75 seconds per hop. Hence, when a packet radio receives a packet addressed to a specific mobile device, the device currently attached to the packet radio is known and the appropriate next hop-packet radio is chosen to forward the packet.

2.6.1 Principles of Packet Forwarding

In PRNETs, a packet traverses over a chosen path hop-by-hop and is acknowledged at every packet radio along the path. Forwarding is accomplished via information read from the device and tier tables, and from the packet headers.

• ETE Header

The end-to-end header (ETE) is created by the source mobile device/terminal, not the packet radio. It includes the source device ID/address, which is used to update the packet radio's device-to-packet radio mapping information, and the destination device ID/address, which is used in packet forwarding. The ETE header remains intact as the packet transits toward the destination device.

• Routing Header

In contrast to the ETE header, the routing header (see Table 2.3) is created by the source packet radio. The routing header encapsulates the ETE header, since it is the routing header that the packet radio will use to forward the packets. Note that the source packet radio ID, sequence number, and destination packet radio ID remain intact throughout the packet's journey toward the destination packet radio. The remaining fields are updated by every intermediate packet radio in the route. Eventually, the routing header is stripped at the destination packet radio and the ETE header is exposed. The packet is then delivered to the destination device.

Table 2.3. The Structure of a Routing Header

Routing Header Field	Purpose
Source PR[a] ID	Acknowledgment
Sequence Number	Acknowledgment
Previous PR ID	Acknowledgment
Previous PR's transmit count	Pacing
Transmitting PR ID	Acknowledgment
Transmitting PR's transmit count	Pacing
Next PR ID	Forwarding/Pacing/Acknowledgment
Tier	Alternate Routing
Destination PR ID	Forwarding

[a]PR stands for packet radio.

2.7 Pacing Techniques

The time at which a packet is selected for transmission is determined by a three-component packing protocol. Transmission parameters are chosen based on measure *link quality* and the *type of service* desired by the user.

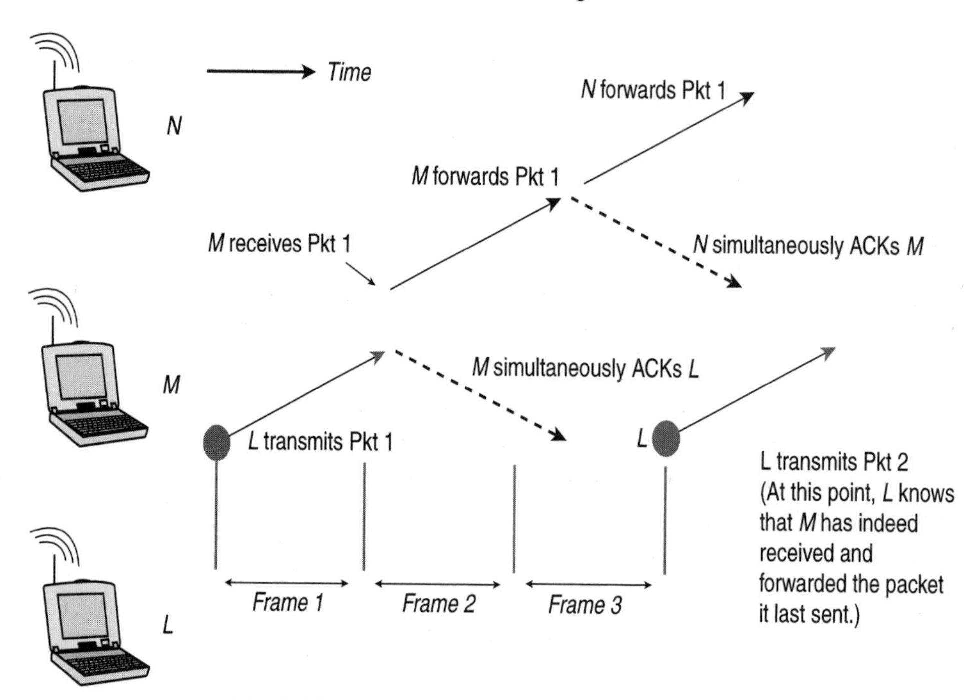

Figure 2.3. Three-frame packet forwarding in PRNETs.

The pacing protocol provides flow and congestion control while ensuring fair use of the radio channel. A *single threading* technique is employed, which requires that a packet transmitted to a certain next packet radio be acknowledged (or discarded) before another packet is sent to the same packet radio. Recall earlier that acknowledgment is provided passively. Hence, the effect is that acknowledgment is not transmitted until the packet radio is ready to accept another packet from the same previous packet radio. Consequently, packet congestion is reflected as much as possible away from a network bottleneck back to the source packet radio. In addition, there is a limit of two buffers allocated per packet radio for packets received over the wired interface for entry into the PRNET.

To estimate the amount of delay needed between packet transmissions, each packet radio measures the forwarding delay of each packet that it forwards. A packet radio records the time at which its transmission completes and when it receives the acknowledgment from the next packet radio. This difference yields the forwarding delay, assuming the propagation time is negligible compared to the transmission time. The forwarding delay, therefore, encompasses all the processing, queuing, carrier sensing/randomization, and transmission delays.

Figure 2.3 shows three packet radios engaged in forwarding packets. The transmitting packet radio $\mathcal{L}$ must allow time for the next packet radio $\mathcal{M}$ not only to receive $\mathcal{L}$'s transmission and to forward it on, but also to receive the acknowledgment from its next packet radio $\mathcal{N}$. In essence, the source packet radio must ensure that not only its immediate neighbor has succesfully received and forwarded the packet, but also that the following neighbor has received and eventually forwarded the same packet. Hence, the source packet radio must wait for a three-frame period. If it waits for only a two-frame period, it has no way of knowing if packet radio $\mathcal{N}$ has successfully received the packet from $\mathcal{M}$. This means that no packet radio can transmit more than one third of the time. Packet radio $\mathcal{L}$ must, therefore, wait to transmit Packet 2 until three times the forwarding delay has elapsed since its transmission of the first packet.

To summarize: By using the pacing mechanism, a node can adapt its transmission time to provide congestion control when faced with a bottleneck at a packet radio; when there is no bottleneck, it can transmit in a manner that uses the channel efficiently.

2.8 Media Access in PRNETs

PRNETs employ the Carrier Sense Multiple Access (CSMA) protocol to coordinate communications among mobile hosts. CSMA prevents a packet radio from transmitting at the same time when a neighboring packet radio is using the medium. A packet radio is aware if a neighbor is transmitting by reading its hardware indication bit-synchronization-in-the-lock. Basically, this bit, when set, implies that the channel is busy and a carrier is being sensed. Whenever a carrier is being sensed, a packet radio will refrain from transmitting.

While carrier sensing reduces the probability of channel contention, it cannot eliminate *hidden terminal* and *exposed nodes* problems. The former is a result of a node lying within the radio range of the receiver, but not another transmitter that is two hops away. The latter results in neighboring nodes of a transmitter being blocked from transmission. These issues will be further discussed in the chapter on media access.

2.9 Flow Acknowledgments in PRNETs

Packets are forwarded via a single communication route through a PRNET. Each packet radio examines the information contained in the packet headers and in its own device and tier tables. Each packet radio must decide if it should be the one to transmit the packet, if it should update the routing header before transmitting, and if it should update its own tables.

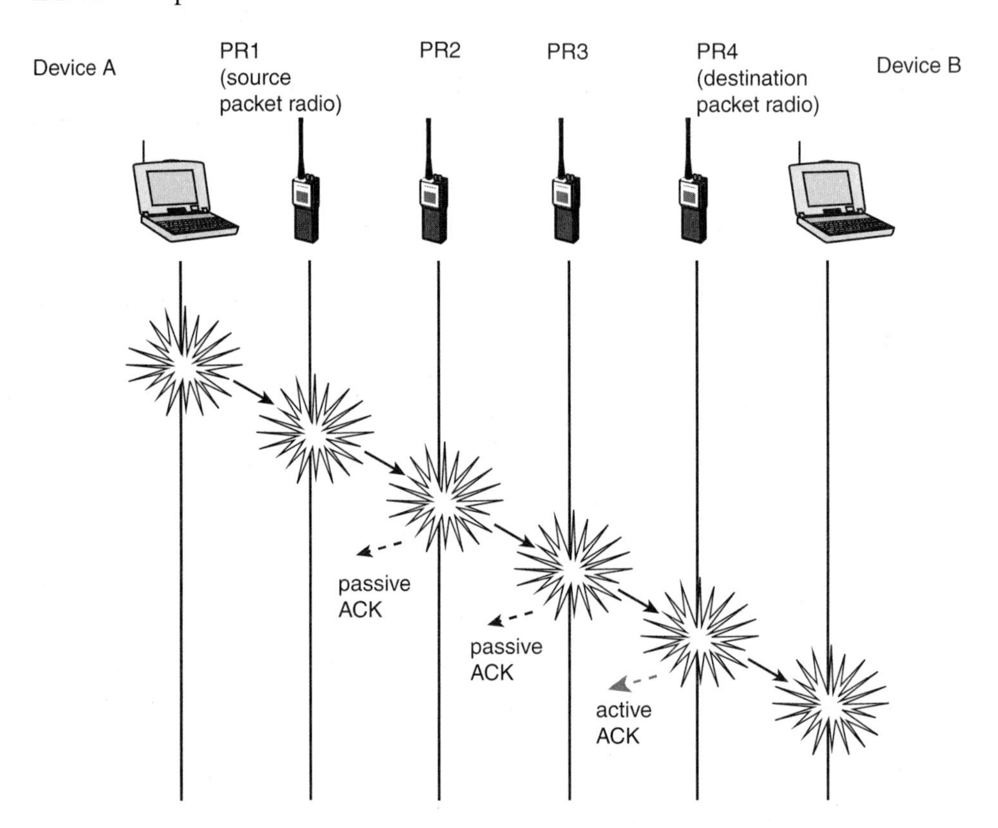

Figure 2.4. The principles of passive and active packet acknowledgments in PRNETs.

Other packet radios within the radio range will also each receive the transmitted packet. If these neigbors are not part of the route, they will discard the overheard packets. The downstream node that receives the packet will process the packet and proceed with issuing a *passive* acknowledgment. The single transmission, therefore, not only forwards the packet on to the next packet radio but also acknowledges the previous packet radio that the packet was successfully received and is being forwarded. This principle of passive acknowledgment will proceed until the packet reaches the destination node. Since the destination node does not have a downstream node and it is the terminating point, an *active* acknowledgment is sent by the destination node to its upstream node to confirm successful reception of the packet.

2.10 Conclusions

The packet radio was the first implementation of an infrastructureless network, where nodes are mobile, including the mobile device. Field trials associated with packet radio performed by DARPA revealed its practicality and also its limitations. Packet radio communications technology was well ahead of computer technology. At that time, only LSI (large scale integration) electronics were realizable, resulting in bulky packet radios that were not easily movable given their size and weight. In addition, the use of tier information propagation similar to classic Bellman-Ford routing resulted in slow network route convergence and the presence of transient loops.

Packet radio selects a route with good quality links that result in the shortest path. It has the ability to measure forwarding delay and it controls packet transmission and forwarding by using pacing techniques. The channel access protocol provides randomization delay to account for transmit-to-receive switchover time and also to avoid collisions. The use of active and passive acknowledgments provide some form of flow control without the excessive use of control packets. These advanced protocols were considered too good for the 1970s.

With the progression of time, we have entered into a new age of advanced microelectronics technology, where devices can be made with a very small form factor. This means that the user terminal, terminal network controller, and the radio itself can all be integrated onto a single device. The repeater can also be a user terminal. This will greatly change the traditional view of packet radio into one where each node is highly mobile, light, and can act as both a repeater and user terminal. We call the wireless interconnection of such devices an ad hoc wireless network. The next few chapters will present information about architectures and protocols associated with ad hoc wireless networks in greater depth.

Chapter 3

AD HOC WIRELESS NETWORKS

3.1 What Is an Ad Hoc Network?

An ad hoc wireless network is a collection of two or more devices equipped with wireless communications and networking capability. Such devices can communicate with another node that is immediately within their radio range or one that is outside their radio range. For the latter scenario, an intermediate node is used to *relay* or *forward* the packet from the source toward the destination.

An ad hoc wireless network is *self-organizing* and *adaptive*. This means that a formed network can be de-formed on-the-fly without the need for any system administration. The term "ad hoc" tends to imply "can take different forms" and "can be mobile, standalone, or networked." Ad hoc nodes or devices should be able to detect the presence of other such devices and to perform the necessary handshaking to allow communications and the sharing of information and services.

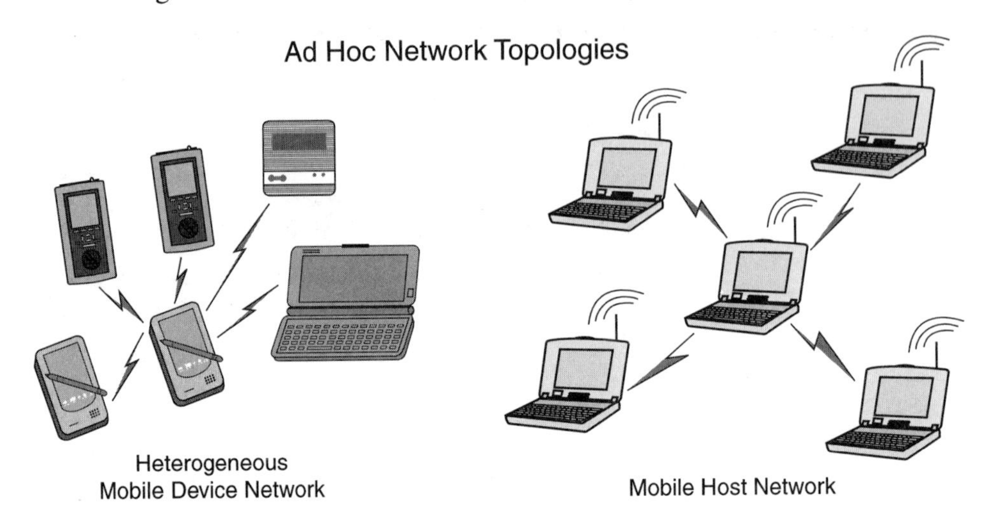

Figure 3.1. (a) Heterogeneous mobile device ad hoc networks, and (b) homogeneous ad hoc network comprising powerful laptop computers.

Since ad hoc wireless devices can take different forms (for example, palmtop, laptop, Internet mobile phone, etc.), the computation, storage, and communications capabilities of such devices will vary tremendously. Ad hoc devices should not only detect the presence of connectivity with neighboring devices/nodes, but also identify what *type* the devices are and their corresponding *attributes*. Since an ad hoc wireless network does not rely on any fixed network entities, the network itself is essentially infrastructureless. There is no need for any fixed radio base stations, no wires or fixed routers. However, due to the presence of mobility, routing information will have to change to reflect changes in link connectivity.

The diversity of ad hoc mobile devices also implies that the battery capacity of such devices will also vary. Since ad hoc networks rely on forwarding data packets sent by other nodes, *power consumption* becomes a critical issue.

3.2 Heterogeneity in Mobile Devices

As shown in Figure 3.2, mobile devices can exist in many forms. There are great differences among these devices, and this heterogeneity can affect communication performance and the design of communication protocols.

Table 3.1 shows some system specifications of existing mobile devices. It is evident that there are differences in size, computational power, memory, disk, and

(a)

(b)

Figure 3.2. (a) Heterogeneous mobile handheld devices (source: COMPAQ), and (b) an Active Badge (source: Olivetti Research).

battery capacity. The ability of an ad hoc mobile device to act as a server or service provider will depend on its computation, memory, storage, and battery life capacity. The presence of heterogeneity implies that some devices are more powerful than others, and some can be *servers* while others can only be *clients*. In addition, relaying packets for others can result in a device expelling its own energy. Hence, a mobile node should examine its own "well-being" before committing to forwarding packets on the behalf of others.

Figure 3.1 shows that despite the differences in ad hoc mobile devices, they can still be networked together wirelessly. The information and processing needs may not be the same, however. A palm pilot with a smaller display may not be capable of displaying large word processing-based applications. An active badge [14] has a limited user interface for word processing. It acts more like a sensor and actuator rather than a user terminal.

Table 3.1. The characteristics of some existing mobile devices.

Device Type	**Form Factor**	**CPU (MPIS)**	**Memory (MB)**	**Disk**	**Battery (Watt-hr)**
Palm Pilot	3.5x4.7 cm	2.7	2-8	None	3-5.5
Active Badge	3.5x3.5 cm	Philips 87C571	64 bytes	None	5V
Cellular Phone	2.5x5.5 cm	16	1Mbit	None	10-20 mA (3.6V)
Pocket PC	13x7.8 cm	130-224	32-64	16MB Flash ROM	3-5
Laptop Computer	12x9 cm	1306-2123	32-128	5-20 GB	37.44-66.60

3.3 Wireless Sensor Networks

Recently, there has been considerable attention devoted to wireless sensor networks [15] and [16], which are crucial for the digital battlefield. These sensors are minute in size and possess both communication and storage capabilities. Some are so small that they take the form of dust, which means that it would be extremely hard for enemies to detect and destroy them.

Microsensors (see Figure 3.3) are not only used in the military. In the healthcare industry, sensors allow continuous monitoring of life-critical information. In the food industry, biosensor technology applied to quality control can help prevent rejected products from being shipped out, thus enhancing consumer satisfaction levels. In agriculture, sensors can help to determine the quality of soil and moisture level; they can also detect other bio-related compounds. Sensors are also widely used for enviromental and weather information gathering. They enable us to make preparations in times of bad weather and natural disaster.

A wireless sensor network is one form of an ad hoc wireless network. Sensors

Figure 3.3. A microsensor.

are wirelessly connected and they, at appropriate times, relay information back to some selected nodes. These selected nodes then perform some computation based on the collected data (a process commonly known as *data fusion*) to derive an ultimate statistic (that reflects an assessment of the environment and tactical conditions) to allow critical decisions to be made.

There are a variety of sensors, including acoustic, seismic, image, heat, direction, smoke, and temperature sensors. The construction of these sensors requires highly integrated electronics, such as MEMS (microelectromechanical systems). Some of the technical challenges behind the realization of wireless sensor networks include: (a) device fabrication, (b) power life conservation, (c) energy-efficient protocols, (d) distributed computation, (e) scalability, (f) data dissemination path derivation, and (g) security.

3.4 Traffic Profiles

Ad hoc wireless communications can occur in several different forms. For a pair of ad hoc wireless nodes, communications will occur between them over a period of time until the session is finished or one of the nodes has moved away. This resembles a peer-to-peer communication scenario.

Another form occurs when two or more devices are communicating among themselves and they are migrating in groups. The traffic pattern is, therefore, one where communications occur over a longer period of time. This resembles the scenario of remote-to-remote communication.

Finally, we can have a scenario where devices communicate in a non-coherent fashion and their communication sessions are, therefore, short, abrupt, and undeterministic. These scenarios are illustrated by Figures 3.4(a), (b), and (c), respectively.

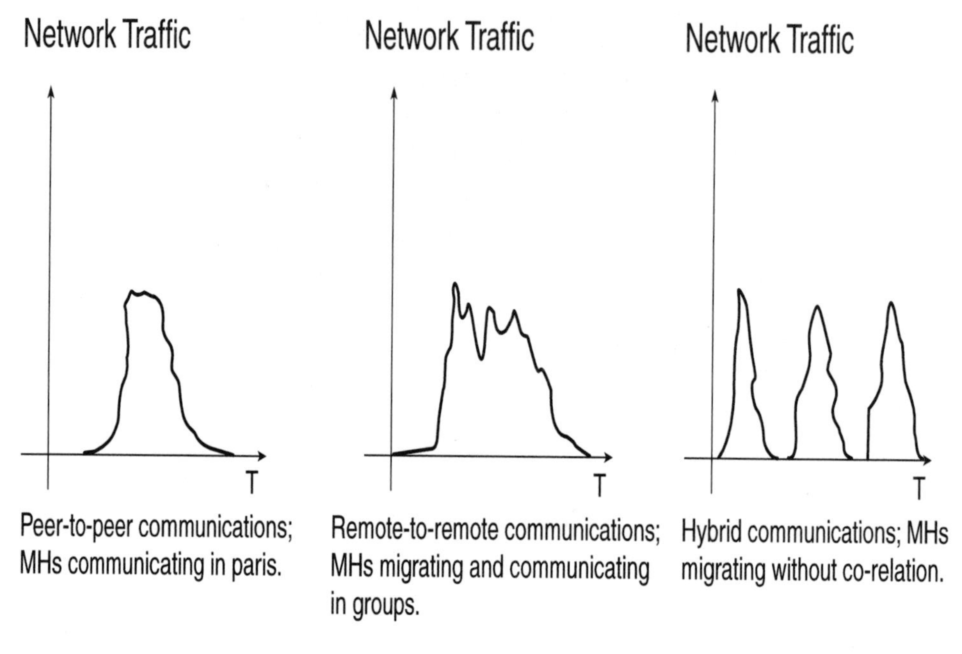

Figure 3.4. Types of traffic patterns for an ad hoc wireless environment. (MH stands for mobile host.)

3.5 Types of Ad Hoc Mobile Communications

Mobile hosts in an ad hoc mobile network can communicate with their immediate peers, that is, peer-to-peer, that are a single radio hop away. However, if three or more nodes are within range of each other (but not necessarily a single hop away from one another), then remote-to-remote mobile node communications exist. Typically, remote-to-remote communications are associated with group migrations. Different types of ad hoc communications result in different traffic characteristics, too.

3.6 Types of Mobile Host Movements

This section examines the types of mobile host movements that can affect the validity of routes directly.

3.6.1 Movements by Nodes in a Route

An ad hoc route comprises the source (SRC), destination (DEST), and/or a number of intermediate nodes (INs). Movement by any of these nodes will affect the validity of the route. An SRC node in a route has a downstream link, and when it moves out of its downstream neighbor's radio coverage range, the existing route will immediately become invalid. Hence, all downstream nodes may have to be informed so they can erase their invalid route entries. Likewise, when a DEST node moves out of the radio coverage of its upstream neighbor, the route becomes invalid. However, unlike the earlier case, here, the upsteam nodes will have to be informed so they can erase their invalid route entries. Lastly, similar to SRC and DEST node movement, any movement by an IN supporting an existing route may cause the route to be invalid.

All these movements cause many conventional distributed routing protocols to respond in sympathy with the link changes. This results in updating all the remaining nodes within the network so that consistent routing information can be maintained. However, the updating process involves broadcasting over the wireless medium, which results in wasteful bandwidth and an increase in overall network control traffic. Hence, new routing protocols are needed.

3.6.2 Movements by Subnet-Bridging Nodes

In addition to the above-mentioned mobility scenarios, any movement by a node that is performing a subnet-bridging function between two mobile subnets can fragment the mobile subnet into smaller subnets. The *property* of a mobile subnet states that if both the SRC and DEST nodes are elements of the subnet, a route or routes should exist unless the subnet is partitioned by some subnet-bridging mobile nodes. On the other hand, movements by certain nodes can also result in subnets merging, yielding bigger subnets. This is illustrated in Figure 3.5.

When mobile subnets merge to form bigger subnets, the routing algorithm may typically accept the new subnet by updating all the nodes' routing tables. This is, however, very inefficient. An efficient routing scheme should forgo this process and choose to update only the affected nodes' association tables, which is already an inherent part of the mobile node's radio data-link layer functions. Likewise, this applies to the partitioning of subnets.

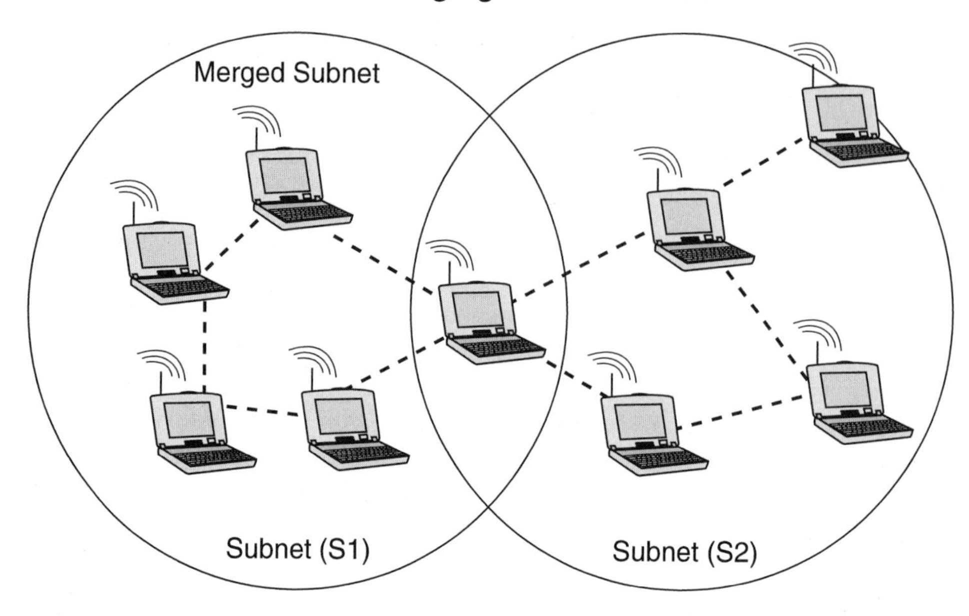

Figure 3.5. Mobile ad hoc subnets merging and fragmenting.

From an application's perspective, mobile subnets can be used to support nomadic collaborative computing, and collaboration partners can grow in size when two collaborating groups join or when new mobile users join by coming into radio range.

3.6.3 Concurrent Node Movements

In reality, concurrent movements by nodes (be they SRC, DEST, or INs) exist. Hence, rules are needed to ensure there is consistency when multiple route reconfiguration or repair processes are invoked. Such processes should ultimately converge where the most appropriate route reconfiguration is performed.

3.7 Challenges Facing Ad Hoc Mobile Networks

3.7.1 Spectrum Allocation and Purchase

Regulations regarding the use of radio spectrum are currently under the control of the FCC. Most experimental ad hoc networks are based on the ISM band. To prevent interference, ad hoc networks must operate over some form of allowed or specified spectrum range. Most microwave ovens operate in the 2.4GHz band, which can therefore interfere with wireless LAN systems. Frequency spectrum is not only tightly controlled and allocated, but it also needs to be purchased. With ad hoc networks capable of forming and deforming on-the-fly, it is not clear who should pay for this spectrum.

3.7.2 Media Access

Unlike cellular networks, there is a lack of centralized control and global synchronization in ad hoc wireless networks. Hence, TDMA and FDMA schemes are not suitable. In addition, many MAC (Media Access Control) protocols do not deal with host mobility. As such, the scheduling of frames for timely transmission to support QoS is difficult.

In ad hoc wireless networks, since the same media are shared by multiple mobile ad hoc nodes, access to the common channel must be made in a distributed fashion, through the presence of a MAC protocol. Given the fact that there are no static nodes, nodes cannot rely on a centralized coordinator. The MAC protocol must contend for access to the channel while at the same time avoiding possible collisions with neighboring nodes. The presence of mobility, hidden terminals, and exposed nodes problems must be accounted for when it comes to designing MAC protocols for ad hoc wireless networks.

3.7.3 Routing

The presence of mobility implies that links make and break often and in an indeterministic fashion. Note that the classical distributed Bellman-Ford routing algorithm is used to maintain and update routing information in a packet radio network. While distance-vector-based routing is not designed for wireless networks, it is still applicable to packet radio networks since the rate of mobility is not high. The bulky and heavy construction of these radios make them less mobile once deployed. However, as mentioned in the previous chapter, advances in microelectronics technology have enabled the construction of small, portable, and highly integrated mobile devices. Hence, ad hoc mobile networks are different from packet radio networks since nodes can move more freely, resulting in a dynamically changing topology. Existing distance-vector and link-state-based routing protocols are unable to catch up with such frequent link changes in ad hoc wireless networks, resulting in poor route convergence and very low communication throughput. Hence, new routing protocols are needed.

3.7.4 Multicasting

The explosion in the number of Internet users is partly attributed to the presence of video and audio conference tools. Such multiparty communcations are enabled through the presence of multicast routing protocols. The multicast backbone (MBone) comprises an interconnection of multicast routers that are capable of tunneling multicast packets through non-multicast routers. Some multicast protocols use a broadcast-and-prune approach to build a multicast tree rooted at the source. Others use core nodes where the multicast tree originates. All such methods rely on the fact that routers are static, and once the multicast tree is formed, tree nodes will not move. However, this is not the case in ad hoc wireless networks.

3.7.5 Energy Efficiency

Most existing network protocols do not consider power consumption an issue since they assume the presence of static hosts and routers, which are powered by mains. However, mobile devices today are mostly operated by batteries. Battery technology is still lagging behind microprocessor technology. The lifetime of an Li-ion battery today is only 2-3 hours. Such a limitation in the operating hours of a device implies the need for power conservation. In particular, for ad hoc mobile networks, mobile devices must perform both the role of an end system (where the user interacts and where user applications are executed) and that of an intermediate system (packet forwarding). Hence, forwarding packets on the behalf of others will consume power, and this can be quite significant for nodes in an ad hoc wireless network.

3.7.6 TCP Performance

TCP is an end-to-end protocol designed to provide flow and congestion control in a network. TCP is a connection-oriented protocol; hence, there is a connection establishment phase prior to data transmission. The connection is removed when data transmission is completed. In the current Internet, the network protocol (Internet Protocol, or IP) is essentially connectionless; therefore, having a connection-oriented, reliable transport protocol over an unreliable network protocol is desirable. However, TCP (Transmission Control Protocol) assumes that nodes in the route are static, and only performs flow and congestion activities at the SRC and DEST nodes.

TCP relies on measuring the round-trip time (RTT) and packet loss to conclude if congestion has occurred in the network. Unfortunately, TCP is unable to distinguish the presence of mobility and network congestion. Mobility by nodes in a connection can result in packet loss and long RTT. Hence, some enhancements or changes are needed to ensure that the transport protocol performs properly without affecting the end-to-end communication throughput.

3.7.7 Service Location, Provision, and Access

While protocols are important for the proper operation of an ad hoc wireless network, service location, provision, and access are equally important. Should we continue to assume that the traditional client/server RPC (remote procedure call) paradigm is appropriate for ad hoc networks? Ad hoc networks comprise heterogeneous devices and machines and not every one is capable of being a server. The concept of a client initiating task requests to a server for execution and awaiting results to be returned may not be attractive due to limitations in bandwidth and power. Perhaps the concept of remote programming as used in mobile agents is more applicable since this can reduce the interactions exchanged between the client and server over the wireless media. Also, how can a mobile device access a remote service in an ad hoc network? How can a device that is well-equipped advertise its desire to provide services to the rest of the members in the network? All these issues demand research.

3.7.8 Security & Privacy

Ad hoc networks are intranets and they remain as intranets unless there is connectivity to the Internet. Such confined communications have already isolated attackers who are not local in the area. Note that this is not the case for wired and wireless-last hop users. Through neighbor identity authenication, a user can know if neighboring users are friendly or hostile. Information sent in an ad hoc route can be protected in some way but since multiple nodes are involved, the relaying of packets has to be authenicated by recognizing the originator of the paket and the flow ID or label.

3.8 Conclusions

This chapter has provided insights into today's ad hoc wireless networks. It was revealed that ad hoc mobile devices can be highly mobile, powerful (in terms of computation and memory capacity), and heterogeneous. Ad hoc wireless networks are not limited to a homogeneous group of laptops. They can consist of a group of minute sensors, communicating to each other and gathering environmental information.

Ad hoc networks need to possess self-organizing characteristics, and they must perform routing and packet-forwarding functions. The topology of an ad hoc wireless network is dynamically changing since devices are not tied down to specific locations over time. The fact that nodes are not static implies that centralized media access is not entirely applicable. Routing protocols in ad hoc networks need to deal with the mobility of nodes and constraints in power and bandwidth. Multicasting in ad hoc wireless networks needs to be efficient since using flooding will only result in a massive consumption of available bandwidth and degrade battery life. Ad hoc devices rely on batteries to operate; hence, any inefficiency in communication protocols can drastically shorten the uptime of these devices. Current transport protocols are not designed for wireless ad hoc networks. In particular, TCP is an end-to-end protocol that cannot distinguish the presence of mobility from congestion. Finally, new methods are needed to faciliate *service location*, *provision*, and *access* in ad hoc wireless networks. In summary, many challenging technical issues have arisen that demand our attention and investigation.

Chapter 4

AD HOC WIRELESS MEDIA ACCESS PROTOCOLS

4.1 Introduction

Wireless media can be shared and any nodes can transmit at any point in time. This could result in possible contention over the common channel. If channel access is probabilistic, then the resultant attainable throughput is low. In an ad hoc wireless network, every node can possibly move, and hence, there is no fixed network node to act as the central controller.

A MAC protocol is a set of rules or procedures to allow the efficient use of a shared medium, such as wireless. We define a *node* as any host that is trying to access the medium. The *sender* is a node that is attempting to transmit over the medium. The *receiver* is a node that is the recipient of the current transmission. The MAC protocol is concerned with per-link communications, not end-to-end.

4.1.1 Synchronous MAC Protocols

In synchronous MAC protocols, all nodes are synchronized to the same time. This is achieved by a timer master broadcasting a regular beacon. All nodes listen for this beacon and synchronize their clocks to the master's time. Central coordination is, therefore, needed to synchronize time events.

4.1.2 Asynchronous MAC Protocols

In asynchronous MAC protocols, nodes do not necessarily follow the same time. A more distributed control mechanism is used to coordinate channel access. As such, access to the channel tends to be contention-based.

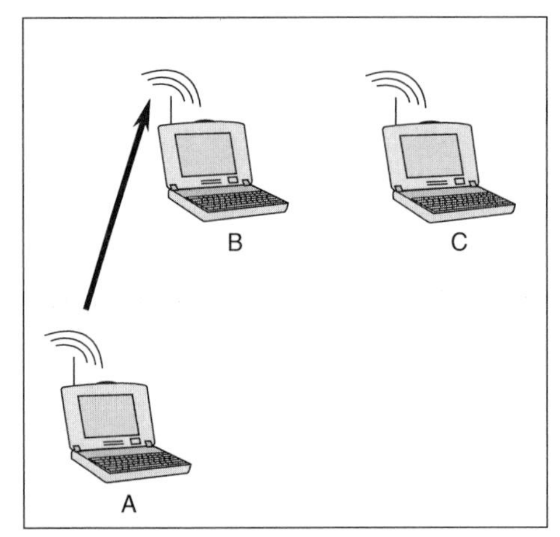

1. A transmits to B. (C does not hear this.)

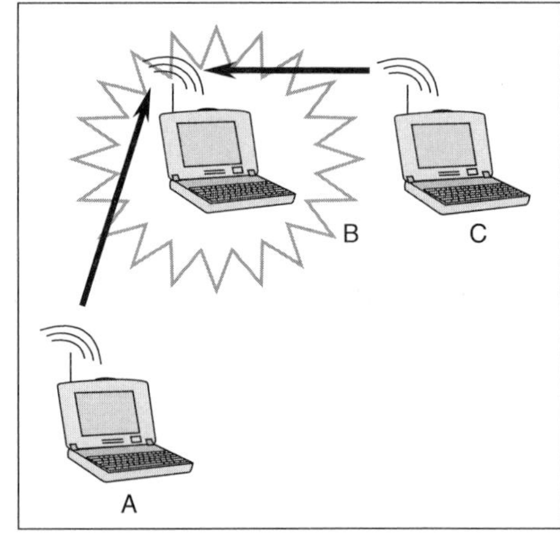

2. C transmits to B ... Collision!

Figure 4.1. The hidden terminal problem.

4.2 Problems in Ad Hoc Channel Access

4.2.1 Hidden Terminal Problem

This is a well-known problem found in contention-based protocols, such as pure ALOHA, slotted ALOHA, CSMA, IEEE 802.11, etc. Two nodes are said to be hidden from one another (out of signal range) when both attempt to send information to the same receiving node, resulting in a collision of data at the receiver node (see Figure 4.1).

To avoid collision, all of the receiver's neighboring nodes need to be informed

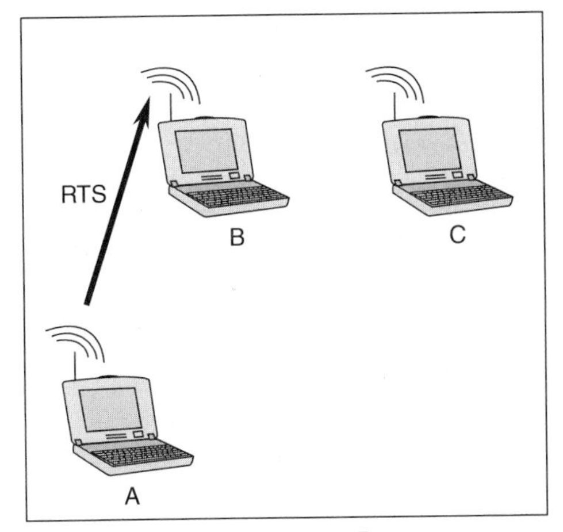

1. A transmits request (RTS) to B.

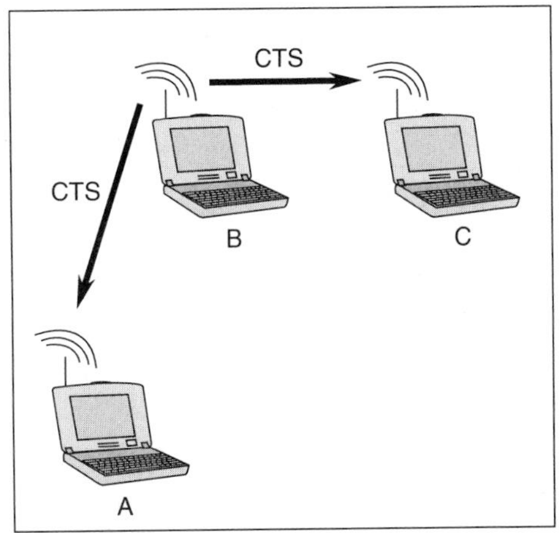

2. B replies that the channel is clear (CTS).
Both A & C overhear the broadcast.

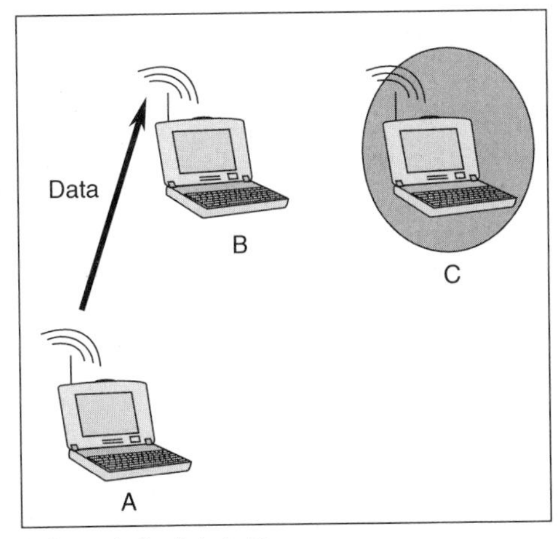

3. A sends its data to B.
C is blocked from transmitting.

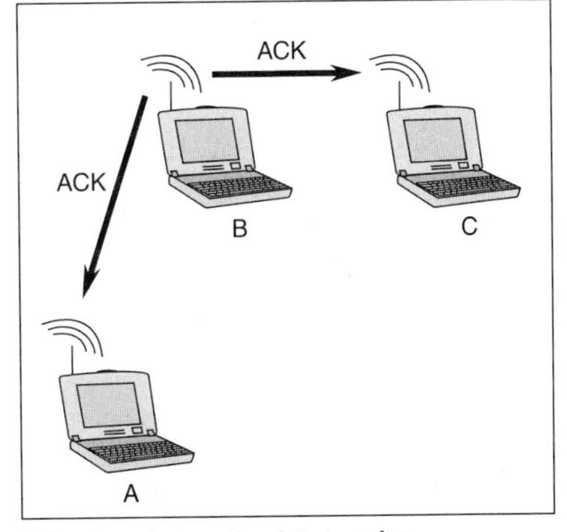

4. B acknowledges the data transfer.

Figure 4.2. Using an RTS-CTS handshake to resolve hidden node problems.

that the channel will be occupied. This can be achieved by reserving the channel using control messages, that is, using a handshake protocol. An RTS (Request To Send) message can be used by a node to indicate its wish to transmit data. The receiving node can allow this transmission by sending a grant using the CTS (Clear To Send) message. Because of the broadcast nature of these messages, all neighbors of the sender and receiver will be informed that the medium will be busy, thus preventing them from transmitting and avoiding collision. Figure 4.2 illustrates the concept of the RTS-CTS approach.

4.2.2 Shortcomings of the RTS-CTS Solution

The RTS-CTS method is not a perfect solution to the hidden terminal problem. There will be cases when collisions occur and the RTS and CTS control messages are sent by different nodes. As shown in Figure 4.3, node B is granting a CTS to the RTS sent by node A. However, this collides with the RTS sent by node D at node C. Node D is the hidden terminal from node B. Because node D does not receive the expected CTS from node C, it retransmits the RTS. When node A receives the CTS, it is not aware of any collision at node C and hence it proceeds with a data transmission to node B. Unfortunately, in this scenario, it collides with the CTS sent by node C in response to node D's RTS.

Another problematic scenario occurs when multiple CTS messages are granted to different neighboring nodes, causing collisions. As shown in Figure 4.4, two nodes are sending RTS messages to different nodes at different points in time. Node A sends an RTS to node B. When node B is returning a CTS message back to node A, node C sends an RTS message to node B. Because node C cannot hear the CTS sent by node B while it is transmitting an RTS to node D, node C is unaware of the communication between nodes A and B. Node D proceeds to grant the CTS message to node C. Since both nodes A and C are granted transmission, a collision will occur when both start sending data.

4.2.3 Exposed Node Problem

Overhearing a data transmission from neighboring nodes can inhibit one node from transmitting to other nodes. This is known as the exposed node problem. An exposed node is a node in range of the transmitter, but out of range of the receiver. This is illustrated in Figure 4.5.

A solution to the exposed node problem is the use of separate control and data channels or the use of directional antennas. The former will be discussed in the PAMAS and DBTMA sections. Figure 4.6a shows that a mobile node using an omni-directional antenna can result in several surrounding nodes being "exposed," thus prohibiting them from communicating with other nodes. This lowers network availability and system throughput. Alternatively, if directional antennas are employed, this problem can be mitigated. As shown in Figure 4.6b, node C can continue communicating with the receiving palm pilot device without impacting the communication between nodes A and B. The directivity provides spatial and connectivity isolation not found in omni-directional antenna systems.

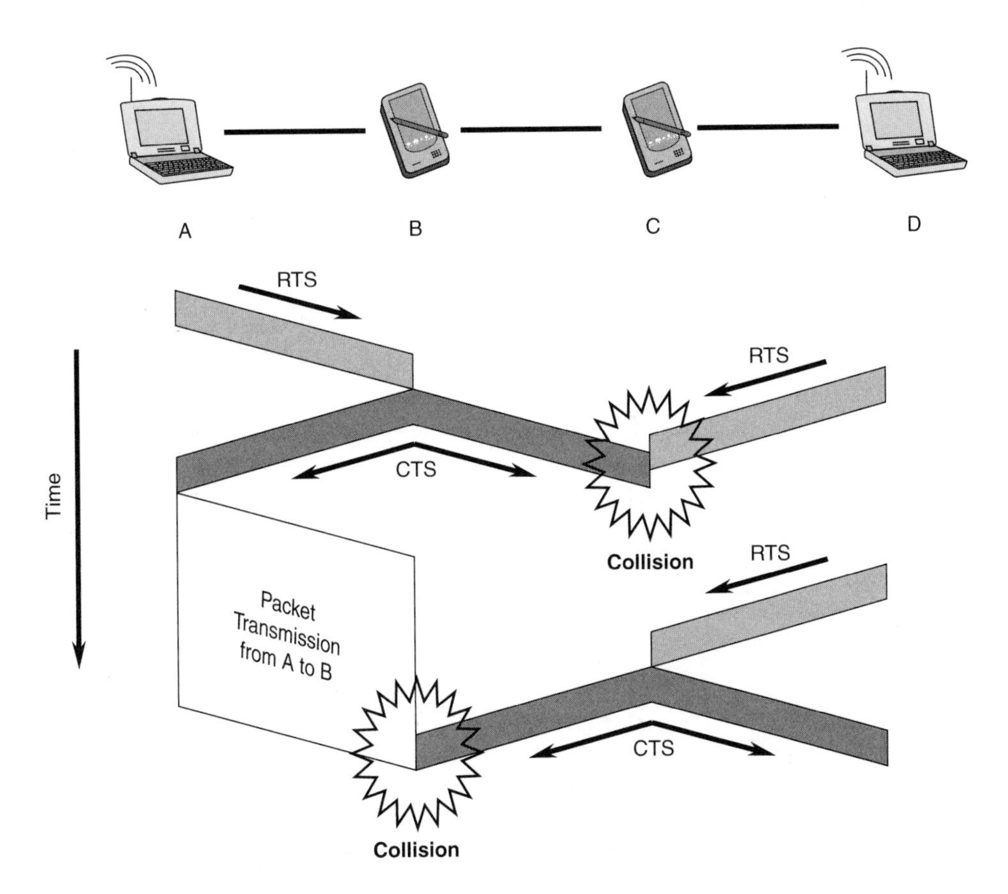

Figure 4.3. The incompleteness of the RTS-CTS method.

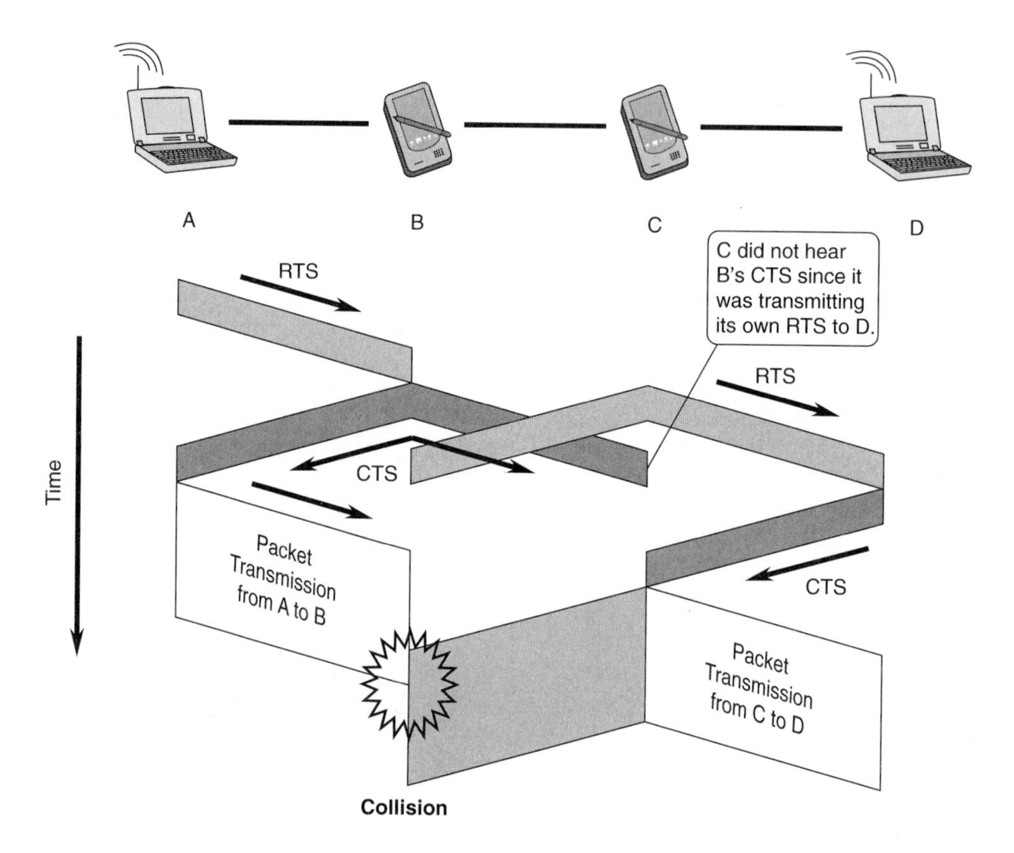

Figure 4.4. Another illustration of the RTS-CTS problem.

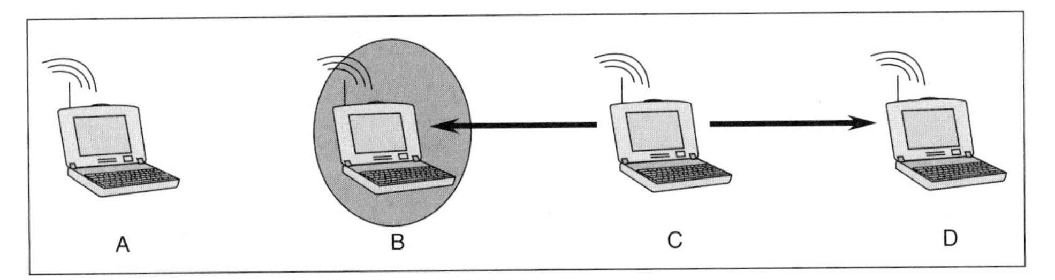

- C is transmitting to D.
- B overhears this, and is blocked.
- B wants to transmit to A, but is being blocked by C.
- Wasted bandwidth!

Figure 4.5. The exposed node problem.

4.3 Receiver-Initiated MAC Protocols

While a MAC protocol can be categorized as synchronous or asynchronous in operation, it can also be distinguished by who initiates a communication request. As shown in Figure 4.7, the receiver (node B) first has to contact the sender (node A), informing the sender that it is ready to receive data. This is a form of polling, as the receiver has no way of knowing for sure if the sender indeed has data to send.

This is also a passive form of initiation since the sender does not have to initiate a request. In addition, there is only one control message used, compared to the RTS-CTS approach. We will discuss MACA-BI, an example of a receiver-initiated MAC protocol, in a subsequent section.

4.4 Sender-Initiated MAC Protocols

Contrary to receiver-initiated MAC protocols, sender-initiated MAC protocols require the sender to initiate communications by informing the receiver that it has data to send. Examples of these sender-initiated protocols include MACA (Multiple Access with Collision Avoidance), MACAW (MACA with Acknowledgment), and FAMA (Floor Acquisition Multiple Access).

As shown in Figure 4.8, node A sends an explicit RTS message to node B (the receiver) to express its desire to communicate. Node B can then reply if it is willing to receive data from node A. If positive, it returns a CTS message to node A. Node A then subsequently proceeds to send data.

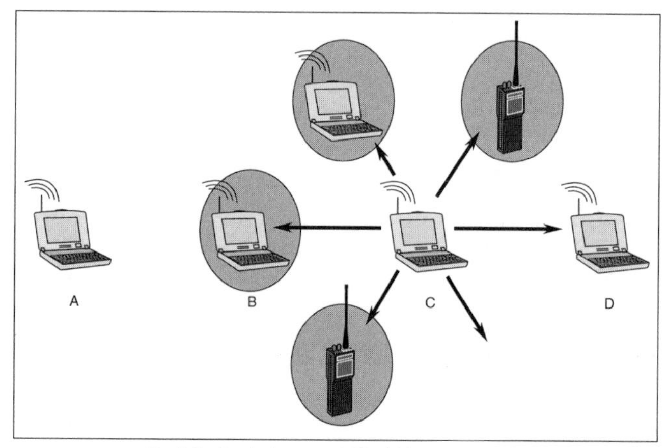

Omni-directional antenna used. All neighbors are exposed.

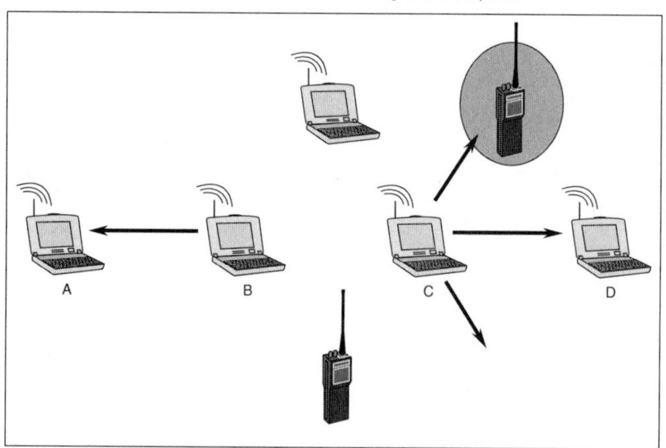

Directional antenna reduces this problem! B is not blocked from sending to A.

Figure 4.6. Using a directional antenna to resolve the exposed node problem.

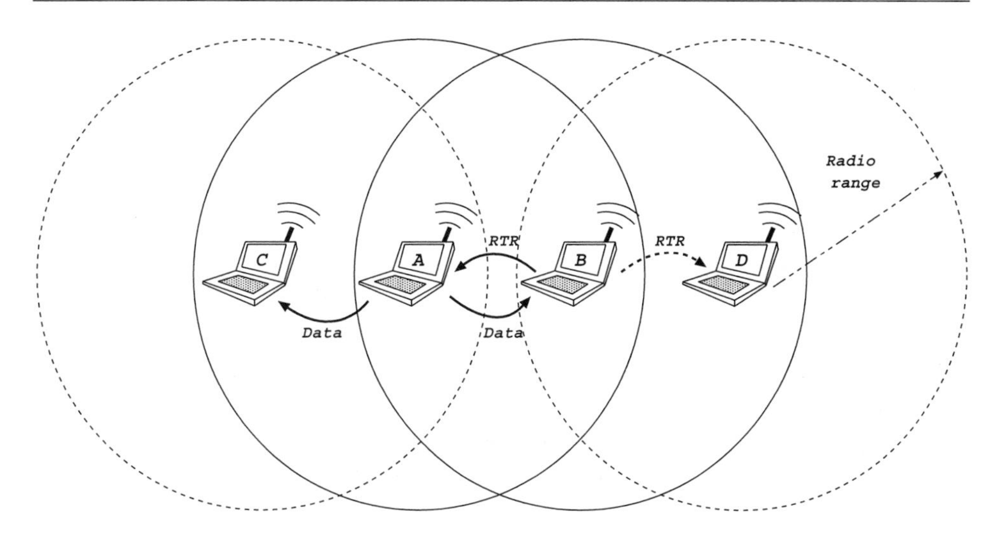

Figure 4.7. An illustration of the receiver-initiated MAC protocol.

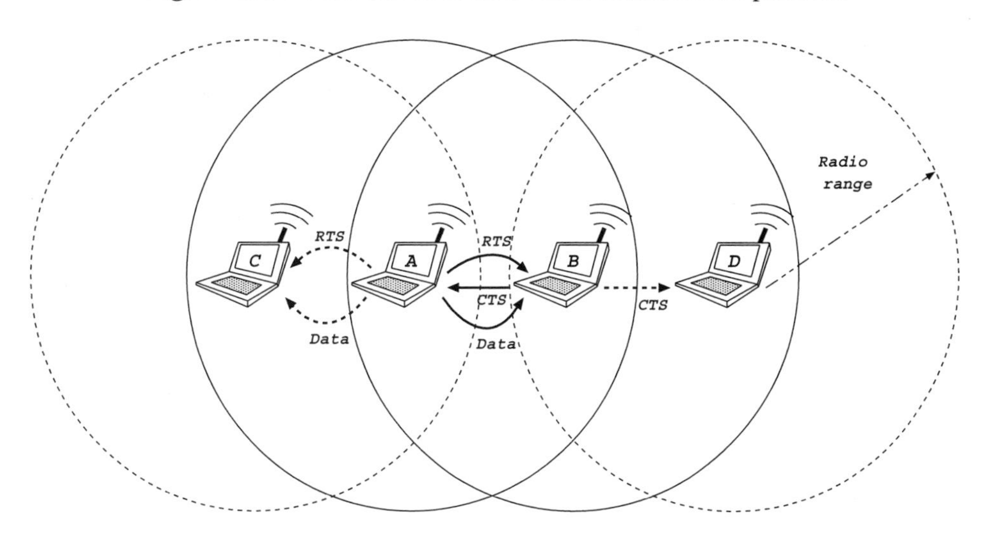

Figure 4.8. Sender-initiated MAC protocols.

4.5 Existing Ad Hoc MAC Protocols

4.5.1 Multiple Access with Collision Avoidance (MACA)

MACA was originally suggested by Phil Karn for amateur packet radio networks. MACA aims to create usable, ad hoc, single-frequency networks. MACA was proposed to resolve the hidden terminal and exposed node problems. It also has the ability to perform per-packet transmitter power control, which can increase the carrying capacity of a packet radio.

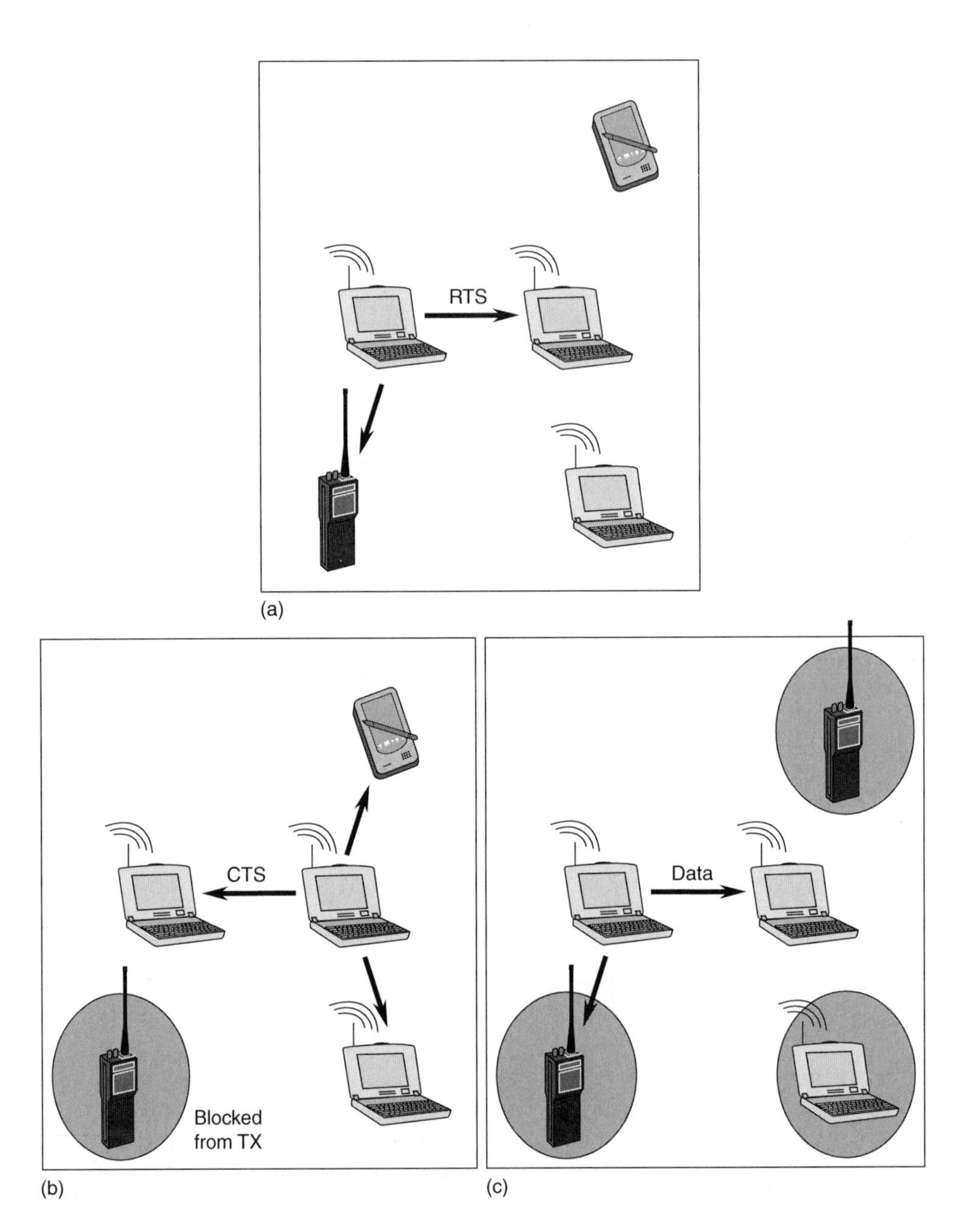

Figure 4.9. An illustration of the control handshake used in MACA.

As shown in Figure 4.9, MACA uses a three-way handshake, RTS-CTS-Data. The sender first sends an RTS to the receiver to reserve the channel. This blocks the sender's neighboring nodes from transmitting. The receiver then sends a CTS to the sender to grant transmission. This results in blocking the receiver's neighboring nodes from transmitting, thereby avoiding collision. The sender can now proceed with data transmission.

MACA has power control features incorporated. The key characteristic of MACA is that it *inhibits a transmitter when a CTS packet is overheard* so as to temporarily limit power output when a CTS packet is overheard. This allows geographic reuse of channels. For example, if node A has been sending data packets to node B, after some time, A would know how much power it needs to reach B. If node A overhears node B's response to an RTS (i.e., a CTS) from a downstream node C, A need not remain completely silent during this time. By lowering its transmission power from the level used to reach node B, node A can communicate with other neighboring nodes (without interfering with node B) during that time with a lower power.

Collisions do occur in MACA, especially during the RTS-CTS phase. There is no carrier sensing in MACA. Each mobile host basically adds a random amount of time to the minimum interval required to wait after overhearing an RTS or CTS control message. In MACA, the *slot time* is the duration of an RTS packet. If two or more stations transmit an RTS concurrently, resulting in a collision, these stations will wait for a randomly chosen interval and try again, doubling the average interval on every attempt. The station that wins the competition will receive a CTS from its responder, thereby blocking other stations to allow the data communication session to proceed.

Compared to CSMA, MACA reduces the chances of data packet collisions. Since control messages (RTS and CTS) are much smaller in size compared to data packets, the chances of collision are also smaller. To summarize: The RTS-CTS period is the *contention period*; after that, data transmission occurs over a *contention-free* period.

4.5.2 MACA-BI (By Invitation)

A shift from the classic three-way handshake MAC protocol is MACA-BI (By Invitation). Invented by Fabrizio Talucci, MACA-BI uses only a two-way handshake, as shown in Figure 4.10. There is no RTS. Instead, the CTS message is renamed as RTR (Ready To Receive). In MACA-BI, a node cannot transmit data unless it has received an invitation from the receiver. Note that the receiver node does not necessarily know that the source has data to transmit. Hence, the receiver needs to predict if indeed the node has data to transmit to it. The timeliness of the invitation will, therefore, affect communication performance.

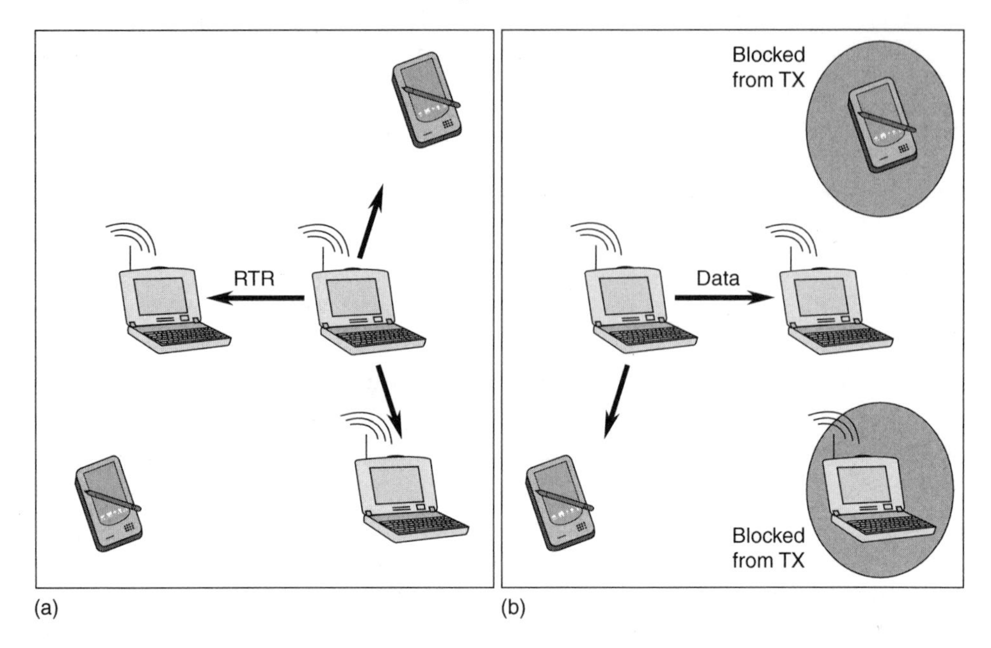

Figure 4.10. An illustration of MACA-BI control handshake.

The author suggested the estimation of packet queue length and arrival rate at the source to regulate the transmission of invitations. One possible way to accomplish this is to piggyback such information into each data packet so that the receiver is aware of the transmitter's backlog. Hence, for constant bit rate (CBR) traffic, the efficiency of MACA-BI will be high since the prediction scheme will work fine. However, for bursty traffic, MACA-BI performance will be no better than MACA.

To enhance the communication performance of MACA-BI under non-stationary traffic situations, a node may still transmit an RTS if the transmitter's queue length or packet delay exceeds a certain acceptable threshold before an RTR is issued. This means that MACA-BI now reverts back to MACA. Figure 4.11 clearly shows the differences between MACA and MACA-BI.

In summary, MACA-BI results in reduced transmit/receive turn around time. Every transmission should be delayed by the transmit-to-receive turn-around time (i.e., up to 25 microseconds) to allow the previous transmitter to switch to receive mode. Because MACA-BI only uses a single control message, this turn around limitation is reduced. Additionally, MACA functionality is preserved in MACA-BI. This includes the collision-free feature of MACA. In fact, MACA-BI is less

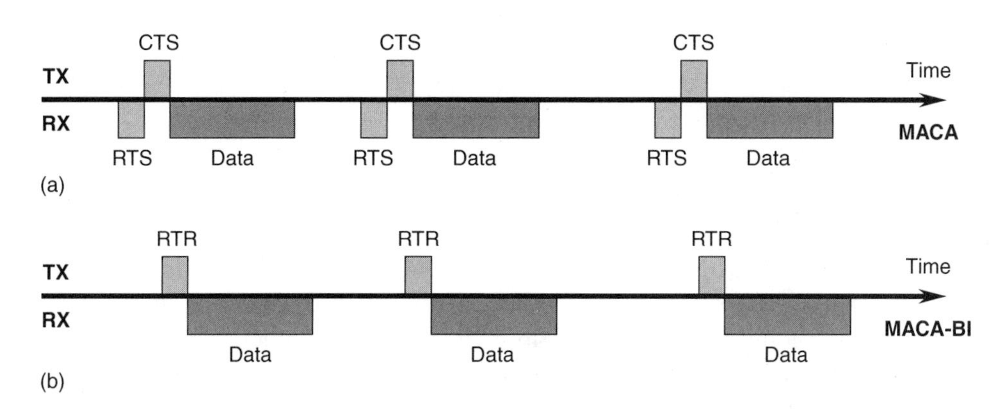

Figure 4.11. Comparing MACA and MACA-BI MAC protocols.

likely to suffer from control packet collision since it uses half as many control packets as MACA.

4.5.3 Power-Aware Multi-Access Protocol with Signaling (PAMAS)

The Power-Aware Multi-Access Protocol with Signalling for ad hoc networks (PAMAS)[17] is based on the MACA protocol with the addition of a separate signalling channel. RTS-CTS dialogue exchanges occur over this channel. PAMAS conserves battery power by selectively powering off nodes that are not actively transmitting or receiving packets.

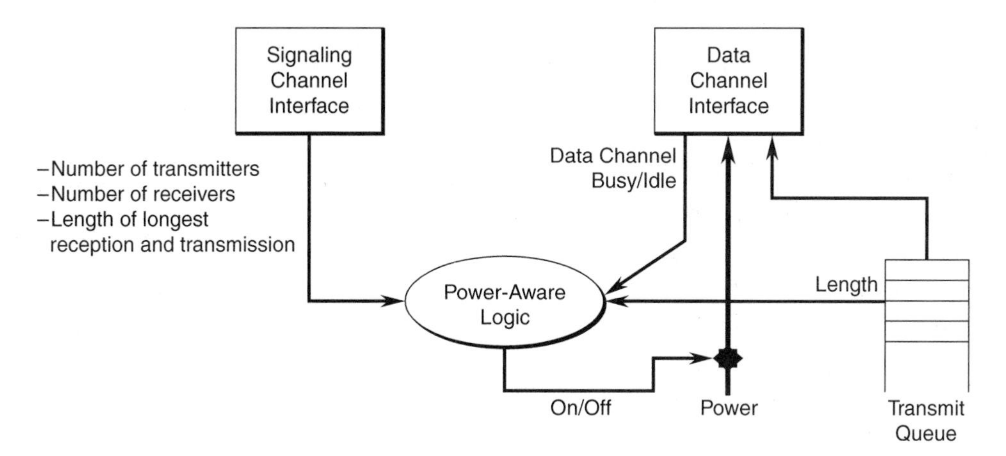

Figure 4.12. PAMAS MAC interfaces and power-aware logic.

When node A wishes to send data to node B, it first sends an RTS message and enters the wait-for-CTS state. If the CTS message does not arrive, node A enters into a binary exponential backoff state and attempts to send the RTS later. If a CTS message does arrive, node A enters the transmit-data state. As for the receiver node B, upon sending the CTS message, it enters the await-data state. If the data indeed begins to arrive, node B will start transmitting a busy tone over the signalling channel and enter the receive-data state.

In PAMAS, nodes are required to shut themselves off if they are overhearing other transmissions not directed to them. In addition, each node makes an independent decision about whether to power off its transceiver. The conditions that force a node to power off include:

a. If a node has no packets to transmit, it should power off if one of its neighboring nodes is transmitting.

b. If a node has packets to transmit, but at least one of the neighboring nodes is transmitting and another is receiving, then it should power off its transceiver.

Note that when a node's transceiver is powered off, it can neither receive nor transmit packets. The author suggested the use of probing to detect when a node should appropriately power up. The duration of power-off is critical since it affects delay and throughput performance. In addition, the author assumes that a node can selectively power down only its data interface and leave the signalling interface power on, as shown in Figure 4.13. Note that some special circuit design may be required to ensure that such control is supported in a radio transceiver.

4.5.4 Dual Busy Tone Multiple Access (DBTMA)

The use of a busy tone [18] was first proposed by Professor Fouad Tobagi from Stanford University. He proposed Busy Tone Multiple Access (BTMA) to solve the hidden terminal problem. However, BTMA relies on a wireless last-hop network architecture, where a centralized base station serves multiple mobile hosts. When the base station is receiving packets from a specific mobile host, it sends out a busy tone signal to all other nodes within its radio cell. Hence, hidden terminals sense the busy tone and refrain from transmitting.

Zygmunt Haas from Cornell applied this concept further for use in ad hoc wireless networks. In DBTMA (Dual Busy Tone Multiple Access)[19] [20], two out-of-band busy tones are used to notify neighboring nodes of any on-going transmission. In addition, the single shared channel is further split into data and control channels. Data packets are sent over the data channel, while control packets (such as RTS and

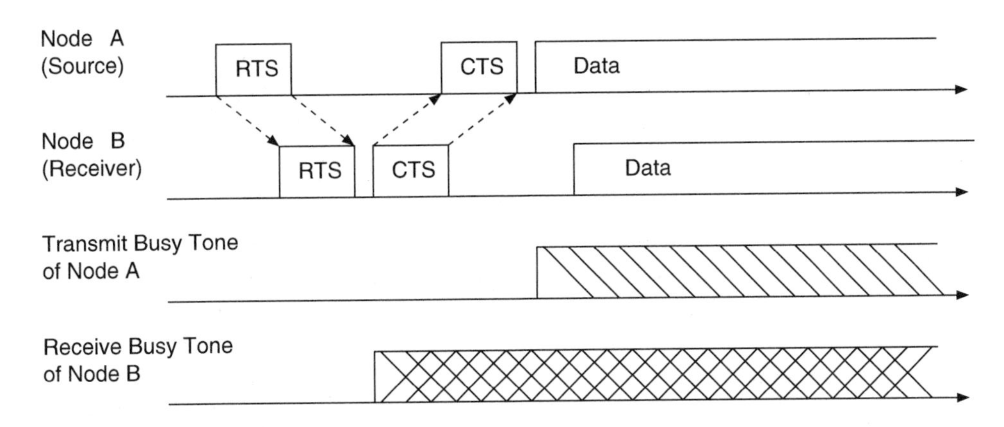

Figure 4.13. The principle of Dual Busy Tone Multiple Access (DBTMA).

CTS) are sent over the control channel. Specifically, one busy tone signifies transmit busy, while another signifies receive busy. These two busy tones are spatially separated in frequency to avoid interference.

The principle of operation of DBTMA is relatively simple. An ad hoc node wishing to transmit first sends out an RTS message. When the receiver node receives this message and decides that it is ready and willing to accept the data, it sends out a receive busy tone message followed by a CTS message. All neighboring nodes that hear the receive busy tone are prohibited from transmitting. Upon receiving the CTS message, the source node sends out a transmit busy tone message to surrounding nodes prior to data transmission. Neighboring nodes that hear the transmit busy tone, are prohibited from transmitting and will ignore any transmission received. Analytical and simulation work performed by the proposers reveal superior performance compared to pure RTS-CTS MAC schemes.

4.6 MARCH: Media Access with Reduced Handshake

In the explanation of PAMAS, it was revealed that overhearing could result in unnecessary power consumption. In addition, we learned that most radios today use omni-directional antennnas. A new MAC protocol that exploits the overhearing characteristic associated with an ad hoc mobile network employing omni-directional antenna is presented in [21]. This is the Media Access with Reduced Handshake (MARCH) protocol.

As shown in Figure 4.14, MARCH improves communication throughput in wireless multi-hop ad hoc networks by reducing the amount of control overhead.

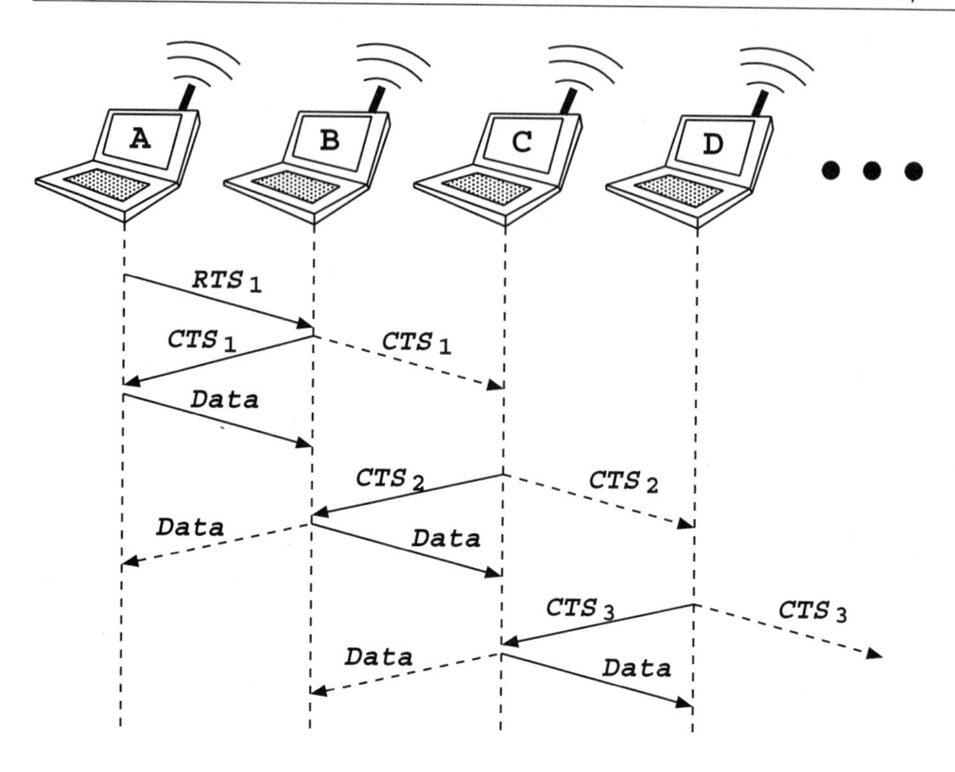

Figure 4.14. The handshake mechanism used in the MARCH protocol.

Unlike other receiver-initiated protocols, MARCH operates without resorting to any traffic prediction. In fact, MARCH exploits the broadcast characteristic of omni-directional antennas to reduce the number of required handshakes. In MARCH, a node has knowledge of data packet arrivals at its neighboring nodes from the overheard CTS packets. It can then initiate an *invitation* for data to be relayed.

Figure 4.14 reveals the broadcast characteristics of omni-directional antennas. Node C will receive the CTS_1 message sent by node B. This characteristic implies that the overheard CTS_1 packet can also be used to convey the infomation of a data packet arrival at node B to node C. Subsequently, after the data packet has been received by node B, node C can invite node B to forward that data via the CTS_2 packet. Hence, the RTS_2 packet can be suppressed here.

The figure also shows that the RTS-CTS handshake is now reduced to a single CTS (CTS-only) handshake after the first hop, and the reduction in the control overhead is hence a function of route length. For an ad hoc route of l hops[1], the

[1]There are $l - 1$ intermediary nodes between the source and destination.

number of handshakes needed to send a data packet from the source to the destination is $2l$ in MACA, l in MACA-BI, but only $(l + 1)$ in MARCH. Hence, if l is large, MARCH will have a very similar number of handshakes as in MACA-BI.

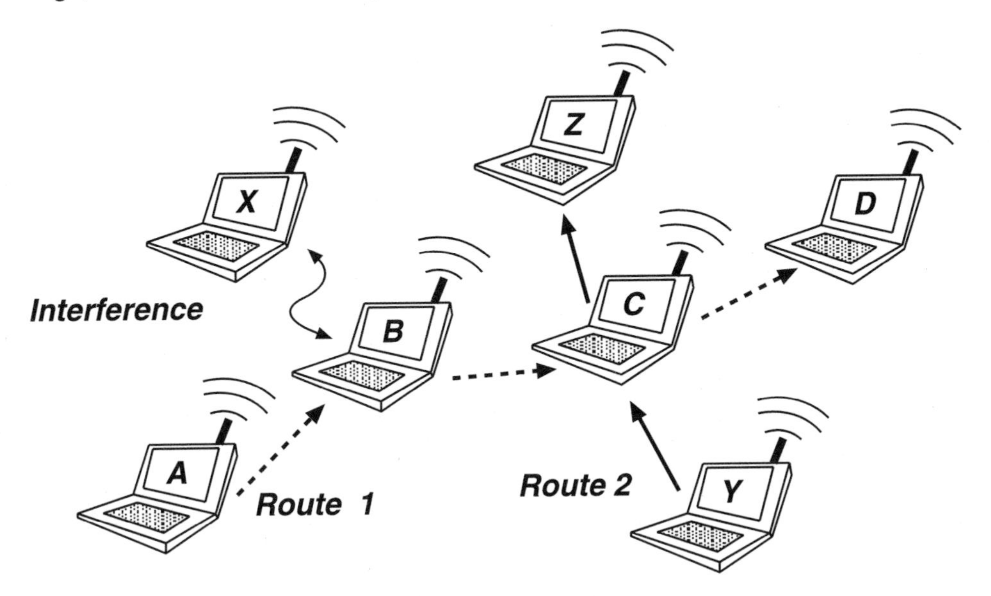

Figure 4.15. Presence of two overlapping routes in an ad hoc mobile network.

MARCH can be viewed as a request-first, pull-later protocol since the subsequent nodes in the path just need to send invitations to pull the data toward the destination node. The RTS-CTS message in MARCH contains:

- The MAC addresses of the sender and receiver
- The route identification number (RT_{ID})

Figure 4.15 further illustrates the operation of MARCH. There are two routes that intersect at a common node C. Route 1 consists of nodes A, B, C, and D, while route 2 consists of nodes Y, C, and Z. These routes can be established through an appropriate routing protocol. In associativity-based routing, the desired route is discovered and activated by the creation of routing entries at the nodes in the route. During route discovery, packets are broadcast. When it comes to data communication, nodes in the route use an underlying protocol such as MARCH.

To begin data transmission in route 1, an RTS_1 packet is first sent from node A to node B. If this packet is successfully received by node B, node B will reply with a CTS_1 packet to grant the data transmission. Meanwhile, CTS_1 is also overheard by node C. According to the MAC address and RT_{ID}, node C knows that the packet was sent by its upstream node B in route 1. A timer T_w is then invoked at node C. T_w is set to a value long enough for node B to receive and process the new data packet. Upon timeout, if the channel is free, node C sends a CTS_2 packet to node B to acquire the data packet. Similarly, node D will overhear the CTS_2 sent by node C and will subsequently invite node C to relay the data packet via CTS_3 once its T_w timer expires. Note that node Z, the downstream node of node C in route 2, will also overhear the CTS_2 packet. To avoid node Z misinterpreting it and initiating an unnecessary CTS-only handshake, the RT_{ID} method is applied.

In MARCH, the MAC layer has access to tables that maintain information on the routes the node participates in, as well as its upstream and downstream neighbors in those routes. This, however, does not mean that MARCH performs any Layer 2 routing. In fact, it just consults those tables to determine if it should respond to a control message (RTS-CTS) particularly to a certain route. If a node is going to initiate a CTS-only handshake in route 1, it encloses its RT_{ID} for that route in the CTS packet. Hence, by checking the RT_{ID} in CTS_2, only node D will react appropriately to the control packet, and may initiate a CTS-only handshake after its T_w timer expires. In short, MARCH does not participate in routing, nor does it make any decisions about the data packets exchanged. It does, however, provide a high-speed, fast relay of MAC frames from the source of a route to the destination without performing an RTS-CTS handshake at every hop. Performance comparison of MARCH versus other MAC protocols can be found in [21].

4.7 Conclusions

In this chapter, we discussed MAC protocols and the issues behind supporting channel access for ad hoc wireless networks. The problems of hidden terminals and exposed nodes are still prevalent in ad hoc networks, and they must be considered in the design of MAC protocols. However, not many of the newly proposed protocols address channel access under the presence of mobility. There is also a need to address high-performance ad hoc mobile systems.

Chapter 5

OVERVIEW OF AD HOC ROUTING PROTOCOLS

Since the advent of the DARPA packet radio networks in the early 1970s, numerous protocols have been developed for ad hoc mobile networks. Such protocols must deal with the typical limitations of these networks, which include high power consumption, low bandwidth, and high error rates. As shown in Figure 5.1, these routing protocols may generally be categorized as: (a) table-driven, and (b) source-initiated on-demand-driven. Solid lines in this figure represent direct descendants, while dotted lines depict logical descendants. Despite being designed for the same type of underlying network, the characteristics of each of these routing protocols are quite distinct. The following sections describe the protocols and categorize them according to their characteristics.

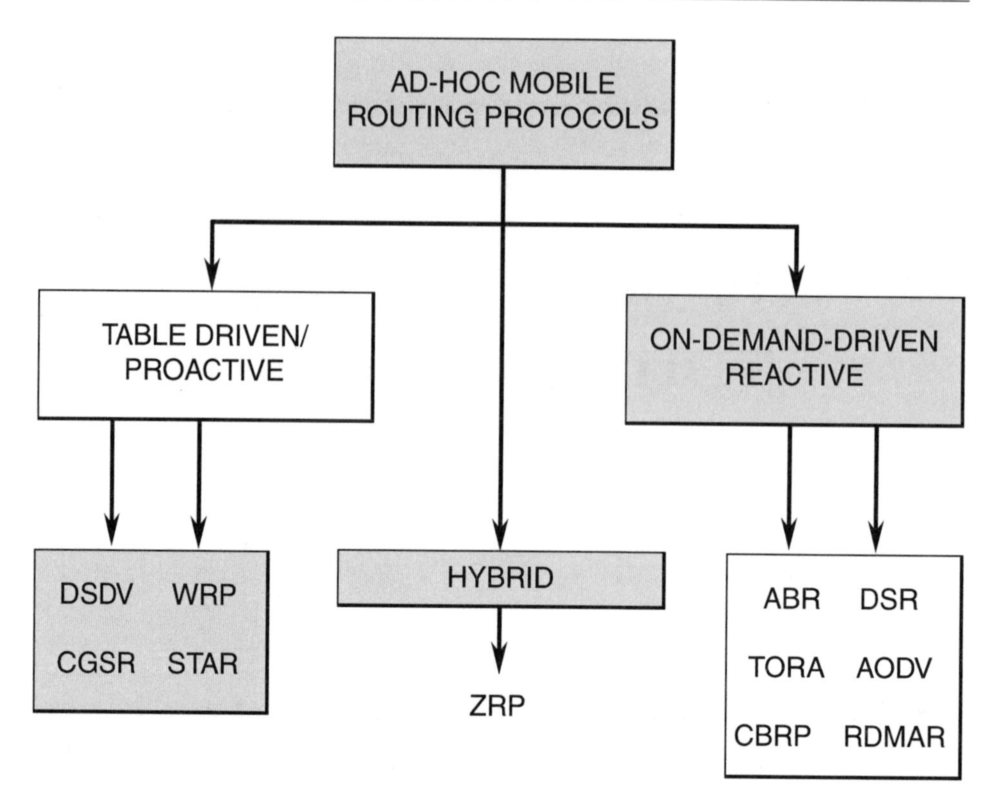

Figure 5.1. Categorization of ad hoc routing protocols.

5.1 Table-Driven Approaches

Table-driven routing protocols attempt to maintain consistent, up-to-date routing information from each node to every other node in the network. These protocols require each node to maintain one or more tables to store routing information, and they respond to changes in network topology by propagating route updates throughout the network to maintain a consistent network view. The areas where they differ are the number of necessary routing-related tables and the methods by which changes in network structure are broadcast.

5.2 Destination Sequenced Distance Vector (DSDV)

Destination Sequenced Distance Vector (DSDV) [22] routing is a table-driven routing protocol based on the classical distributed Bellman-Ford routing algorithm. The improvement made here is the avoidance of routing loops in a mobile network of routers. Each node in the mobile network maintains a routing table in which all of the possible destinations within the non-partitioned network and the number of routing hops (in this case, number of radio hops) to each destination are recorded. Hence, routing information is always made readily available, regardless of whether the source node requires a route or not.

A sequence numbering system is used to allow mobile hosts to distinguish stale routes from new ones. Routing table updates are sent periodically throughout the network to maintain table consistency. This can, therefore, generate a lot of control traffic in the network, rendering an inefficient utilization of network resources. To alleviate this problem, DSDV uses two types of route update packets. The first is known as *full dump*. This type of packet carries all available routing information and can require multiple network protocol data units (NPDUs). During periods of occasional movement, these packets are transmitted infrequently. Smaller *incremental* packets are used to relay only information that has changed since the last full dump.

New route broadcasts will contain the address of the destination node, the number of hops to reach the destination, the sequence number of the information received regarding the destination, as well as a new sequence number unique to the broadcast. The route labeled with the most recent sequence number (in increasing order) is always used. In the event that two updates have the same sequence number, the route with the smaller hop count is used.

5.3 Wireless Routing Protocol (WRP)

A novel part of WRP[23][24] stems from the way in which it achieves loop freedom. In WRP, routing nodes communicate the distance and second-to-last hop information for each destination in the wireless network. WRP belongs to the class of path-finding algorithms with an important exception. It avoids the *count-to-infinity* problem by forcing each node to perform consistency checks of predecessor information reported by all its neighbors. This ultimately eliminates looping situations and provides faster route convergence when a link failure event occurs.

In WRP, nodes learn about the existence of their neighbors from the receipt of acknowledgments and other messages. If a node is not sending packets, it must send a HELLO message within a specified time period to ensure connectivity information is properly reflected. Otherwise, the lack of messages from the node can indicate the failure of that wireless link and this may cause a false alarm. When a mobile receives a HELLO message from a new node, that new node information is added to the mobile's routing table, and the mobile sends the new node a copy of its routing table information.

WRP must maintain four tables, namely: (a) distance table, (b) routing table, (c) link-cost table, and (d) message retransmission list (MRL) table. The distance table indicates the number of hops between a node and its destination. The routing table indicates the next-hop node. The link-cost table reflects the delay associated with a particular link. The MRL contains the sequence number of the update message, a retransmission counter, an acknowledgment required flag vector, and a list of the updates sent in the update message. The MRL records which updates in an update message need to be retransmitted and which neighbors should acknowledge the retransmission.

To ensure that routing information is accurate, mobiles send update messages periodically to their neighbors. The update message contains a list of updates (the destination, the distance to destination, the predecessor of the destination), as well as a list of responses indicating which mobile should acknowledge the update. A mobile sends update messages after processing updates from neighbors or when a link change is detected. In the event of a link failure, nodes detecting the failure will send update messages to their neighbors, and those neighbors will modify their distance table entries and check for new possible paths through other nodes.

5.4 Cluster Switch Gateway Routing (CSGR)

Cluster Switch Gateway Routing (CSGR) [25] is a table-driven-based routing protocol where mobile nodes are grouped into clusters and each cluster has a cluster head. This grouping also introduces a form of hierarchy. A cluster head can control a group of ad hoc hosts, and clustering provides a framework for code separation (among clusters), channel access, routing, and bandwidth allocation. To elect a cluster head, a distributed cluster head selection algorithm is used. Although using a cluster head allows some form of control and coordination, it does impose a reliance from other nodes within the cluster. When a cluster head moves away, another new cluster head must be selected. This can be problematic if a cluster head is changing frequently and nodes will be spending a lot of time converging to a cluster head instead of forwarding data toward their intended destinations. To avoid invoking cluster head reselection every time the cluster membership changes, a least cluster change (LCC) algorithm is introduced. Using the LCC algorithm, cluster heads only change when two cluster heads come into contact, or when a node moves out of all other cluster heads.

CSGR uses DSDV as the underlying routing scheme. However, it modifies DSDV by using a hierarchical cluster-head-to-gateway routing approach to route

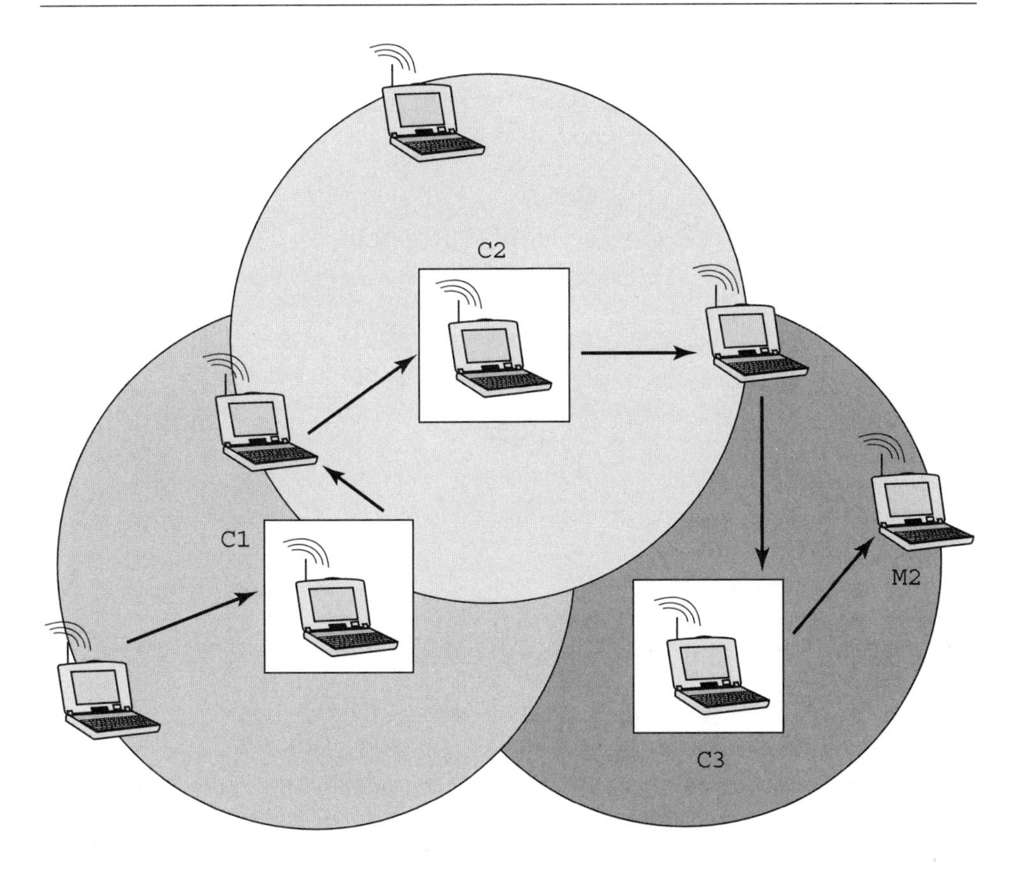

Figure 5.2. A CSGR path is constrained to cluster heads in the path and hence the route so formed is not necessarily optimal.

traffic from source to destination. Gateway nodes are nodes that are within communication range of two or more cluster heads. As shown in Figure 5.2, a packet sent by a node is first routed to its cluster head, and then the packet is routed from a cluster head to a gateway to another cluster head, and so on until the cluster head of the destination node is reached. The packet is then transmitted to the destination.

In CSGR, each node must keep a *cluster member table*, where it stores the destination cluster head for each mobile host in the network. These cluster member tables are broadcast periodically by each node using the DSDV protocol. Nodes receiving this update will refresh their cluster member tables to ensure their validity. In addition to the cluster member table, each node must also maintain a routing table, which is used to determine the next hop to reach the destination. On receiving a packet, a node will consult its cluster member and routing tables to determine the nearest cluster head along the route to the destination. The node then checks its routing table to determine the next hop node to use reach the cluster head. In summary, updates are needed for both routing and cluster member tables in CSGR.

5.5 Source-Initiated On-Demand Approaches

An approach that is different from table-driven routing is source-initiated on-demand routing. This type of routing creates routes only when desired by the source node. When a node requires a route to a destination, it initiates a route discovery process within the network. This process is completed once a route is found or all possible route permutations have been examined. Once a route has been discovered and established, it is maintained by some form of route maintenance procedure until either the destination becomes inaccessible along every path from the source or the route is no longer desired.

5.6 Ad Hoc On-Demand Distance Vector Routing (AODV)

The Ad Hoc On-Demand Distance Vector (AODV) routing protocol described in [26] builds on the DSDV algorithm previously described. AODV is an improvement on DSDV because it typically minimizes the number of required broadcasts by creating routes on an on-demand basis, as opposed to maintaining a complete list of routes as in the DSDV algorithm. The authors of AODV classify it as a *pure on-demand route acquisition* system, as nodes that are not on a selected path do not maintain routing information or participate in routing table exchanges. However, the distance vector function of AODV seems to be absent.

When a source node wants to send a message to some destination node and does not already have a valid route to that destination, it initiates a *path discovery* process to locate the other node. It broadcasts a route request (RREQ) packet to its neighbors, which then forward the request to their neighbors, and so on, until either the destination or an intermediate node with a "fresh enough" route to the destination is located (Figure 5.3a). AODV uses destination sequence numbers to ensure that all routes are loop-free and contain the most recent route information. Each node maintains its own sequence number, as well as a broadcast ID. The broadcast ID is incremented for every RREQ the node initiates, and together with the node's IP address, uniquely identifies an RREQ. Along with its own sequence number and the broadcast ID, the source node includes in the RREQ the most recent sequence number it has for the destination. Intermediate nodes can reply to the RREQ only if they have a route to the destination whose corresponding destination

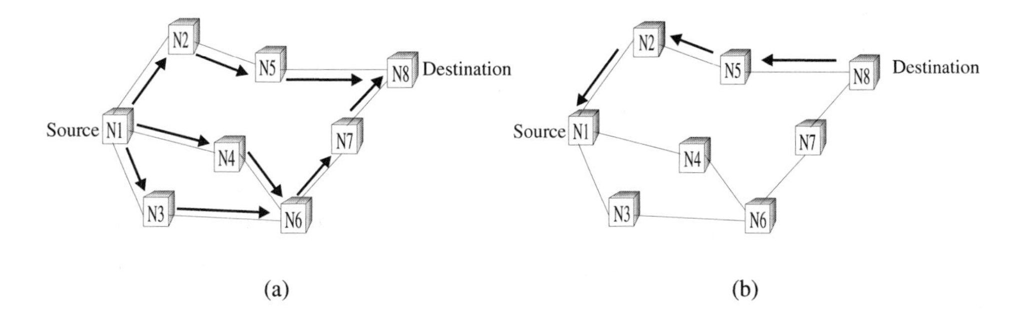

Figure 5.3. AODV route discovery.

sequence number is greater than or equal to that contained in the RREQ.

During the process of forwarding the RREQ, intermediate nodes record in their route tables the addresses of neighbors from which the first copy of the broadcast packet was received, thereby establishing a reverse path. If additional copies of the same RREQ are later received, these packets are silently discarded[1]. Once the RREQ has reached the destination or an intermediate node with a "fresh enough" route, the destination/intermediate node responds by unicasting a route reply (RREP) packet back to the neighbor from which it first received the RREQ (Figure 5.3b). As the RREP is routed back along the reverse path, nodes along this path set up forward route entries in their route tables that point to the node from which the RREP came[2] These forward route entries indicate the active forward route. Associated with each route entry is a route timer, which causes the deletion of the entry if it is not used within a specified lifetime. Because an RREP is forwarded along the path established by an RREQ, AODV only supports the use of symmetric links.

In AODV, routes are maintained as follows: If a source node moves, it has to reinitiate the route discovery protocol to find a new route to the destination. If a node along the route moves, its upstream neighbor notices the move and propagates a *link failure notification* message (an RREP with an infinite metric) to each of its active upstream neighbors to inform them of the erasure of that part of the route [26]. These nodes in turn propagate the *link failure notification* to their upstream neighbors, and so on, until the source node is reached. The source node may then choose to re-initiate route discovery for that destination if a route is still desired[3].

[1]Hence, consideration for other better routes is absent in AODV.

[2]This approached was first proposed in Associativity Based Routing (ABR) in 1994 and protected by the ABR US patent.

[3]Hence, AODV does not exploit the fast and localized partial route recovery method as proposed

An additional aspect of the protocol is the use of *hello* messages which are periodic local broadcasts made by a node to inform each mobile node of other nodes in its neighborhood. Hello messages can be used to maintain the local connectivity of a node. However, the use of hello messages is not required. Nodes listen for retransmissions of data packets to ensure that the next hop is still within reach. If such a retransmission is not heard, the node may use any one of a number of techniques, including the reception of hello messages. Hello messages may list the other nodes from which a mobile has heard, thereby yielding a greater knowledge of network connectivity.

5.7 Dynamic Source Routing (DSR)

The Dynamic Source Routing (DSR) protocol presented in [27] is an on-demand routing protocol that is based on the concept of source routing. Mobile nodes are required to maintain route caches that contain the source routes of which the mobile is aware. Entries in the route cache are continually updated as new routes are learned.

The protocol consists of two major phases: (a) route discovery, and (b) route maintenance. When a mobile node has a packet to send to some destination, it first consults its route cache to determine whether it already has a route to the destination. If it has an unexpired route to the destination, it will use this route to send the packet. On the other hand, if the node does not have such a route, it initiates route discovery by broadcasting a *route request* packet. This *route request* message contains the address of the destination, along with the source node's address and a unique identification number. Each node receiving the packet checks whether it knows of a route to the destination. If it does not, it adds its own address to the *route record* of the packet and then forwards the packet along its outgoing links. To limit the number of *route requests* propagated on the outgoing links of a node, a mobile only forwards the *route request* if the request has not yet been seen by the mobile and if the mobile's address has not already appeared in the *route record.* A *route reply* is generated when either the *route request* reaches the destination itself, or when it reaches an intermediate node that contains in its route cache an unexpired route to the destination [27]. By the time the packet reaches either the destination or such an intermediate node, it contains a *route record* yielding the sequence of hops taken.

Figure 5.4a illustrates the formation of the *route record* as the *route request* message propagates through the network. If the node generating the *route reply*

in Associativity Based Routing.

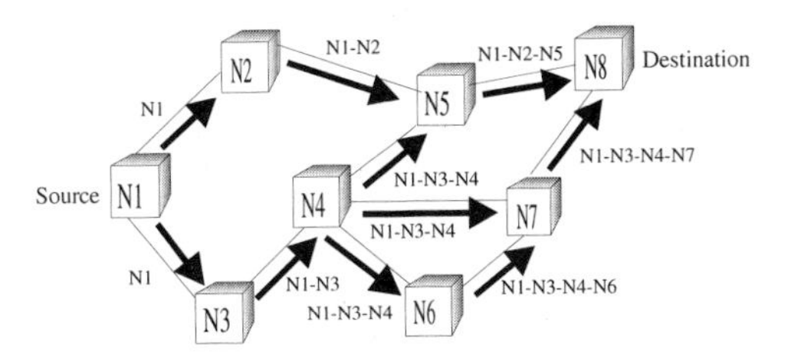

(a) Building of the route record during route discovery

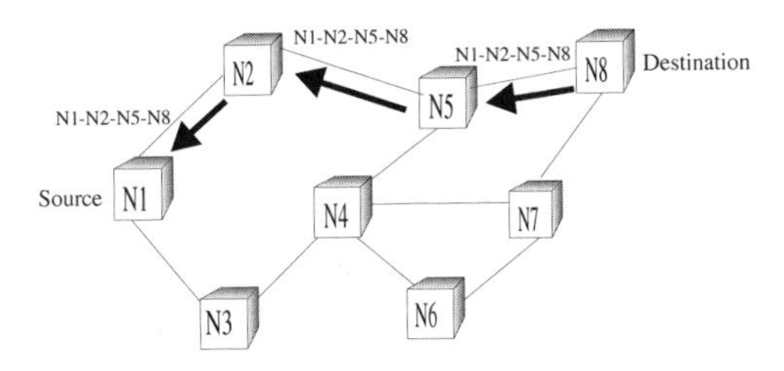

(b) Propagation of the route reply with the route record

Figure 5.4. The creation of source route record in DSR. Note that each data packet sent by the source has to contain complete source route information. This adds to the overhead per data packet.

is the destination, it places the *route record* contained in the *route request* into the *route reply*. If the responding node is an intermediate node, it appends its cached route to the *route record* and then generates the *route reply*. To return the *route reply*, the responding node must have a route to the initiator. If it has a route to the initiator in its route cache, it may use that route. Otherwise, if symmetric links are supported, the node may reverse the route in the *route record*. If symmetric links are not supported, the node may initiate its own route discovery and piggyback the *route reply* on a new *route request*[4]. Figure 5.4b shows the transmission of the *route reply* with its associated *route record* back to the source node.

Route maintenance is accomplished through the use of *route error* packets and acknowledgments. *Route error* packets are generated at a node when the data link

[4]This is a recent addition and is not part of the original DSR protocol. See RODA [28] on handling asymmetric links in ad hoc wireless networks.

layer encounters a fatal transmission problem. The source is always interrupted when a route is truncated. When a *route error* packet is received, the hop in error is removed from the node's route cache and all routes containing the hop are truncated at that point. In addition to *route error* messages, acknowledgments are used to verify the correct operation of the route links. Such acknowledgments include passive acknowledgments[5] (where a mobile is able to hear the next hop forwarding the packet along the route).

5.8 Temporally Ordered Routing Algorithm (TORA)

TORA is a highly adaptive, loop-free, distributed routing algorithm based on the concept of link reversal [29]. TORA is proposed to operate in a highly dynamic mobile networking environment. It is source-initiated and provides multiple routes for any desired source/destination pair. The key design concept of TORA is the localization of control messages to a very small set of nodes near the occurrence of a toplogical change. To accomplish this, nodes need to maintain routing information about adjacent (one-hop) nodes. The protocol performs three basic functions: (a) route creation, (b) route maintainence, and (c) route erasure.

During the route creation and maintenance phases, nodes use a "height" metric to establish a DAG (directed acyclic graph) rooted at the destination. Thereafter, links are assigned a direction (upstream or downstream) based on the relative height metric of neighboring nodes, as shown in Figure 5.5a. This process of establishing a DAG is similar to the query/reply process proposed in LMR (Lightweight Mobile Routing) [30].

In times of node mobility, the DAG route is broken and route maintenance is necessary to re-establish a DAG rooted at the same destination. As shown in Figure 5.5b, upon failure of the last downstream link, a node generates a new reference level, which results in the propagation of that reference level by neighboring nodes, effectively coordinating a structured reaction to the failure. Links are reversed to reflect the change in adapting to the new reference level. This has the same effect as reversing the direction of one or more links when a node has no downstream links. Timing is an important factor for TORA because the "height" metric is dependent on the logical time of a link failure; TORA assumes all nodes have synchronized clocks (accomplished via an external time source such as the global positioning system (GPS)). Hence, it is unclear if TORA would function properly in an environment where GPS is not available or is not reliable (i.e., positional errors are

[5] Again, this is a subsequent proposal. The original DSR protocol introduced in 1996 does not have this feature, unlike Associativity-Based Routing.

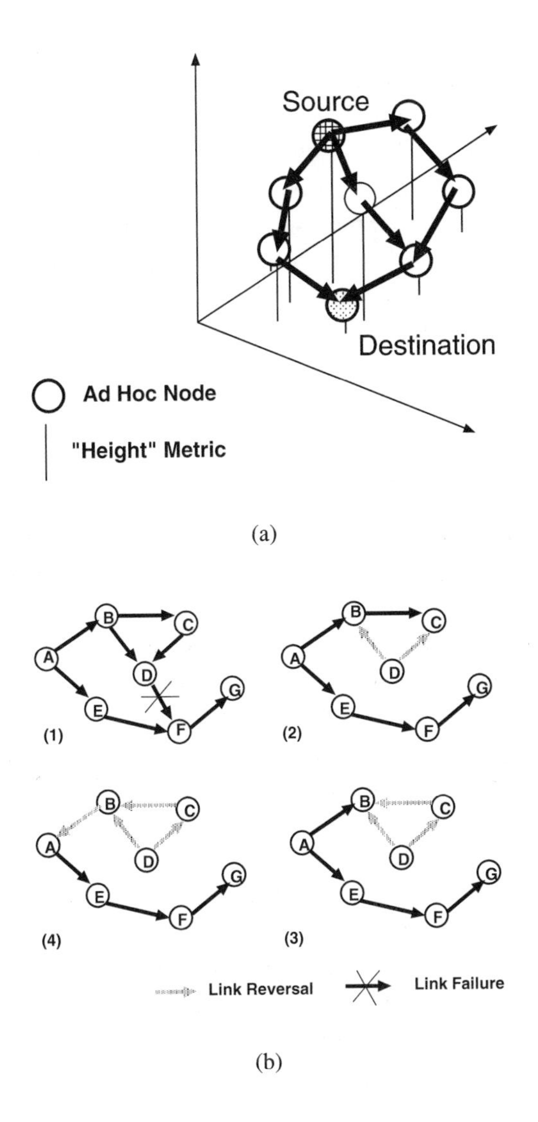

Figure 5.5. (a) Route creation (showing link direction assignment), and (b) route maintenance (showing link reversal phenonemon) in TORA.

large).

TORA's metric is a quintuple comprising five elements, namely: (a) logical time of link failure, (b) the unique ID of the node that defined the new reference level, (c) a reflection indicator bit, (d) a propagation ordering parameter, and (e) the unique ID of the node. The first three elements collectively represent the reference level. A new reference level is defined each time a node loses its last downstream link due to a link failure. TORA's route erasure phase essentially involves flooding a broadcast "clear packet" (CLR) throughout the network to erase invalid routes.

In TORA, there is a potential for oscillations to occur, especially when multiple sets of coordinating nodes are concurrently detecting partitions, erasing routes, and building new routes based on each other. Because TORA uses internodal coordination, its instability problem is similar to the "count-to-infinity" problem in distance-vector routing protocols, except that such oscillations are temporary and route convergence will ultimately occur.

5.9 Signal Stability Routing (SSR)

Another on-demand protocol is the Signal Stability-Based Adaptive Routing (SSR) protocol presented in [31]. SSR is a descendent of Associativity-Based Routing (ABR), and ABR predates SSR. Similar to ABR, SSR selects routes based on the *signal strength* between nodes and on a node's location stability. Hence, SSR offers little novelty. SSR route selection criteria has the effect of choosing routes that have "stronger" connectivities. SSR can be divided into two cooperative protocols: (a) the Dynamic Routing Protocol (DRP), and (b) the Static Routing Protocol (SRP).

The DRP is responsible for the maintenance of the signal stability table (SST) and the routing table (RT). The SST records the signal strength of neighboring nodes, which is obtained by periodic beacons from the link layer of each neighboring node. The signal strength may be recorded as either a strong or weak channel. All transmissions are received by, and processed in, the DRP. After updating all appropriate table entries, the DRP passes a received packet to the SRP.

The SRP processes packets by passing them up the stack if they are the intended receivers, or looking up their destination in the RT and then forwarding them if they are not. If no entry is found in the RT for the destination, a *route-search* process is initiated to find a route. Route requests are propagated throughout the network, but are only forwarded to the next hop if they are received over strong channels and have not been previously processed (to prevent looping). The destination chooses the first arriving *route-search* packet to send back because it is most probable that the packet arrived over the shortest and/or least congested path[6]. The DRP then reverses the selected route and sends a *route-reply* message back to the initiator. The DRP of the nodes along the path update their RTs accordingly.

[6]SSR route selection method, therefore, does not consider other QoS factors and the first arriving route search packet may not be long-lived.

The assumption made in SSR is that *route search* packets arriving at the destination might have chosen the path of strongest signal stability, as the packets are dropped at a node if they have arrived over a weak channel. If there is no *route-reply* message received at the source within a specific timeout period, the source changes the *PREF* field in the header to indicate that weak channels are acceptable, as these may be the only links over which the packet can be propagated.

When a failed link is detected within the network, intermediate nodes will send an error message to the source indicating which channel has failed. The source then initiates another *route search* process to find a new path to the destination. Thereafter, the source sends an erase message to notify all nodes of the broken link. Hence, similar to DSR and AODV, the source is always interrupted when the route is broken due to mobility, which is undesirable.

5.10 Location-Aided Routing (LAR)

Compared to other ad hoc routing schemes, LAR utilizes location information (via, say, the GPS) to improve the performance of ad hoc wireless networks.

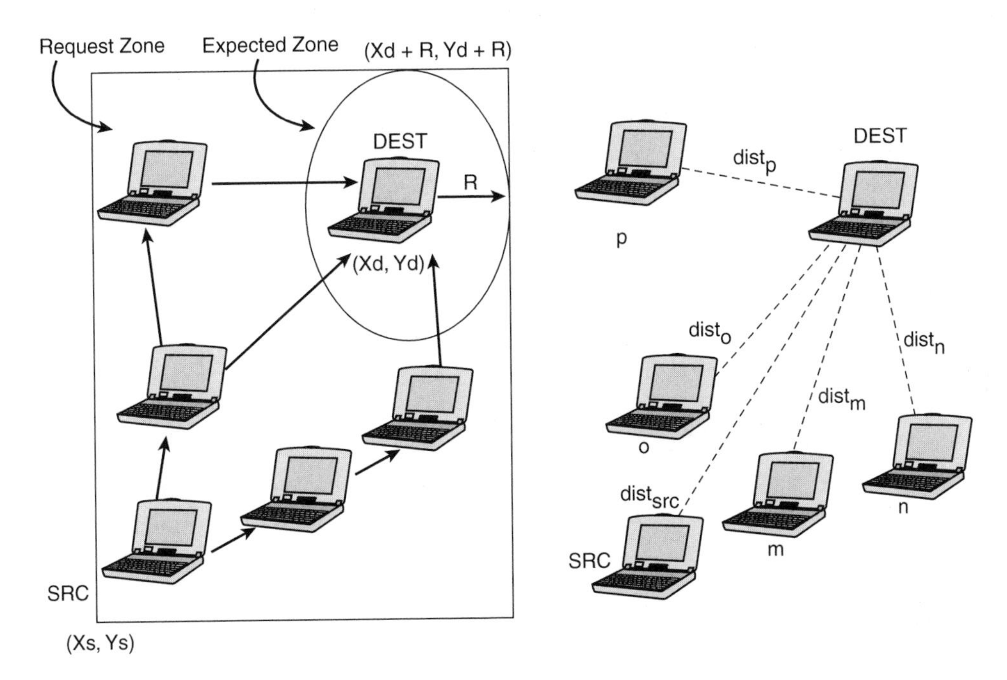

Figure 5.6. (a) Concepts of request zone and expected zone in LAR, and (b) consideration of route physical distance.

LAR limits the search for a new route to a smaller request zone, thereby resulting in reduced signaling traffic. LAR defines two concepts: (a) *expected zone*, and (b) *request zone*. However, LAR makes several assumptions. First, it assumes that the sender has advanced knowledge of the destination location and velocity. Based on the location and velocity, the expected zone can be defined. The request zone, however, is the smallest rectangle that includes the location of the sender and the expected zone. This is illustrated in Figure 5.6. With LAR, the sender explicitly specifies the request zone in its route request message. Nodes that receive the request message but do not fall within the request zone discard the packet. This, therefore, acts as a limiting boundary on the propagation of the route request message. This is known as the LAR1 scheme. The other scheme is to consider a route that has a shorter physical distance from the source to the destination node.

5.10.1 Shortcomings of LAR

The use of positioning information has been widely used for location tracking and navigation. For example, GPS is commonly used in cars to assist drivers in navigating through foreign land. It is also used for militrary and defense operations. However, GPS availability is not yet worldwide and positional information does come with deviation, i.e., errors. In addition, it is not necessarily true that all future mobile devices will be equipped with GPS receivers. Heterogeneous devices will exist and hence some devices will not have GPS receivers. In such situations, location based routing will suffer and fail to operate. Positional errors can also cause problems in routing. In addition, without considering signal stength, power life, and connectivity information, communication performance will suffer if data are routed based on location information alone. Lastly, prior and advance information about the positional information of the destination node may not be readily available at the source.

5.11 Power-Aware Routing (PAR)

Mobile ad hoc devices can take many different forms and some will have higher power demands than others. In Power-Aware Routing (PAR) [32], battery life is taken as the routing metric. PAR advocates for:

- Minimizing the energy consumed per packet
- Maximizing the time before the network is partitioned
- Minimizing the variance in node power levels

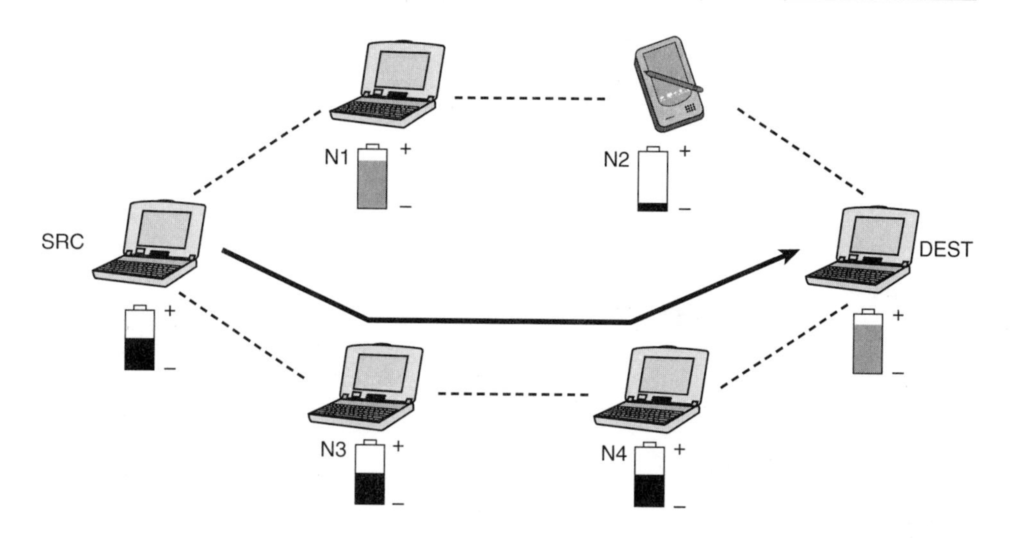

Figure 5.7. In power-aware routing, remaining battery life is used as the primary routing metric.

- Minimizing the cost per packet
- Minimizing the maximium node cost

The protocol selects routes that have a longer overall battery life. As shown in Figure 5.7, there are two possible routes from SRC to DEST. However, $\mathcal{N}2$ in route { SRC, $\mathcal{N}1$, $\mathcal{N}2$, DEST } is powering down sooner compared to $\mathcal{N}3$ and $\mathcal{N}4$ in route { SRC, $\mathcal{N}3$, $\mathcal{N}4$, DEST } . Hence, it makes sense to select the latter route.

5.12 Zone Routing Protocol (ZRP)

The Zone Routing Protocol (ZRP) [33] is a hybrid protocol incorporating the merits of *on-demand* and *proactive* routing protocols. A routing zone is similar to a cluster with the exception that every node acts as a cluster head and a member of other clusters. Zones can overlap. Each node specifies a zone radius in terms of radio hops. The size of a chosen zone can, therefore, affect ad hoc communication performance.

In ZRP, a routing zone comprises a few mobile ad hoc nodes within one, two, or more hops away from where the central node is formed. Within this zone, a table-driven-based routing protocol is used. This implies that route updates are performed for nodes within the node. Each node, therefore, has a route to all other

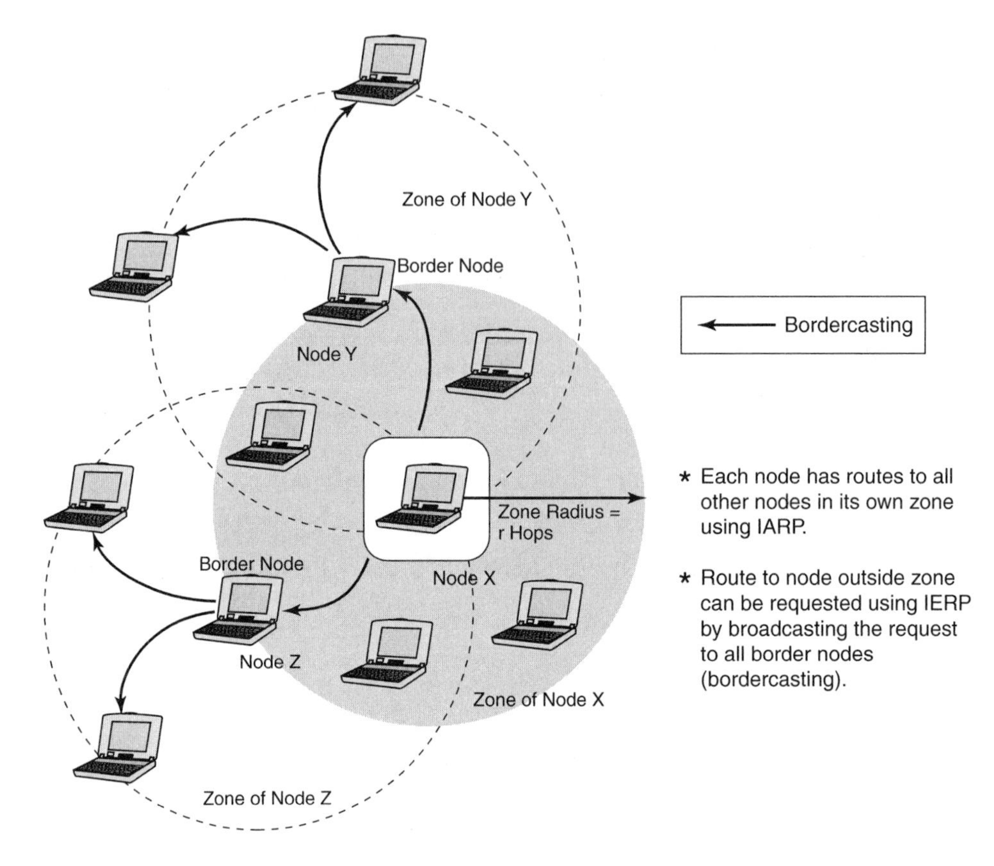

Figure 5.8. Hybrid Approach - Zone routing protocol.

nodes within the zone. If the destination node resides outside the source zone, an on-demand search-query routing method is used.

ZRP itself has three sub-protocols: (a) the proactive (table-driven) Intrazone Routing Protocol (IARP), (b) the reactive Interzone Routing Protocol (IERP), and (c) the Bordercast Resolution Protocol (BRP). IARP can be implemented using existing link-state or distance-vector routing. Unlike OSPF [34] or RIP [35], the propagated routing information is propagated to the border of the routing zone.

ZRP's IARP relies on an underlying neighbor discovery protocol to detect the presence and absence of neighboring nodes, and therefore, link connectivity to these nodes. Its main role is to ensure that each node within the zone has a consistent routing table that is up-to-date and reflects information on how to reach all other nodes in the zone.

IERP, however, relies on border nodes to perform on-demand routing to search for routing information to nodes residing outside its current zone. Instead of allowing the query broadcast to penetrate into nodes within other zones, the border nodes in other zones that receive this message will not propagate it further. IERP uses the bordercast resolution protocol.

Because parts of an ad hoc route are running different routing protocols, their characteristics will therefore be different. Some parts of the route is dependent on proper routing convergence, while the other part is dependent on how accurate the discovered interzone route is. This can make assurance of routing stability very difficult. Without proper query control, ZRP can actually perform worse than standard flooding-based protocols [36].

ZRP's route discovery process is, therefore, route table lookup and/or interzone route query search. When a route is broken due to node mobility, if the source of the mobility is within the zone, it will be treated like a link change event and an event-driven route updates used in proactive routing will inform all other nodes in the zone. If the source of mobility is a result of the border node or other zone nodes, then route repair in the form of a route query search is performed, or in the worst case, the source node is informed of route failure.

5.13 Source Tree Adaptive Routing (STAR)

The Source Tree Adaptive Routing (STAR) [37] [38] protocol is a proactive routing protocol that does not require periodic routing updates, nor does it attempt to maintain optimum routes to destinations. It was a protocol developed in the SPARROW [39] project, which was part of the DARPA GloMo program.

STAR examines the updating strategies used in table-driven routing for wireless networks. In the optimum routing approach (ORA), a routing protocol must perform routing updates fast enough so as to provide optimum paths with regard to the defined metric. A major problem with ORA-based routing is that if topology changes occur frequently, the rate of routing updates increases dramatically. This results in a scenario where more control messages are present in the network than user data, which is undesirable.

By contrast, the least overhead routing approach (LORA) is commonly found in on-demand routing protocols. It attempts to minimize control overhead by:

(a) Maintaining path information only for the destinations the router needs to support, that is, *active* routes

(b) Using the paths found after a flood search as long as the paths are still valid, even if the paths are not optimum

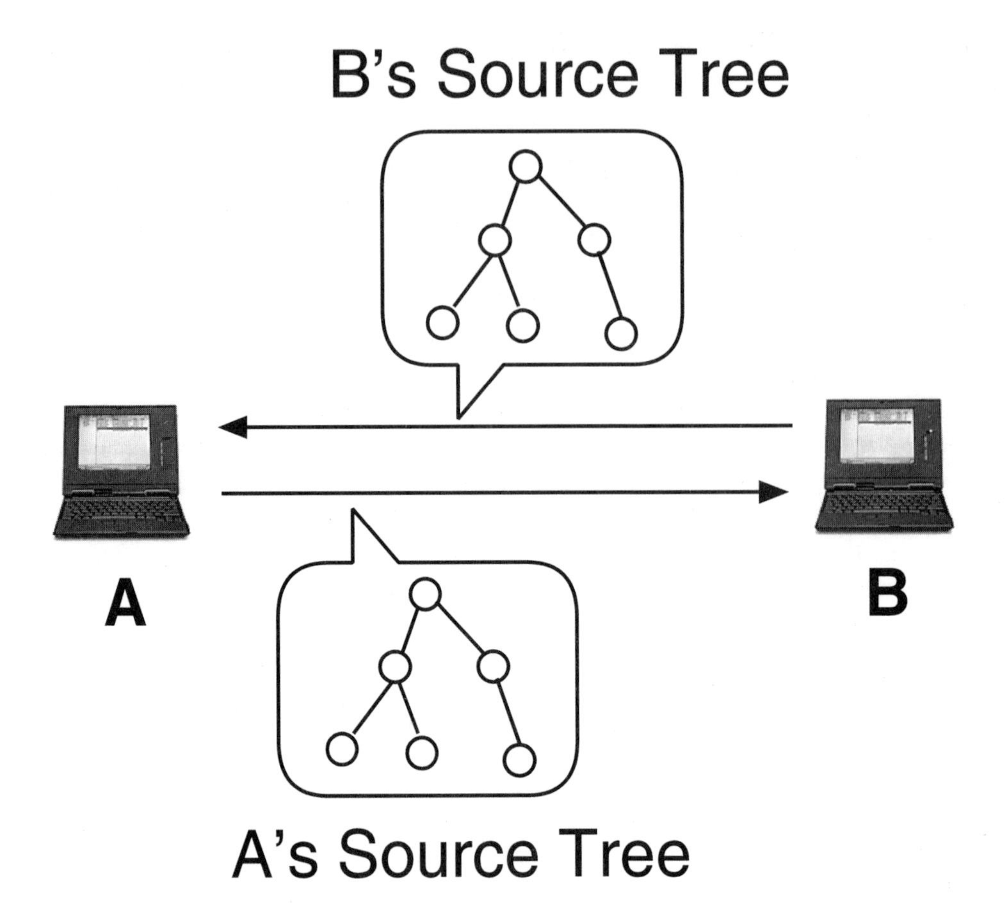

Figure 5.9. STAR - Source tree adaptive routing.

STAR is a table-driven routing protocol, but it aims to implement LORA, which can reduce the number of routing updates. STAR uses an underlying neighbor discovery protocol to allow it to discover the presence and mobility of neighboring nodes. In STAR, each ad hoc node maintains a source tree. The set of links used by a router (ad hoc host) in its preferred path to a destination is called the *source tree* of the router. In STAR, a node knows its adjacent links and the source trees reported by its neighbors. The aggregation of a router's adjacent links and the source trees reported by its neighbors constitute a partial topology graph. Each node runs a route selection algorithm on its own source tree to derive a routing table that specifies the successor to each destination.

STAR uses sequence numbers to validate link-state updates (LSUs). A router that receives an LSU will accept it if the received LSU has a larger sequence number than the previously stored value or there is no entry for that link. To reduce the number of updates, only changes to the validity of the source tree are propagated. Such changes include when the source loses all paths to a destination, detects a new neighbor, or encounters a pending long-term routing loop. STAR updates are known as LSUs and each update carries a sequence number to reflect the history of the update. Note that STAR differs from traditional link-state routing since the entire topology information is neither used nor sent. STAR uses Dijkstra's shortest path algorithm to select routes.

In summary, STAR must perform ORA table-driven-based routing to construct source trees. Thereafter, LORA table-driven routing can be used to selectively perform updates.

5.14 Relative Distance Microdiversity Routing (RDMAR)

RDMAR estimates the distance, in radio hops, between two nodes using the relative distance estimation algorithm. It is a source-initiated on-demand routing protocol having features found in ABR [40][41][42]. Based on the distance estimate, it can limit the flooding of route discovery packets to a radius of this estimate, thus reducing the amount of flooding.

The limitations of RDMAR are: (a) the estimated distance is a function of the previously recorded relative distance, (b) the assumption on fixed speed of mobile hosts, and (c) fixed radio transmission range. It is assumed in RDMAR that all ad hoc mobile hosts are migrating at the same fixed speed. This assumption can make good practical estimation of relative distance very difficult.

RDMAR route discovery involves the transmission of route discovery packets in the search for a route to the destination. If a current relative estimate is present, then the search flood is limited to this relative distance. The destination node then returns a route reply message over the reverse path. As the reply message moves back toward the source, intermediate nodes establish the forward route to the destination hop-by-hop, in a manner similar to associativity based routing.

When an existing route is broken, a route maintenance process is used to perform route repair. If a node discovers that its neighbor link has gone away, it invokes a localized route discovery[7] to find a partial route to the destination. If this node detects that link failure is closer to the source node, it sends a route failure message to the source. Nodes along the path that receive this failure message must remove

[7]This is similar to the method first proposed in ABR in 1994.

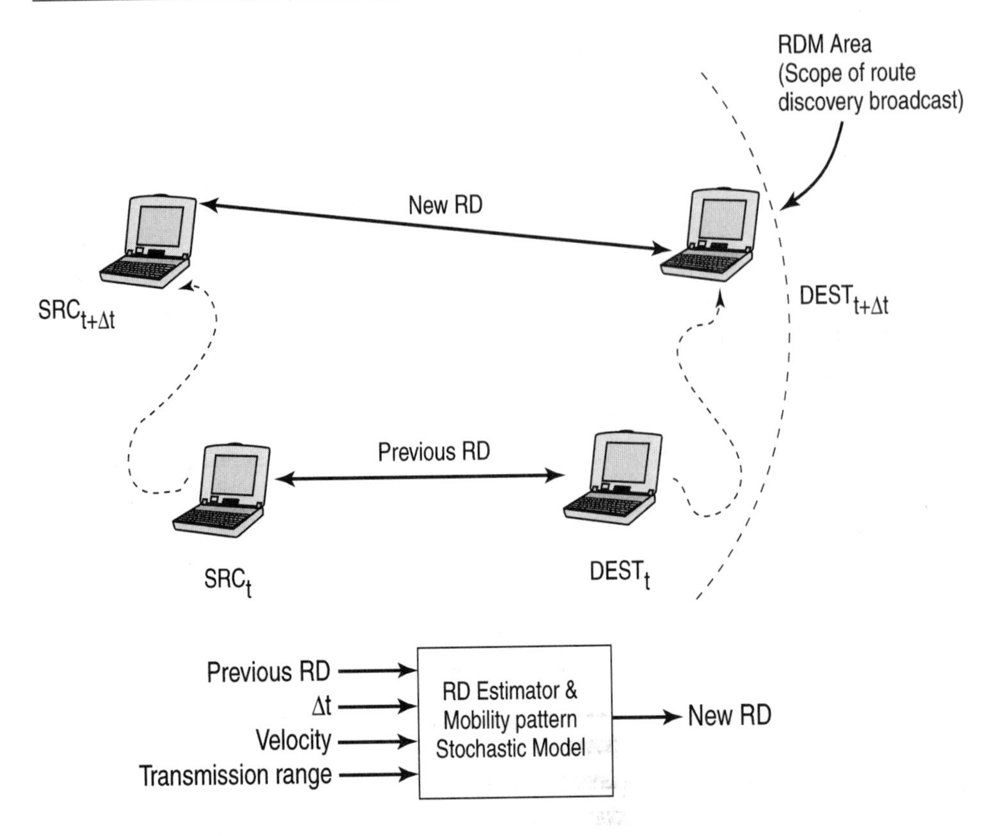

Figure 5.10. RDMAR - Relative distance microdiversity routing.

the routing table entry for the destination that is currently not reachable.

Although the current RDMAR specification states that unidirectional links are supported, the route discovery process assumes that all links are bidirectional. It has not yet addressed how unidirectional links can be handled. Although RDMAR claims to consider route stability, in actuality, it uses shortest route as the routing metric.

RDMAR does not employ beaconing. It does not use a route cache. Each node has a routing table that lists all available destinations and the number of hops to each. For an ad hoc wireless network with $\mathcal{N}$ nodes, $\mathcal{N}$-1 row entries are needed. In addition to the list of reachable destinations, the routing table contains the estimated relative distance (RD) between the node and destination concerned. A "Time_Last_Update" (TLU) field indicates the time elapsed since the node last received routing information for the destination. An "RT_Timeout" field records the remaining amount of time before the route is considered invalid. A "Route Flag" is used to indicate that the route to the destination is active.

In addition to the routing table, a data retransmission table (DRT) is employed. Before a source node transmits a data packet to its destination, it first ascertains if it has a route to the destination. If this is true, it keeps a copy of the packet in its DRT. This packet is kept in the DRT buffer until the packet is acknowledged by overhearing the forwarding done by the node concerned. If this acknowledgement is not received, retransmission occurs. If the node has not had a route to the destination, then this buffering allows the route discovery process to derive a route to the destination.

Unlike the ABR protocol, RDMAR route discovery and reply packets are fixed in length. There is no way to choose the best path since the routing metric does not take into account route stability. RDMAR aims to support QoS, but it is unknown how such routes can be discovered and differentiated.

In summary, RDMAR does not take into account signal stability, but concentrates more on relative distance. However, relative distance cannot reflect route stability. RDMAR also incorporates features from other existing on-demand routing protocols.

5.15 Conclusions

While several existing routing protocols are discussed here, it is essential that readers perform their own reasoning of the advantages, disadvantages, novelty, originality, and impact of these protocols. Some of them may have little original contributions and are extensions of existing methods. Source routing, link-state, and distance-vector routing have been in existence for years. Extensions are acceptable if they can satisfy the requirements properly and yield good communication performance. In addition to the issue about suitability, there are also concerns about protocol correctness, complexity, practicality, and communication performance. Utlimately, if an approach cannot be realized or cannot provide reasonable communication performance to support useful applications, then it is unlikely that the protocol can be accepted, deployed, and used widely.

Chapter 6

ASSOCIATIVITY-BASED LONG-LIVED ROUTING

6.1 A New Routing Paradigm

Research on routing for communication networks has been on-going for many years. The significance of a routing algorithm is to compute a path from one node to another within a finite time. When multiple paths exist, then some form of selection criteria is needed. A common and widely used route selection criteria is shortest path. A path that results in the shortest end-to-end route is usually preferred. Shortest path can sometime refer to minimum-hop, that is, the least number of links used in the route. This has strong implications on total link utilization and the state of network congestion.

Many routing algorithms developed over the years are variants of shortest path routing. Progress in routing research has been made toward supporting QoS and addressing scalability. However, in mobile networks, routing protocols for wired networks are not applicable since they are capable of handling host and router mobility.

Several researchers still attempt to propose variants of existing wired routing protocols based on shortest path. However, **shortest path-based routing cannot be appropriately applied to wireless ad hoc networks**. A shortest ad hoc route derived at time $\mathcal{T}$ may no longer be valid at time $\mathcal{T}$+1 since any nodes in the route could have moved or the characteristics of the links in the route could have changed. Hence, shortest path routing is not directly applicable nor appropriate for use in ad hoc wireless networks. A new routing paradigm is needed.

6.2 Associativity-Based Long-Lived Routing

Associativity[1] is related to the *spatial*, *temporal*, and *connection* stability of a mobile host (MH). Specifically, associativity is measured by one node's connectivity relationship with its neighboring nodes. A node's association with its neighbors changes as it is migrating, and its transition period can be identified by associativity ticks or counts. Migration is such that after this unstable period, there exists a period of stability (i.e., a node is constantly associated with certain neigbors over time without losing connectivity with these neighbors), where the MH will spend some dormant time[2] within a wireless cell before it starts to break its connectivity relationship with its surrounding neigbors and move outside the boundary of the existing wireless cell. The threshold where associativity transitions take place is defined by $A_{threshold}$, as shown in Figure 6.2.

Figure 6.2 shows the spatial and temporal representation of a mobile node with its neighbors. The mobile node hears beacons from its neigbors as it moves from one point in space to another. Its affiliation with neighboring nodes also changes with time and space. Specifically, the mobile node collects more beacons from neighboring node C since its mobility profile is such that it continues to maintain good connectivity with C, despite the fact that it is moving over time and space. Good neighbors, therefore, are those that yield high associativity ticks.

In ABR, each MH periodically transmits short beacons identifying itself (more like hello messages) and constantly updates its associativity ticks/counts in accordance to the MHs sighted (i.e., hearing others' beacons) in the neighborhood. For

[1]Associativity-based long-lived routing (ABR) was invented by the author at Cambridge University, England in fall 1994. It was filed for a patent in 1996 and the patent was granted in 1999. It is the first routing protocol that advocates the selection of stable links and routes for ad hoc wireless networks.

[2]This does not imply that the mobile host is not moving.

The United States of America

United States Patent
Granted To

Chai Keong Toh
CAMBRIDGE UNIVERSITY

November 16, 1999 NUMBER: 5,987,011

ROUTING METHOD FOR
AD-HOC MOBILE NETWORKS

The Commissioner of Patents and Trademarks has received an application for a patent for a new and useful invention. The requirements of law have been complied with, and it has been determined that a patent on the invention shall be granted under the law.

Therefore, this

United States Patent

Grants to the person or persons having title to this patent the right to exclude others from making, using or selling the invention throughout the United States of America for the term of this patent, subject to the payment of maintenance fees as provided by law.

Commissioner of Patents and Trademarks

Figure 6.1. Associativity Based Routing (ABR) patent: One the first few patents issued for routing in ad hoc mobile networks.

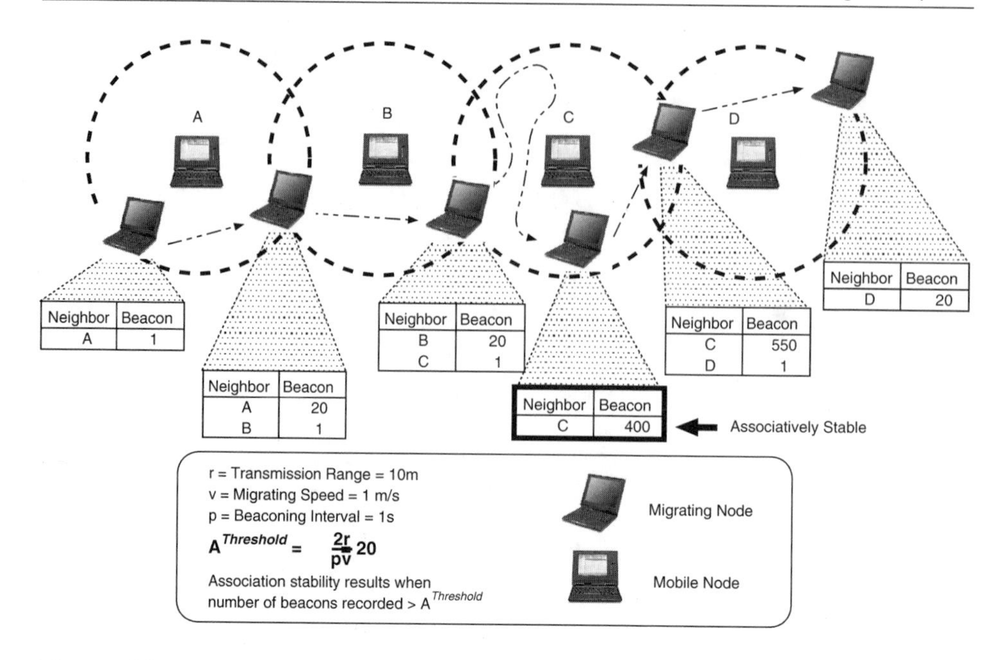

Figure 6.2. Temporal and spatial representation of associativity of a mobile node with its neighbors.

example, consider an ad hoc wireless network where each wireless cell is 10 m in diameter and where each MH is beaconing once every second. Hence, a mobile node migrating at pedestrian speed (2 m/s) across a wireless cell with one or more neighboring MHs will record an associativity tick of no more than 5. This is $A_{threshold}$, and any associativity ticks greater than this threshold imply periods of association stability (i.e., the mobile node concerned is not just transiting through its neighbors). Sending packets to a *transiting* neighbor is useless since link connectivity with this neighbor will be broken very soon. In essence, deriving *stable links* is crucial to obtaining *stable routes* in an ad hoc mobile network.

In ABR, an MH is said to exhibit a high state of mobility when it has low associativity ticks with its neighboring nodes. However, if high (i.e., greater than $A_{threshold}$) associativity ticks are observed, the MH is regarded to be in a stable state[3] and this is the ideal point at which to select it to perform ad hoc routing. Because of the interdependent relationship in associativity, if all the MHs in a route have high associativity ticks, an interlocking phenomenon arises where one host's degree of associativity will be high if others do not move out of reachability and

[3]This does not imply that the MH is not moving.

Neighboring Nodes	Associativity Ticks	Link Delay	Signal Strength	Power Life	Route Relaying Load
N_a	30	100	0.8	0.8	3
N_b	10	80	0.4	0.7	1
N_c	19	60	0.5	0.5	0

Table 6.1. ABR Stability Table.

enter a stable[4] state.

The above discussion concentrates on explaining associativity ticks. However, *stability* in ABR refers to more than just associativity ticks. It also includes *signal strength* and *power life*, as shown in table 6.1. The former defines the quality of the signal propagation channel while the latter describes the current power life of the device. Advances in radio transceiver technology has enabled one to monitor signal strength over time and store this information into memory. Similar, advances in smart battery technology has enabled us to monitor remaining power life of battery-powered devices. Such information, therefore, can be used to govern route stability.

6.2.1 New Routing Metrics

Although using spatial, temporal, and connectivity relationships among mobile nodes is considered to be novel in ABR, there are also other novelties in terms of routing metrics for ad hoc mobile networks. Conventional routing metrics are characterized by: (a) fast adaptability to link changes (i.e., route recovery time), (b) minimum-hop paths to destinations, (c) propagation delay, (d) loop avoidance, and (e) link capacity. However, fast adaptability at the expense of frequent broadcasts and excessive radio bandwidth consumption is undesirable. The qualities of a good route should not only include the number of hops and the round trip propagation delay.

The **longevity** of a route is of paramount importance in ad hoc mobile networks as the merits of a shorter hop but short-lived route will be denigrated due to frequent data flow interruptions and the frequent need for route reconstruction. This new metric indicates that the classic shortest path metric is not necessarily applicable nor useful in ad hoc wireless networks. From another perspective, **fair route relaying load** is also important as no one particular mobile node should be unfairly burdened to support many routes and to perform many packet relaying functions. This is an issue about fairness, and even route relaying load can alleviate the possibility of

[4]"Stable" here refers to maintaining connectivity relationships with surrounding MHs over time.

network congestion in an ad hoc mobile network.

6.2.2 Route Selection Rules

Given a set of possible routes from SRC to DEST, if a route consists of mobile nodes having high associativity ticks (therefore indicating *spatial*, *temporal*, and *connection* stability), then that route will be chosen by the destination, despite other shorter hop routes. However, if the overall degree of association stability of two or more routes is the same, then the route with the least number of hops will be chosen. If multiple routes have the same minimum hop count, then one of the routes will be arbitrarily selected. The ABR route selection algorithm, which is executed at the destination node, is formally stated in Figure 6.2.

6.3 ABR Protocol Description

The ABR protocol is a source-initiated on-demand routing protocol that consists of the following three phases: (a) *route discovery* phase, (b) *route reconstruction* phase, and (c) *route deletion* phase. Initially when an SRC node desires a route, the route discovery phase is invoked. When the links of an established route change due to source/destination/intermediate or subnet-bridging MH migration, the route reconstruction phase is then invoked. When the SRC no longer desires the route, it initiates the route deletion phase. These three phases will be discussed briefly in the sections that follow. A full description of the ABR protocol can be found in [40].

6.3.1 Route Discovery Phase

The route discovery phase consists of a broadcast query (BQ) and an await reply (REPLY) cycle. Initially, all nodes except those of DEST's neighbors have no routes to DEST. A node desiring a route to DEST broadcasts a BQ message, which is propagated throughout the ad hoc mobile network in search of MHs that have a route to DEST. Here, a sequence number is used to uniquely identify each BQ packet, and no BQ packet will be broadcast more than once.

Once the BQ is broadcast by the SRC, all intermediate nodes (INs) that receive the query check if they have processed this packet. If affirmative, the query packet is discarded; otherwise, each node checks if it is the DEST. If it is not the DEST, the IN appends its MH address/identifier at the IN identifiers (IDs) field of the query packet and broadcasts it to its neighbors (if it has any). The associativity ticks with its neighbors will also be appended, along with its route relaying load, link

The ABR Route Selection Algorithm

Let S_i be the set of possible routes from SRC→DEST, where $i = 1, 2, 3....$
Let RL^i_j be the relaying load in each node j of a route in S_i, where $j = 1, 2, 3....$
Let RL_{max} be the maximum route relaying load allowed per MH.
Let $AT_{threshold}$ be the min. associativity ticks required for association stability.
Let AT^i_j represent the associativity ticks in each node j of a route in S_i.
Let H_i represent the aggregate degree of association stability of a route in S_i.
Let L_i represent the aggregate degree of association instability of a route in S_i.
Let $H_{i_{ave}}$ represent the average degree of association stability of a route in S_i.
Let $L_{i_{ave}}$ represent the average degree of association instability of a route in S_i.
Let Y_i represent the number of nodes of a route in S_i having acceptable route relaying load.
Let U_i represent the number of nodes of a route in S_i having unacceptable route relaying load.
Let $Y_{i_{ave}}$ represent the average acceptable route relaying load factor.
Let $U_{i_{ave}}$ represent the average unacceptable route relaying load factor.

Begin

For each route i in S_i
Begin
$a \leftarrow 0$
For each node j in route S_i
Begin
If ($AT^i_j \geq AT_{threshold}$) H_i++ ; else L_i ++;
If ($RL^i_j \geq RL_{max}$) U_i++; else Y_i ++; a ++;
End
$H_{i_{ave}} = H_i/a$; $L_{i_{ave}} = L_i/a$; $U_{i_{ave}} = U_i/a$; $Y_{i_{ave}} = Y_i/a$;
End

Best Route Computation
Let the set of acceptable routes with $U_{i_{ave}} = 0$ and $H_{i_{ave}} \neq 0$ be P_l, where $P_l \subseteq S_i$
Begin
** Find Route With Highest Degree of Association Stability **
Compute a route k with $H_{k_{ave}} > H_{l_{ave}}, \ \forall \ l \neq k$.
or if a set of routes K_n exists such that $H_{K1_{ave}} = H_{K2_{ave}} ... = H_{Kp_{ave}}$
where n = {1,2,3,...,p}
Begin
** Compute Minimum Hop Route Without Violating Relaying Load **
Compute a route K_k with $Min\{K_k\} < Min\{K_m\}, \ \forall \ m \neq k$.
or if a set of routes K_o exists such that $Min\{K_1\} = Min\{K_2\}... = Min\{K_q\}$,
where o = {1,2,3,...,q}
Begin
** Multiple Same Associativity & Minimum Hop Routes Exists **
Arbitrarily select a minimum hop route K_k from K_o
End
End
End
End

Table 6.2. ABR Route Selection Algorithm.

propagation delay, remaining power life, and route hop count. The piggybacking of associativity ticks into the BQ query packet is illustrated in Figure 6.3.

The next succeeding IN will erase its upstream neighbor's associativity ticks

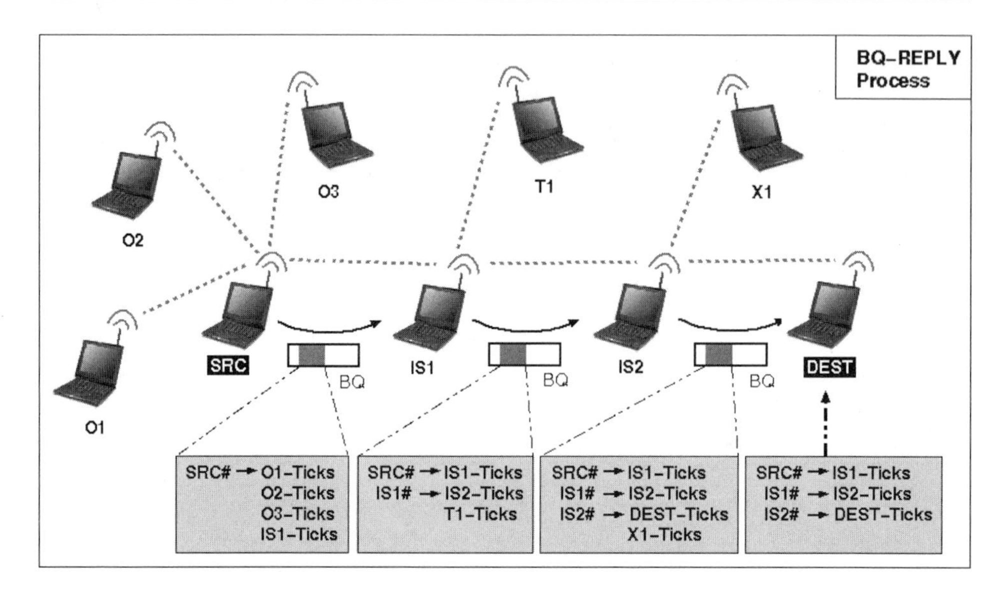

Figure 6.3. Updating the associativity metric during BQ packet propagation.

and retain only those concerned with itself and its upstream neighbor. In this manner, the query packet reaching the DEST will only contain the intermediate MH addresses (hence recording the path taken) and their associativity ticks (hence recording the stability state of INs supporting the route) and relaying loads, together with information on route forwarding delays and hop count. The resulting BQ packet (see Figure 6.4) is, therefore, variable in length.

The DEST will, at an appropriate time after receiving the first BQ packet, know all the possible routes and their qualities. It can then select the best route (based on the selection criteria mentioned earlier) and send a REPLY packet back to the SRC via the route selected, as shown in Figure 6.5. This causes INs in the route to mark their route to the DEST as valid, meaning that all other possible routes will be inactive and will not relay packets destined for DEST, even if they hear the transmission. This, therefore, avoids duplicated packets from arriving at DEST. Similar to BQ, the REPLY packet is also variable in length.

6.3.2 Route Reconstruction (RRC) Phase

Although the route selected using ABR tends to be long-lived, there can still be cases where the association stability relationship is violated. For example, the user of a mobile device may decides to head to the restroom or to go away for another unexpected engagement. When this happens, RRC procedures are invoked to cope

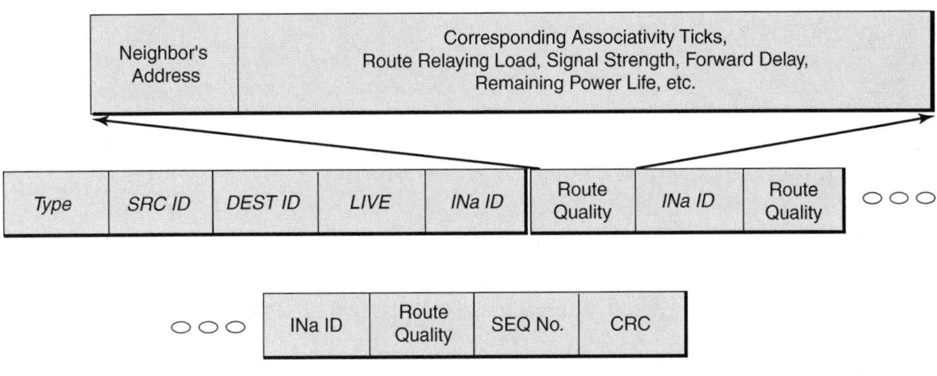

Figure 6.4. ABR route discovery using BQ (broadcast query) control packet. Notice that associativity and other QoS information are piggyback into this packet as it moves from one node to the next.

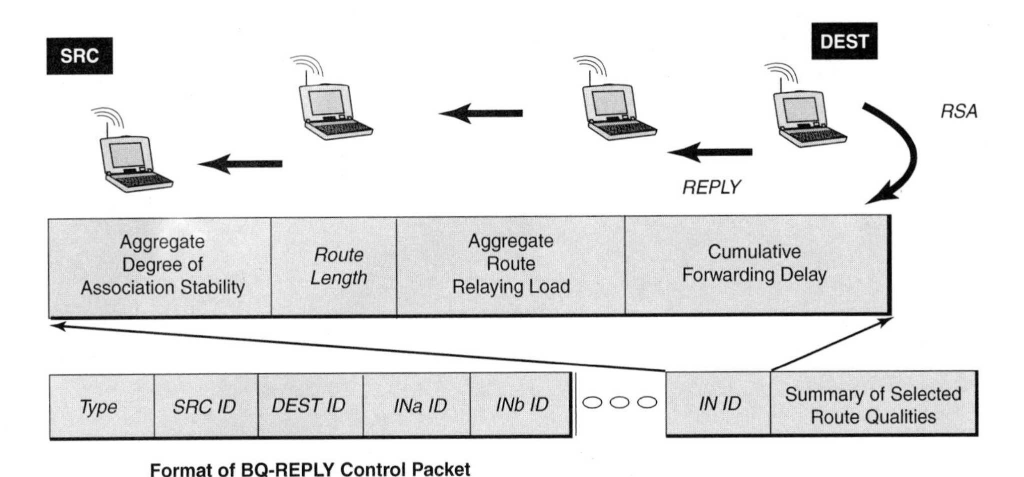

Figure 6.5. Destination node sends a REPLY packet back over the selected route. Nodes in this route will be programmed to perform ad hoc routing.

with mobility.

ABR localizes the route repair operation intelligently to avoid excessive control overhead and disturbing unconcerned nodes. ABR route recovery is fast because it uses the principle of partial route discovery. It repairs the broken route on-the-fly in realtime. The ABR route maintenance phase consists of the following operations, namely:

(a) partial route discovery,

(b) invalid route erasure,

(c) valid route update, and

(d) new route discovery (worst case).

ABR handles unexpected moves by attempting to locate an alternative valid route quickly without resorting to a BQ unless necessary. The following narrations will refer to Figures 6.6(a), (b), and (c), respectively.

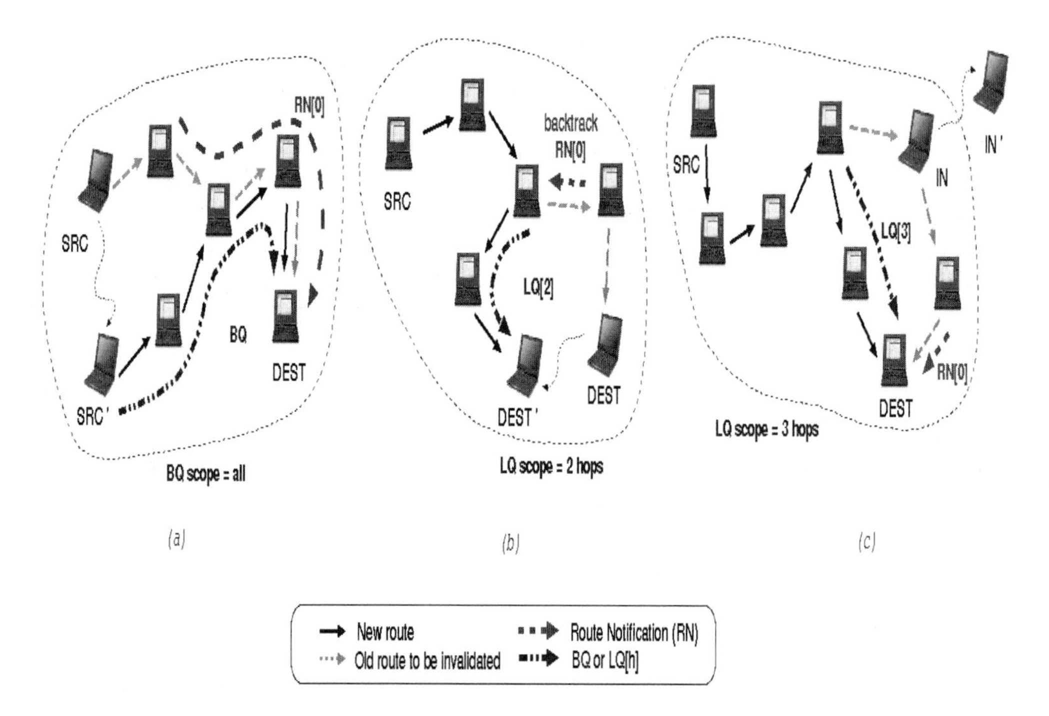

Figure 6.6. ABR route maintenance when SRC, DEST, and IN move.

Since the routing protocol is source-initiated, any moves by SRC will invoke an RRC process equivalent to that of a route initialization, that is, via a BQ_REPLY process. It will be clear later that this avoids multiple RRC conflicts as a result of concurrent node movements[5].

[5]Other source-initiated on-demand protocols do not seem to address this issue of concurrent nodes' movement.

When DEST moves, DEST's immediate upstream neighbor (i.e., the pivoting node) will erase its route. It then performs an LQ[H][6] (localized query) process to ascertain if DEST is still reachable. If DEST receives the LQ, it will select the best partial route (again based on association stability criteria) and send a REPLY; otherwise, the LQ_TIMEOUT period will be reached and the pivoting node will backtrack to the next upstream node. During the backtrack, the new pivoting node will erase the route through that link and perform an LQ[H] process until the new pivoting node is greater than one-half $hop_{src-dest}$ away from DEST or a new partial route is found. If no partial route is found, the pivoting node will send a route notification RN[DIR='1'] packet back to SRC to initiate a BQ process. The format of LQ control packet is shown in 6.7 and similar to BQ, it is variable in length.

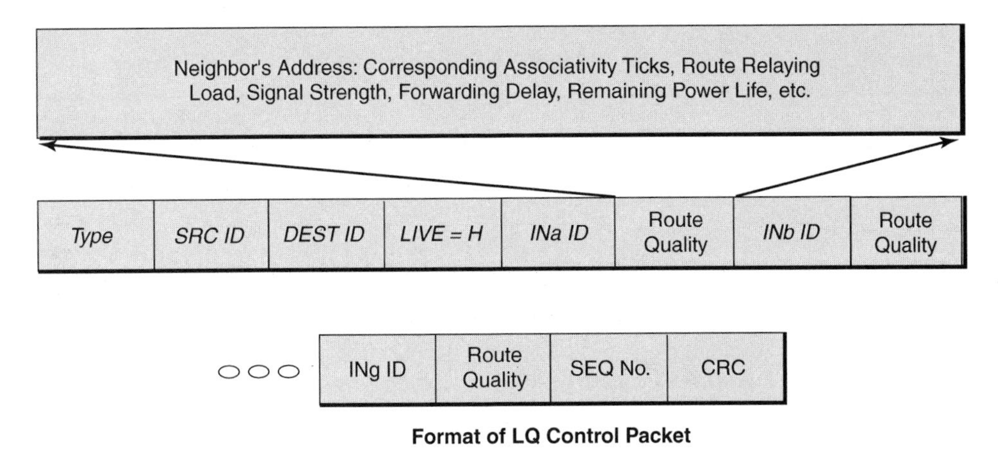

Figure 6.7. Handling mobility and route repair using LQ control packet.

As shown in Figure 6.8, the RN packet comprises an ORG ID field, which stores the pivoting node address and a STEP flag which indicates the type of route notification that is to be carried out. When STEP=0, the backtracking process is to be performed one hop at a time (in the upstream direction), while when STEP=1 this implies that the RN control packet will be propagated straight back to the source node to invoke a BQ-REPLY cyle or to the destination node to erase invalid routes. The RN control packet also comprises a DIR flag which serves to indicate the direction of RN[1] propagation.

If any IN in the route moves and breaks the association stability characteristics, an RRC process is necessary. The immediate upstream node will invoke an LQ[H] process to quickly locate an alternate stable partial route. The LQ[H] process is

[6] "H" here refers to the hop count from the upstream node to DEST.

performed based on a suitable $\mathcal{H}$ value. If the pivoting node is $\mathcal{X}$ hops away from the destination node via the previous active route, then $\mathcal{H}$=$\mathcal{X}$ will be used in the hope that the destination node is still within $\mathcal{X}$ hops range (reachable via other paths) or shorter. The immediate downstream node, however, will immediately send a route erase message toward DEST, i.e., RN control packet with STEP=0 and DIR=0 (downstream). In this manner, invalidated route entries are deleted. Again, multiple partial routes can exist and DEST will select the best possible route. If no partial routes to DEST exist, then the next upstream node will invoke another LQ process. This backtracking process proceeds until: (a) a partial route is found, or (b) the number of backtracks exceeds one-half the route length[7]. If all possible LQs are unsuccessful, the SRC will time out and may invoke a BQ again.

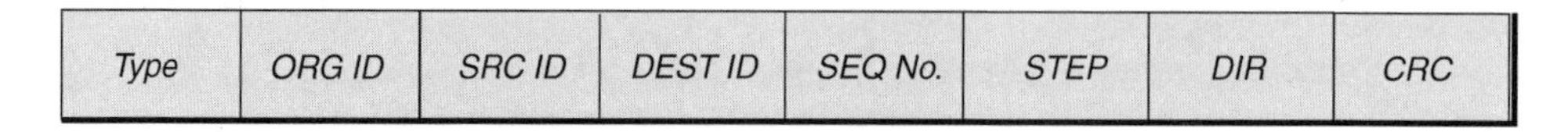

Format of RN Control Packet

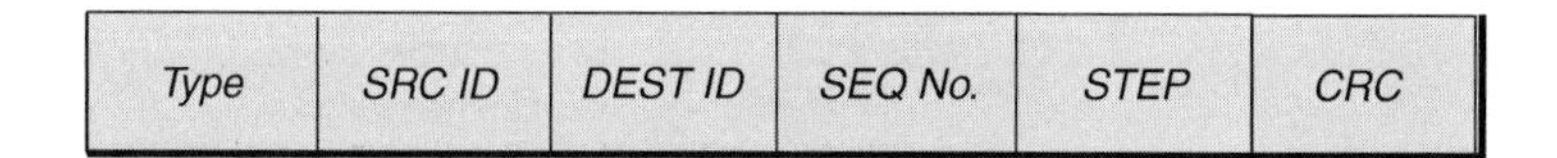

Format of RD Control Packet

Figure 6.8. Internal structure of RN and RD control messages.

Race conditions exist due to multiple invocations of RRC processes as a result of concurrent movements by SRC, DEST, and INs. However, the ABR protocol is able to resolve the conflicts of multiple RRCs by ensuring that ultimately, there is only one RRC that succeeds. Each LQ process is tagged with a sequence number so that earlier LQ processes will terminate when a newer LQ process is invoked. In the same vein, if nodes processing LQs hear a new BQ for the same connection, the LQ process is aborted.

6.3.3 Alternate Routes

In ABR, no attempt is made to retain alternate routes, as maintaining them causes overhead. Only one route will be selected and only one route is valid for a particular route request. The avoidance of using alternate routing information means that problems associated with looping due to INs having stale routes are absent and there

[7]This was introduced to avoid excessive delay in route reconfiguration, especially when the route length is long.

is no need for periodic network-wide broadcast and route updates. Any alternate route will have to be discovered via an LQ or BQ process, which may give rise to better (shorter hop and long-lived) routes. The original ABR patent does not advocate the use of route caches since there is little point of caching discovered routes if they are going to be changed soon. Maintaining the "validity' of cached routes will also result in additional control overhead, which is undesirable.

6.3.4 Route Deletion Phase

When a discovered route is no longer desired, a route delete (RD) broadcast will be initiated by the SRC so that all INs will update their routing table entries. The format of RD control messasge is shown in Figure 6.8. A full broadcast is used compared to a directed broadcast since the nodes supporting an active route would have changed during route reconstructions. Similar to a BQ, the RD control packet has a LIVE field of infinity to achieve a full wave-like broadcast. In addition to this *hard state* approach, a *soft state* approach is possible where the route entries are invalidated upon time out, when there is no traffic activity related to the route over a period of time. This approach, therefore, is performed at each node in the route.

6.3.5 ABR Headers and Tables

Since a long packet header results in low channel utilization efficiency, in ABR, each data packet header will only contain the neighboring node routing information, not all nodes in the route[8]. Each IN will renew the next hop information contained in the header before propagating a packet upstream or downstream. The individual fields in the packet header are summarized in Table 6.3.

The ABR routing table of a node supporting existing routes is shown in Table 6.4. The table reveals that every node supporting on-going routes will map incoming packets from a particular upstream node to the corresponding out-going downstream node. Every node will also keep track of its distance (hop count) to DEST and record the total routes that it is currently supporting, that is, its route relaying load.

The ABR neighboring table is usually updated by the data-link layer protocol, which generates, receives, and interpret beacons from the neighboring MHs and passes this information up to the higher protocol layers. Associativity tick entries are updated according to the number of times the same beacon was heard from a

[8]Source routing used in DSR will have to include addresses of all nodes in the discovered route in every data packet. The control overhead per data packets grows significantly as the route path gets longer.

Table 6.3. ABR Packet Header

Routing Header Field	Function
SRC ID	Packet Forwarding
DEST ID	Route Identification
Sequence No.	Duplicates Prevention, Uniqueness
Service Type	Packet Priority
Last IN ID	Passive Acknowledgement
Next IN ID	Duplicates Prevention, Routing
Current IN ID	Acknowledgement, Routing

Table 6.4. ABR Routing Table

Destination	Source	Incoming IN	Out-Going IN	Distance
N_a	N_x	N_z	N_j	4
N_k	N_y	N_i	N_o	3
Total No. of Active Routes Supported (Relay Load):				2

respective neighboring node. The structure of a neighboring table is shown in Table 6.5.

To avoid mobile hosts from processing and relaying the same BQ, RD, or LQ packet twice, BQ, RD and LQ "seen" tables are needed. If the received control packet type, route identifier, and sequence number match an entry in the "seen" table list, the packet is discarded. The contents of these "seen" tables will be erased after a certain timeout period or when a route is no longer desired by the SRC. However, the timeout period must be long enough to allow the neighboring nodes to forward the control packets to their neighbors. As shown in Figure 6.9, mobile node $\mathcal{B}$ is the source node. It sends the first BQ message to mobile nodes $\mathcal{A}$, $\mathcal{C}$, and $\mathcal{D}$. These neighboring nodes then forward the BQ packet to their neighbors,

Table 6.5. ABR Neighboring Table

Neighboring Nodes	Associativity Ticks (units)	Link Delay (msecs)
N_a	5	100
N_b	15	50

as shown by the arrows labelled with BQ[2a] and BQ[2b]. Mobile node $\mathcal{B}$ should ignore BQ[2a] and BQ[2b] packets since they are considered echo packets. Hence, the BQ entry in the seen table should not be erased until at least this period of receiving echos from neighors.

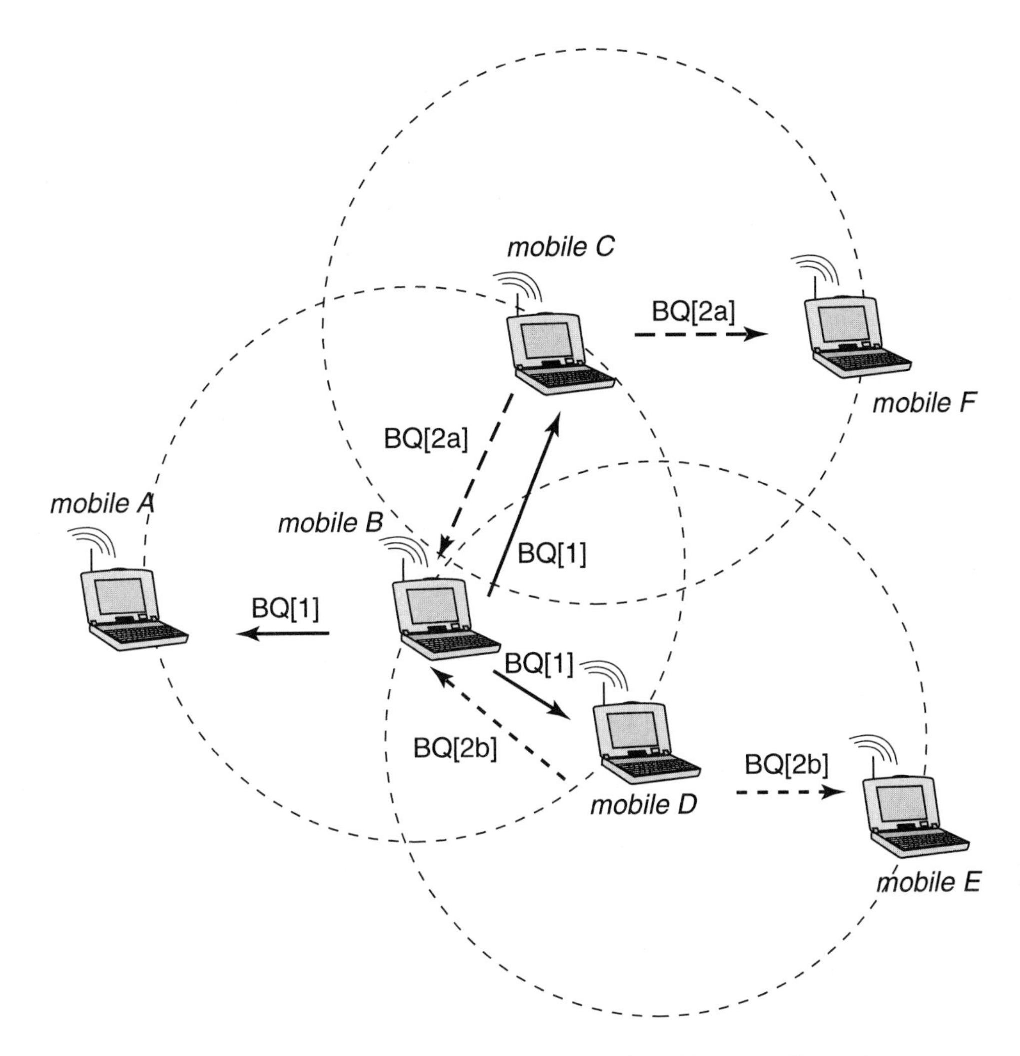

Figure 6.9. Erasure of BQ seen table entry in ad hoc mobile nodes.

Finally, since REPLY and RN control packets use "directed" broadcast (since intended recipients' addresses are contained in the control packet), "seen" tables for these packets are not necessary.

6.3.6 ABR Protocol Summary

The outstanding feature of the ABR protocol is that no RRCs are needed so long as the property of associativity interlock remains valid. When this property is violated, the appropriate RRC procedure will be invoked to handle mobility. For INs and destination node moves, an LQ-REPLY cycle is used. If this cycle does not successfully locate an alternate route, the BQ-REPLY cycle is eventually invoked. This BQ-REPLY cycle is also initiated when the SRC node moves. For movements by subnet-bridging MHs, there are two scenarios. If existing ad hoc routes are all within the subnets, then no RRC is necessary. However, if routes span across subnets, migration of a subnet-bridging MH can partition the network and a BQ-REPLY cycle will be necessary. Finally, for concurrent node movements, only one RRC cycle will be ultimately valid. Table 6.6 summarizes the ABR protocol procedure for each scenario.

Table 6.6. Summary of ABR Protocol Procedures

<table>
<tr><th>Associativity Valid</th><th colspan="6">Associativity Violated</th></tr>
<tr><td rowspan="3">No Route Reconstructions Needed</td><td colspan="2">INs & DEST Moves</td><td>SRC Moves</td><td colspan="2">Subnet-Bridging MH Moves</td><td>Concurrent Moves</td></tr>
<tr><td>Normal Case</td><td>Worst Case</td><td rowspan="2">BQ-REPLY Cycle Success</td><td>Route within Subnet</td><td>Route Spans across Subnets</td><td rowspan="2">Ultimately Only One Route Reconstruction Cycle Is Valid</td></tr>
<tr><td>LQ, REPLY Cycle Success</td><td>BQ, REPLY Cycle Success</td><td>No Route Reconstructions Needed</td><td>Network Is Partitioned; BQ-REPLY Cycle Will Retry Before Aborting</td></tr>
</table>

6.4 Conclusions

Breaking from the traditional routing paradigm based on shortest path is a key feature in ABR. The protocol exploits the *spatial*, *temporal*, *connection*, and *power* characteristics of neighboring nodes to construct a route that is long-lived. Best still, the protocol *programs* nodes in the selected route so that each packet will be forwarded accordingly. There is no need for source routing.

The key argument lies on the fact that selecting a route based on shortest path is useless if the path is going to be broken in the next instance due to mobility or power depletion. Hence, it makes more sense to select nodes to form a route that is likely to last over time and at least at the lifetime of a connection. If one can transmit all information quickly and reliably throughout the lifetime of an ad hoc mobile connection, then performance would be satisfying.

ABR employs this key concept to derive long-lived routes. In addition, ABR is a source-initiated protocol, meaning that there is no need for periodic route updates and await for route convergence. Recall that transient loops are present in table-driven protocols. ABR does not employ route caches since maintaining the validity of these caches incurs significant control overhead.

Finally, ABR attempts to localize route repair operations only to the affected region, thus reducing the amount of control overhead introduced during this operation. The protocol does not interrupt the source node each time a route is truncated - a problem commonly found in other protocols. The consideration for QoS requirements and the exploitation of new long-lived partial routes ensure that the repaired route is still likely to be long-lived.

Chapter 7

IMPLEMENTATION OF AD HOC MOBILE NETWORKS

7.1 Introduction

Although there are many protocols proposed for ad hoc wireless networks, few have actually demonstrated practicality of implementation and operation. Some protocols are highly complex and involve the computation of parameters that are not readily available in practical systems. This chapter examines how ABR (Associativity Based Routing) - an self-organizing, on-demand and source-initiated routing protocol, is implemented using current-off-the-shelf (COS) hardware and new innovative software.

7.2 ABR Protocol Implementation in Linux

7.2.1 System Components

The ad hoc wireless networking testbed established comprises several IBM and Compaq laptops running the Lunix operating system and equipped with Lucent Technologies WaveLAN Wireless PCMCIA adapters, as shown in Figure 7.1. The Linux OS has a copy of the TCP/IP/Ethernet protocol suite. This protocol suite is further enhanced with the ABR protocol to support ad hoc wireless networking and communications.

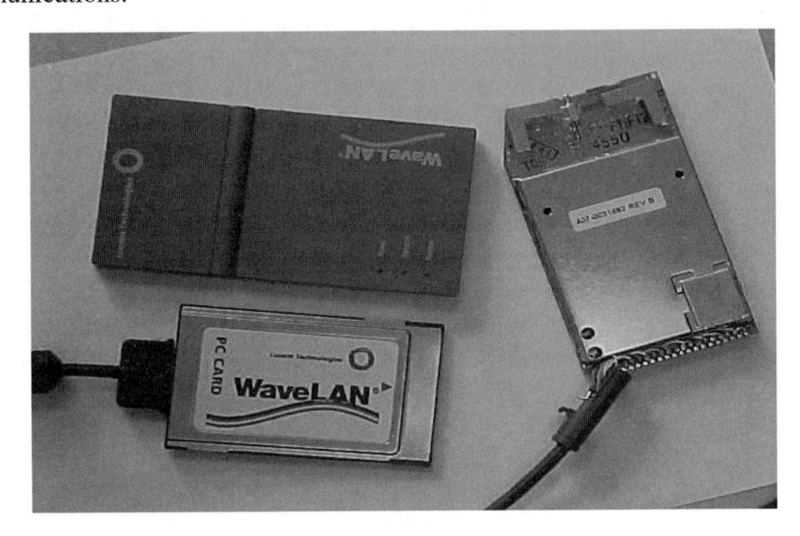

Figure 7.1. The Lucent WaveLAN PCMCIA card and antenna module.

Linux is a variant of UNIX for x86 PC (and a few other architectures). It is capable of multi-tasking and multi-user operation, which allows many users to have access to the same machine, running multiple processes at the same time. More importantly, Linux distributions include most of the required development software (compilers, libraries, tools, etc.), a convenient UNIX and X window system environment, and a full suite of TCP/IP networking software. In addition, its open architecture allows easy incorporation of the ABR protocol into TCP/IP. The widespread use of Linux has also resulted in easy availability of audio, video, and network device drivers for use in various experiments.

The laptop computers used for the testbed have an Intel Mobile Pentium II processor, 32 MB of memory, 2.5inch 5.1 GB of hard disk drive, and an active 13.3inch TFT color display with 1024 by 768 resolution. They are also equipped with standard I/O interfaces (such as serial, parallel, USB, diskette, keyboard/mouse, dock-

Data Communications	Performance
Data Rate	2 Mbps
Media Access	Ethernet CSMA/CA
Bit Error Rate	Better than 10^{-8}

Table 7.1. Data Specifications of WaveLAN PCMCIA Wireless Adapter (source Lucent).

Radio Specifications	915 MHz	2.4GHz
Receiver Sensitivity	-80dBm	-82 dBm
Modulation Technique	Spread Spectrum (DQPSK)	Spread Spectrum (DQPSK)
Output Power	-80 dBm	-82 dBm
Range (open office)	250m	200m
Range (semi-open office)	60m	50m
Range (Closed office)	30m	30m

Table 7.2. RF Specification of WaveLAN PCMCIA Wireless Adapter (source Lucent).

ing interface, audio I/O, and external monitor connector), an internal 56K modem, and two Type I/II PCMCIA card slots.

The radio adapter chosen is the 2.4GHz WaveLAN/PCMCIA card by Lucent Technologies (as shown in Figure 7.1), which implements the CSMA/CA media access protocol. WaveLAN provides a fast and reliable solution for wireless last-hop access to services residing in the wired network. In the testbed to be discussed, WaveLAN base stations are not utilized. Instead, only the wireless adapters are used to provide point-to-point and multihop wireless connectivity. The data and RF specifications related to the wireless adapter are shown in Tables 7.1 and 7.2 respectively.

7.2.2 Software Layering Architecture

As shown in Figure 7.2, ABR was implemented as a sublayer between the IP and Ethernet MAC layers. Since ABR is a sublayer, packets are able to bypass the ABR module if it is not necessary. If ad hoc routing is required, ABR routing will then supersede IP routing to support ad hoc mobile communications. With this architecture, we are able to handle both non-ABR (regular IP) traffic over wired or one-hop, peer-to-peer wireless networks and ABR traffic over multi-hop and peer-to-remote wireless networks with the same kernel. Ideally, the ABR sublayer would sit completely above the Ethernet layer, but due to the Linux implementation of the TCP/IP protocol suite, ABR sits partially in Ethernet (see Figure 7.2). The reason for this is due to how route discovery has been implemented. Route discovery procedures are currently residing at the same level as ARP (Address Resolution Protocol)[43]. Details on implementing route discovery will be discussed later.

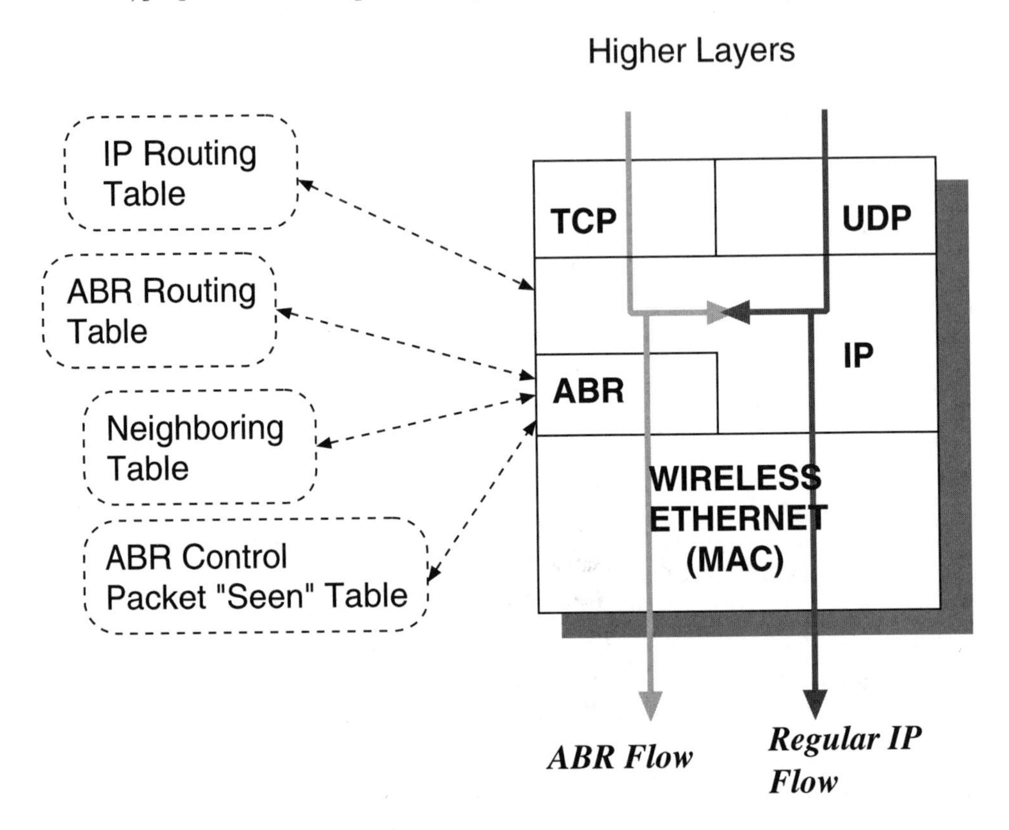

Figure 7.2. The software architecture of TCP/UDP/IP/ABR/Ethernet protocol implementation.

Figure 7.3 shows the ABR packet flow in a typical ad hoc scenario with two wireless hops. Although only a single intermediate node is shown, it should be evident how the inclusion of more intermediate nodes (INs) would behave. For a typical data packet originated at the endpoints, the packet must traverse the entire protocol stack, from the application layer down to the device driver. At the IN, the packet is brought up to the ABR sublayer, examined, and re-sent out. Hence, during ad hoc routing, packets do not go beyond the ABR sublayer. All control packets (such as LQ, BQ, RN, etc.) are originated from and terminated at the ABR

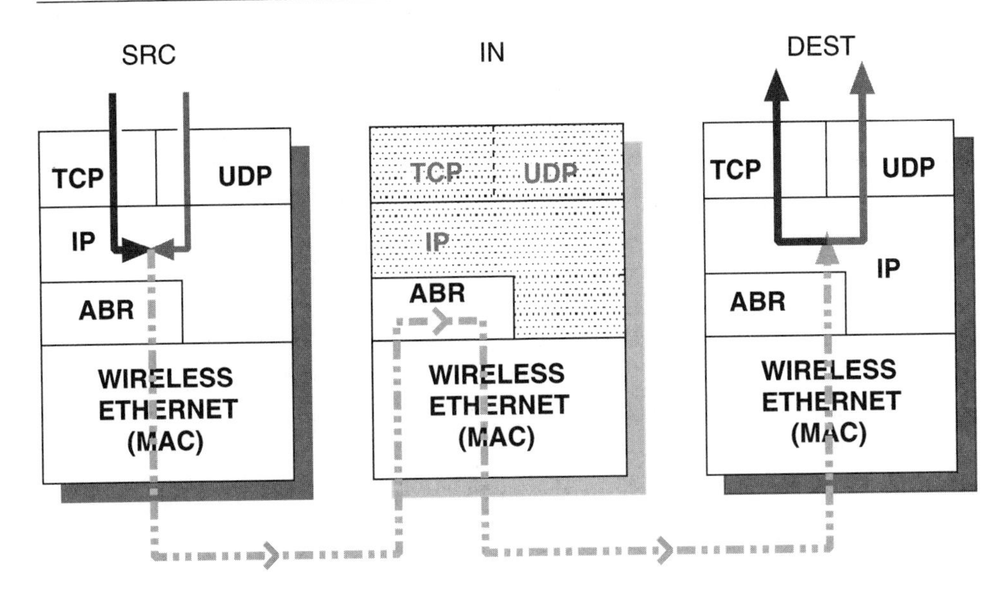

Figure 7.3. The principles of packet relaying in ABR protocol.

sublayer. The relaying of control packets is the same as for data packets.

7.2.3 Implementing ABR Packet Headers and Beaconing

As discussed earlier, ABR control and data packets have the same packet header. Basically, ABR packet headers are broken into two components: (a) the *base header* (or common header), and (b) the *type-specific header*. Every packet type has the same base header, but a different type-specific header. The "type" here refers to BQ, LQ, RD, or RN. The base header contains fields that are common to all ABR packet types. The format of this table is shown in Table 7.3. The complete header format and size, broken down by packet type, is revealed in Table 7.4.

Beacons are implemented to support the associativity ticks in the ABR protocol. Beacons are processed separately from the regular ABR traffic flow of *control* and *data* packets. Periodically, each node will transmit a beacon. Whenever a node receives a beacon from a neighbor, the associativity tick counter corresponding to that neighbor is incremented. ABR beacons are implemented within the ABR sublayer. However, the ABR beacon module could have been implemented at the data-link layer. The decision for implementing the ABR beacon module at the network layer was motivated by the fact that it was simplier to do so than to modify the WaveLAN device driver.

Table 7.3. ABR Base Header Format

Type	Field	Bytes
Base	Type	1/2
	Version	1/2
	Padding	1
	Length	2
	Source Address	4
	Dest Address	4
	CRC	4
Subtotal		16

7.2.4 Implementing ABR Outflow and Inflow

"Outflow" here refers to a source (SRC) wishing to send data to a receiver. During the experimentation, it was realized that trapping output from the IP to the device driver was essential for ABR to process packets. The ABR software checks to see if the destination (DEST) is an ABR host. If not, the packet proceeds as normal, oblivious of the ABR sublayer. Otherwise, for ABR destinations, it redoes the MAC header and inserts an ABR header between the Ethernet MAC and IP headers. A route discovery process is initiated if no route to DEST is available, and this is ascertained by examining the entries of the ABR routing table.

The phenomenon where packets arriving at an ABR host is known as *inflow*. In the implementation, each packet is registered with an ABR packet type, ETH_P_ABR, with the device driver. When a packet of such type arrives, it is passed up the protocol stack rather than being dropped. ABR control packets are processed in the ABR sublayer, invoking the respective protocol routines. ABR data packets, however, have the ABR header stripped and passed up to the IP layer for further processing. Hence, for data packets, the ABR sublayer appears transparent to IP and higher layers. This also implies that TCP/IP-based applications can be supported over ABR in an ad hoc wireless network environment.

7.2.5 Implementing ABR Routing Functions

- **Route Discovery (BQ/REPLY)**

  Route discovery is implemented by following the state machine shown in Figure 7.4. When a route request is initiated by an application, such as TELNET, a check for available routes to DEST is first performed. If the route

Table 7.4. Format of ABR Packet Headers Implemented

Type	Field	Bytes
BQ	Base Header	16
	Sequence No.	4
	No. of Hosts in Route	2
	IN Addresses	*4
	Associativity Ticks	*2
	No. of Neighboring Hosts	2
	Neighboring Host Addresses	*4
	Associativity Ticks with Neighbors	*2
Total		Variable
LQ	Base Header	16
	Sequence No.	4
	Live	2
	Number of Hosts in Route	2
	IN Addresses	*4
	Associativity Ticks	*2
	No. of Neighboring Hosts	2
	Neighboring Host Addresses	*4
	Associativity Ticks with Neighbors	*2
Total		Variable
RN	Base Header	16
	Originator ID	4
	Dir	1/2
	Step	1/2
Total		21
RD	Base Header	16
	Previous Node ID	4
	Sequence No.	4
Total		24
REPLY	Base Header	16
	Number of Hosts in Route	2
	Selected Host Addresses	*4
Total		Variable
DATA	Base Header	16
	Previous Node Address	4
	Next Node Address	4
Total		24

* Indicates a variable-length field.

request is new, then it must be discovered through the use of BQ packets. Note that the BQ/REPLY state machine is implemented in both the source and destination nodes.

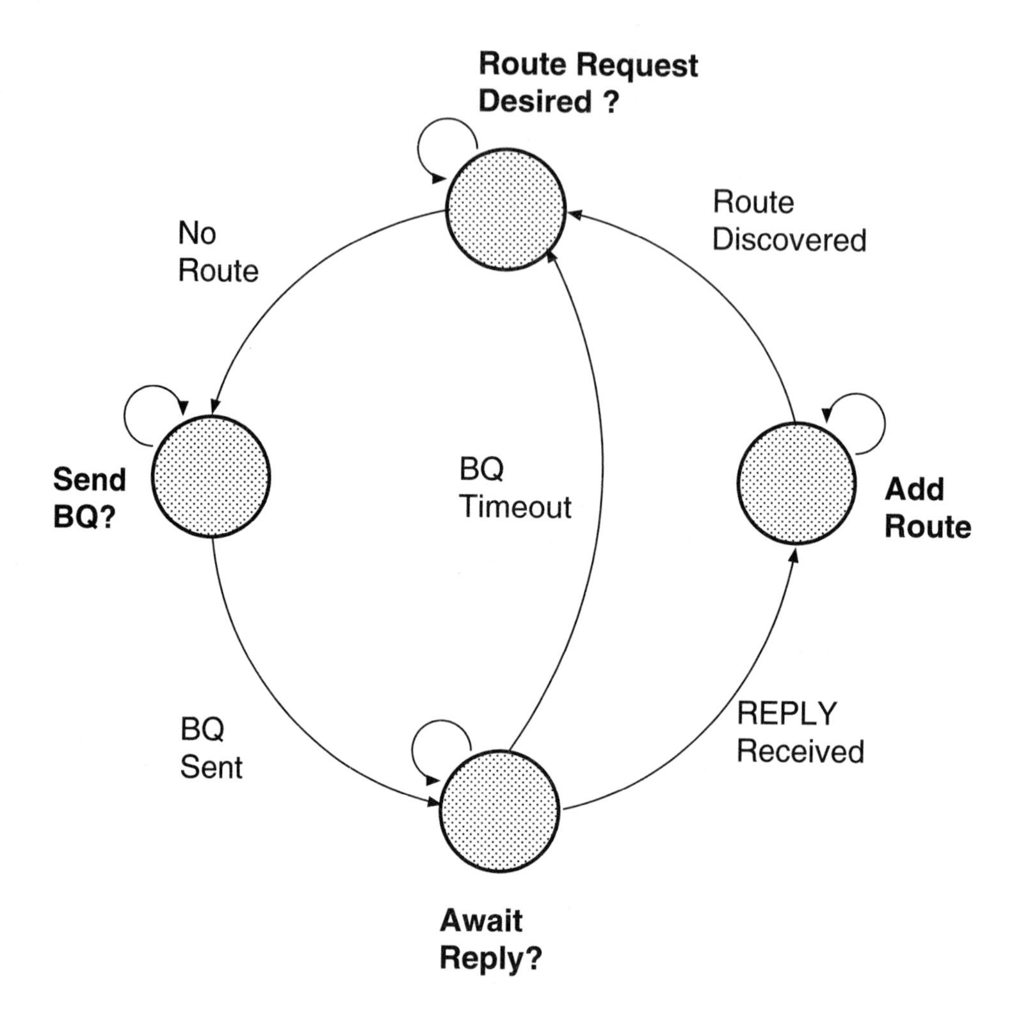

Figure 7.4. The ABR route discovery protocol state machine.

The route discovery phase is delayed until the packet is ready to be handed to the device driver. Another possible implementation when there is no route to the DEST is to perform route discovery in IP. This would mean "sleeping" the packet handler while route discovery is taking place. However, this is impossible since the networking code is not allowed to sleep, that is, it must be returned as soon as possible[1]. Therefore, another solution is necessary.

ARP has a similar dilemma with unresolved IP addresses. The Linux ARP implementation internally queues packets with unresolved addresses, and the networking code is allowed to return as if packets were successfully transmitted. The application is unaware that its packets are queued internally. When an ARP reply comes back, the ARP module checks its internal queue for packets awaiting a reply. Then, packets that have a resolved IP address are finally sent to the device driver.

In the same vein, ABR packets are checked before being sent to the interface. If route discovery is not required, the next-hop address is written into the ABR header and the packet is sent to the device driver. If route discovery is required, the packet is queued internally and the BQ/REPLY cycle is invoked by transmitting a BQ packet. While route discovery is in progress, a new unresolved routing entry for the destination is added to the routing table, and packets awaiting a REPLY packet are queued. When a REPLY packet arrives, a route is added and any queued packets are sent out via the device driver. The BQ/REPLY routine is implemented as a state machine, which is illustrated in Figure 7.4.

- **Route Reconstruction, or RRC (LQ/REPLY)**

  The ABR protocol invokes RRC when a link in the route is invalidated. This phrase is interpreted as:

  > *"A host has moved when the expected beacons do not arrive after a specified time period."*

  After receiving a beacon, a timeout for ABR_BEACON_INTERVAL is set. If the timer expires before receiving another beacon, then the host is assumed to have moved (or we have moved away from it). In an ad hoc wireless network environment, moves are relative. For instance, if host $\mathcal{A}$ moves away from host $\mathcal{B}$, it can also be interpreted as host $\mathcal{B}$ moving away from host $\mathcal{A}$. When movement is detected and the on-going route is invalidated, an RRC process is invoked, which determines what actions are needed:

  (a) Nothing is performed if the host that has moved is not part of any active routes.

[1] In private emails with Alan Cox, a contributor to major portions of the Linux networking code, it was confirmed that sleeping in the protocol stack (e.g., IP) is simply not allowed.

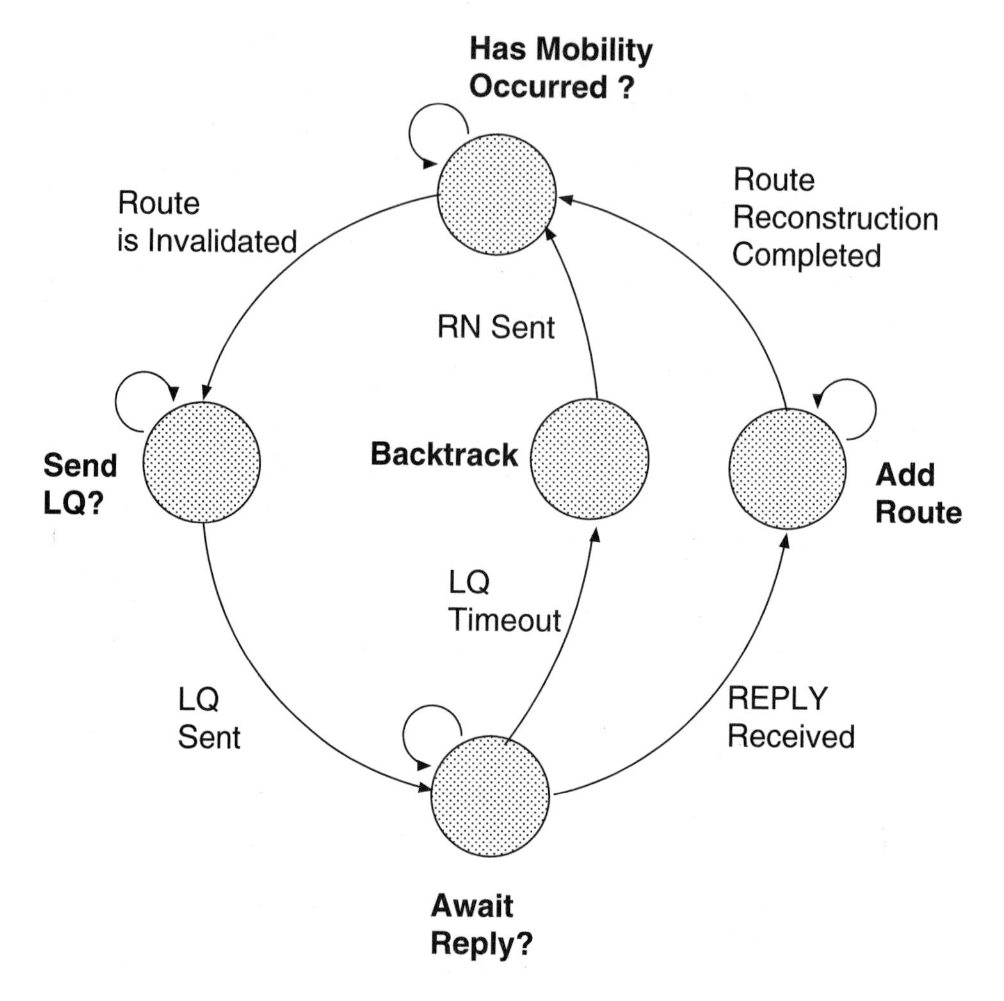

Figure 7.5. The ABR route reconstruction protocol state machine.

(b) A BQ packet is sent if the current host is the SRC and the host that has moved is the next-hop node downstream or a RN[1] message is received.

(c) An LQ packet is sent if the host that has moved is the next-hop node downstream of an active route.

(d) A RN[0] is sent in the downstream direction, if the host that has moved is the previous hop (upstream) of an active route.

When an LQ packet is sent, a timer is set for LQ_TIMEOUT. If a RE-

PLY packet is not received within this time, a backtracking process is initiated. Backtracking here refers to requesting the upstream node to intiate an LQ/REPLY cycle. However, if the current node is at the midpoint of the old route, then the backtrack process stops at this node and an RN[1] packet is sent back to the SRC to invoke a BQ. This is to shorten the RRC time and limit the degree of RRC traffic. The state diagram for the LQ/REPLY cycle is shown in Figure 7.5.

- **Route Deletion (RD)**

  To remove undesired routes, the SRC node can explicitly send an RD packet so that nodes supporting such routes can update their routing tables. In addition to the explicit invocation of an RD, a soft state approach is also available.

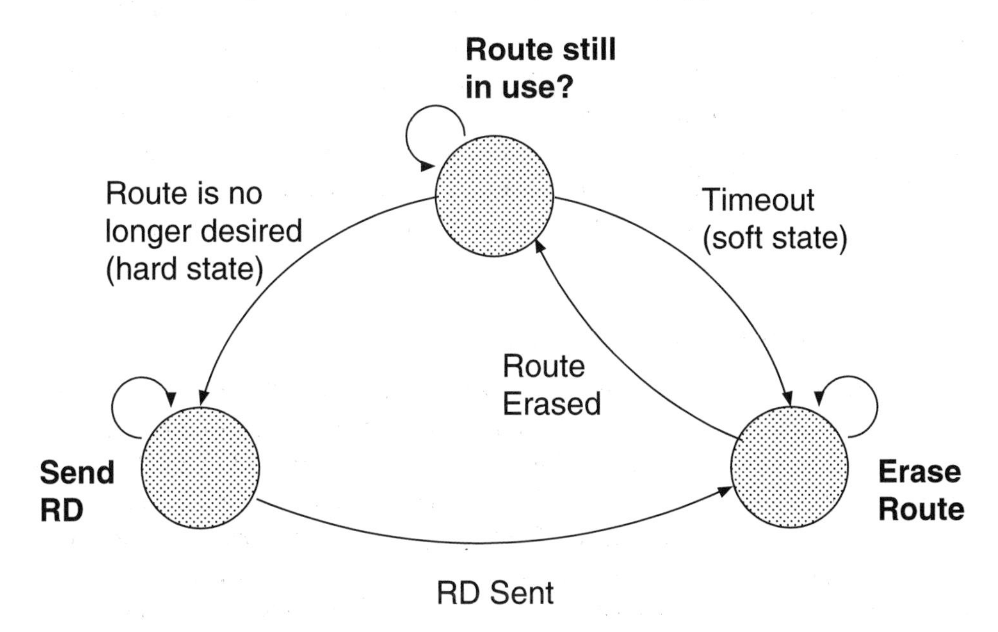

Figure 7.6. The ABR route delete protocol state machine.

Basically, a ROUTE_TIMEOUT is set on each route entry at each node in the route. This allows routes that are no longer in use to be erased. If this timeout (ROUTE_TIMEOUT) occurs at the SRC, an RD packet is broadcast instead of a mere route table update. This is to prevent other nodes from performing RRCs on routes that are no longer desired. Timeouts are useful especially if nodes fail to receive the RD packet. Note that RD packets can only be sent by SRC and INs will never send an RD packet, even if they time out on the route. The RD state diagram is shown in Figure 7.6.

- **Implementing Packet Forwarding**

  When a packet arrives at an ABR host, it is examined and a decision is made as to whether the packet should be passed up the protocol stack (if the host is indeed the DEST) or forwarded. In the latter case, a lookup on the ABR routing table is necessary to determine the next-hop node address. The packet next-hop field is then updated and sent to the device driver for transmission.

7.3 Experimentation and Protocol Performance

After implementing the ABR protocol, it is important to evaluate its performance. Hence, several one-, two-, three-hop route experiments were performed. Laptops were spatially separated, as shown in Figure 7.7a. Because of the large wireless cell size, specially designed metallic shields were used on the WaveLAN antennas, as shown in Figure 7.7b. In addition, most of the laptops were placed inside offices.

Performance parameters of interest include: (a) control packet overhead, (b) route discovery time, (c) data throughput, (d) end-to-end delay, and (e) the effect of beaconing interval on battery life.

7.3.1 Control Packet Overhead

Before proceeding with experimentation results, an examination of the ABR control packet overhead is necessary. An IP header is 20 bytes in size (assuming with no IP options). From Table 7.4, an ABR data packet header is 24 bytes. With Ethernet, the maximum transfer unit (MTU), minus the MAC header, is 1500 bytes. The packet overhead of ABR is therefore: $\frac{24+20}{1500} = 2.9\%$ whereas IP has an overhead of: $\frac{20}{1500} = 1.3\%$. In comparison, the ABR header adds a mere 1.6% overhead. However, as noted earlier, since ABR basically replaces IP to support ad hoc routing, the IP header is not really necessary. If the IP header is replaced with ABR, then the total overhead would be: $\frac{24}{1500} = 1.6\%$, which translates to a 0.3% increase. From these numbers, the overhead incurred by the ABR header is considered negligible. IP is retained so that ad hoc routing is transparent to existing TCP/UDP/IP-based applications.

7.3.2 Route Discovery Time

To measure the route discovery time, three different test scenarios were performed in an indoor environment. Each scenario had a route with a different hop count,

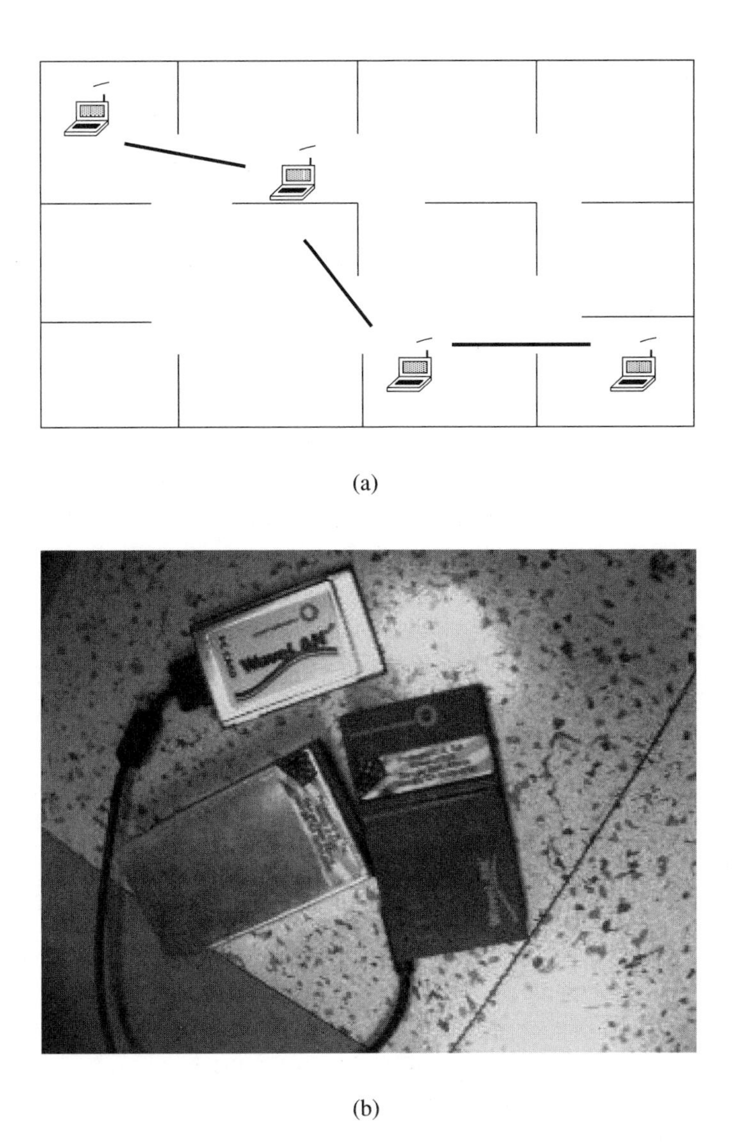

(a)

(b)

Figure 7.7. (a) Experimental setup of an ad hoc mobile network inside a building and (b) Use of metallic shield to limit radio transmission range.

ranging from one to three hops. Using these scenarios, the `ping` application was used to initiate sending data from one ABR host to another.

Route discovery time is the time when a BQ packet is sent by the SRC until the

Table 7.5. Results of route discovery times wrt. route hop count

Route Hop Count	Average Route Discovery Time (ms)	Standard Deviation
1	6.5	0.23
2	13.1	0.35
3	18.6	0.66

time when a route is added to the routing table (after receiving a REPLY from the DEST). It reflects how long a user has wait before data can be transmitted. For each route, the route discovery process is performed 100 times. Basically, when one ping packet is sent, the route is invalidated and route discovery is repeated 99 times. Route discovery time accounts for the entire round-trip time (RTT) from SRC to DEST. The results are shown in Table 7.5. Clearly, as the route hop count increases, the route discovery time also increases. These latencies are relatively short and easily tolerable by mobile users.

7.3.3 End-to-End Delay

In this experiment, data are sent using `ping` from the SRC to the DEST for 100 times and the RTT are recorded. For each set of tests, the `ping` packet size is gradually increased. The default ping packet size was 64 bytes. For every `ping` packet received at the DEST, a `ping` status packet was returned to the SRC, revealing the RTT. The end-to-end delay incurred was, therefore, one-half the RTT.

As shown in Table 7.6, for a single hop, the delay is 3.3 ms. For a larger hop count, this delay increases almost proportionally. However, as shown in Figure 7.8, this delay increases with increasing packet size, signifying that packet transmission time is getting more significant than propagation time. This was true for the one-, two-, and three-hop routes under test. Note that for two- and three-hop routes, there was no response back from ping when the `ping` packet size went beyond 1500 bytes. This suggests that as `ping` packet size increases, more packets need to be transmitted and reassembled. The link layer will perform retransmission if packets are not properly received. All these delays prevent the `ping` reply packet from arriving at the SRC before it expires.

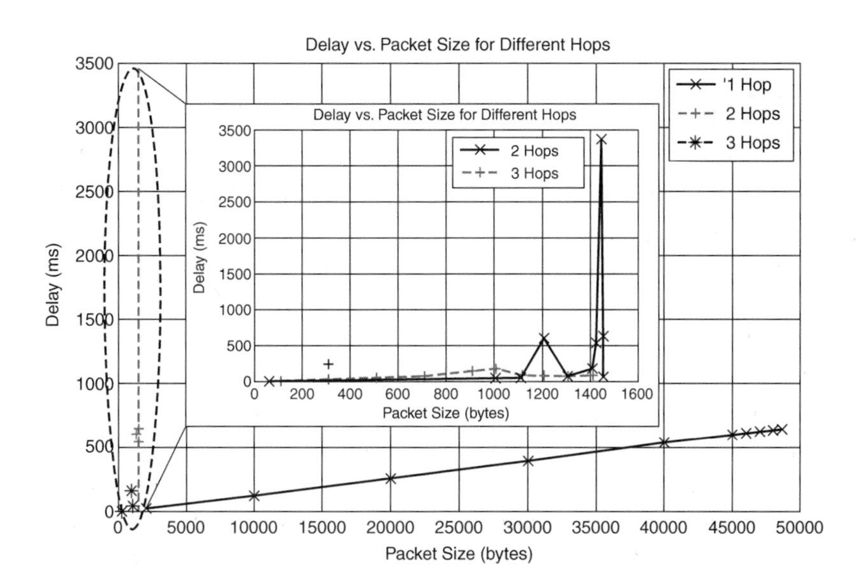

Figure 7.8. End-to-end delay performance wrt. route hop count (at one second beaconing interval).

Table 7.6. Average end-to-end delay wrt. route hop count for 64 Bytes of `ping` transmission.

Route Hop Count	Average End-to-End Delay (ms)
1	3.3
2	7.86
3	10.80

7.3.4 Data Throughput

Based on the statistics provided by performing `ping` 100 times, data throughput was derived and is shown in Figure 7.9a. An evaluation of throughput with respect to packet size, and route hop count is made. The average throughput is computed by dividing the packet size by the average end-to-end delay. The actual throughput is computed by dividing the total number of bytes received by the total end-to-end delay. Note that for one hop, the communication throughput goes as high as 1.17 Mbps. At two hops, the throughput falls by more than 50%, down to 440 Kbps. Extending another radio hop brings this throughput down to about 300 Kbps. It is also observed that as packet size increases, communication throughput increases, and it saturates at larger packet size. The limit in throughput is due to the fact that there is no allowable channel reuse and media access is based on contenting for access over a single and common frequency channel among all ad hoc nodes.

7.3.5 Effects of Beaconing on Battery Life

Since ABR relies on the periodic beaconing of mobile hosts to gather associativity information, it is important to know the effects of beaconing on power consumption and battery life. The laptops that were used in the experiments had NiMh batteries with a battery life of about 2.8 hours. These laptops had several power management features, such as:

(a) **standby mode** - where the LCD backlight is turned off, the hard disk motor is switched off, and the CPU is set to the lowest speed

(b) **suspend mode** - where the LCD backlight, LCD power, and hard disk are powered off, the CPU clock is set to lowest speed, and the CPU is stopped

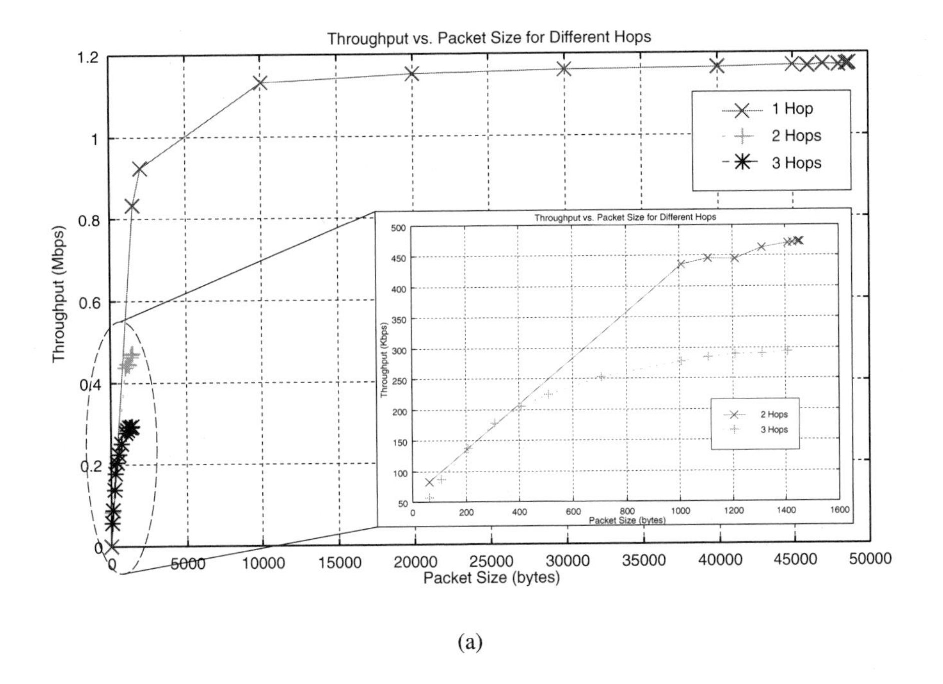

(a)

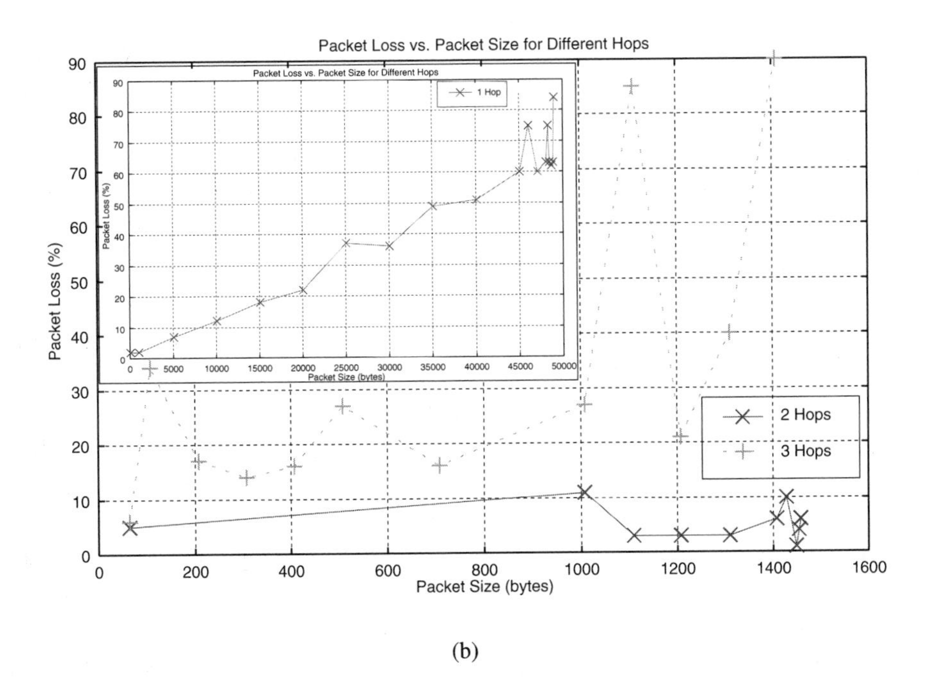

(b)

Figure 7.9. (a) Throughput performance, and (b) Packet loss performance wrt. route hop count (1 second beaconing interval).

(c) **hibernation mode** - where the system status, RAM, VRAM, and setup data are stored on the hard disk and the system is then powered off.

During the experiments, these power management features were not disabled since normal users would have them enabled to prolong the lifetime of the laptop's operation. With the same experimental setup mentioned earlier, the beaconing interval on the mobile hosts was varied and their battery life was noted at regular periods. Figure 7.10b shows the window capture when an ad hoc host is beaconing at a 100ms interval. The percentage of battery life remaining obtained for different beaconing intervals is shown in Figure 7.10a, which reveal that varying the beaconing interval from 0.1 to 120 secs does not significantly affect the gradient of the graphs for more than two hours of continuous operation. This means that increasing the beaconing frequency **does not** necessarily result in significant reduction in battery life. In-depth discussions of beaconing on power life and communication performance will be made in subsequent chapters.

7.4 Important Deductions

The implementation of ABR protocol in Linux is fairly straightforward. Only minor modifications are made to the existing TCP/UDP/IP/Ethernet stack to accommodate the ABR sublayer to support ad hoc wireless networking. The challenge lies in deciding where the ABR sublayer should reside and how to interface the ABR sublayer with the existing IP and Ethernet layers. Comparing the route discovery times (Table 7.5) with the data round trip times (i.e., twice the end-to-end delay), the former is shorter since route discovery occurs at the ABR sublayer, whereas data packets orginate up at the application layer, resulting in a few milliseconds of processing time as the packets traverse the protocol stack.

The implementation results also reveal that route discovery is relatively fast, on the order of a few milliseconds for the range of route hop counts examined. The throughput performance for multi-hop routes is satisfactory. At three hops, a throughput of 485 Kbps can support many existing applications. The RRC times obtained indicate that partial RRC can be established quickly in times of mobility. Extrapolating from the numbers reported, ABR could scale to a larger environment than the one tested. The route discovery time (from Table 7.5) is just about 6.5 ms per hop. Hence, one could possibly discover a 10-hop route in less than a second. The throughput would be lower, however, as a result of longer delays and also the inability of using multiple contention-free channels.

The effects of beaconing on battery life was also examined. Increasing the beaconing interval does not significantly change the degradation characteristic of

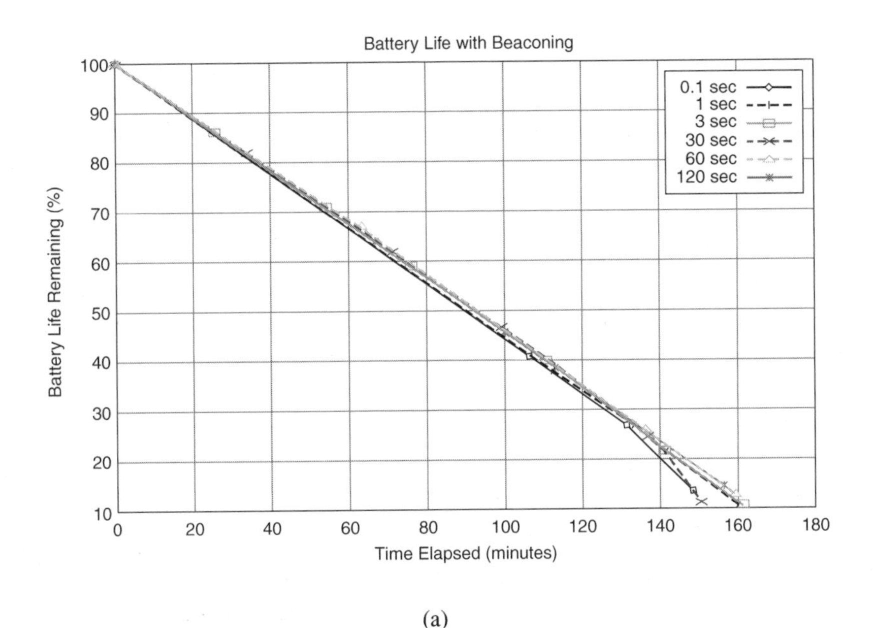

(a)

```
nxterm
Iface   MTU Met  RX-OK RX-ERR RX-DRP RX-OVR   TX-OK TX-ERR TX-DRP TX-OVR Flags
lo     3584   0     26      0      0      0      26      0      0      0 BLRU
eth0   1500   0      0      3      0      0    2020      0      0      0 BRU
Kernel Interface table
Iface   MTU Met  RX-OK RX-ERR RX-DRP RX-OVR   TX-OK TX-ERR TX-DRP TX-OVR Flags
lo     3584   0     26      0      0      0      26      0      0      0 BLRU
eth0   1500   0      0      3      0      0    2030      0      0      0 BRU
Kernel Interface table
Iface   MTU Met  RX-OK RX-ERR RX-DRP RX-OVR   TX-OK TX-ERR TX-DRP TX-OVR Flags
lo     3584   0     26      0      0      0      26      0      0      0 BLRU
eth0   1500   0      0      3      0      0    2040      0      0      0 BRU
Kernel Interface table
Iface   MTU Met  RX-OK RX-ERR RX-DRP RX-OVR   TX-OK TX-ERR TX-DRP TX-OVR Flags
lo     3584   0     26      0      0      0      26      0      0      0 BLRU
eth0   1500   0      0      3      0      0    2050      0      0      0 BRU
Kernel Interface table
Iface   MTU Met  RX-OK RX-ERR RX-DRP RX-OVR   TX-OK TX-ERR TX-DRP TX-OVR Flags
lo     3584   0     26      0      0      0      26      0      0      0 BLRU
eth0   1500   0      0      3      0      0    2060      0      0      0 BRU
Kernel Interface table
Iface   MTU Met  RX-OK RX-ERR RX-DRP RX-OVR   TX-OK TX-ERR TX-DRP TX-OVR Flags
lo     3584   0     26      0      0      0      26      0      0      0 BLRU
eth0   1500   0      0      3      0      0    2070      0      0      0 BRU
```

(b)

Figure 7.10. (a) Effects of beaconing interval on battery life, and (b) Window capture showing beaconing at 100ms interval.

battery life under the condition that power management is turned on. These findings, therefore, allow us to deduce that implementing ABR to support multi-hop neworking is both *feasible* and *practical* while providing reasonable and acceptable communication performance.

7.5 Conclusions

In this chapter, the implementation of the ABR protocol using COS hardware and newly enhanced communicatons software was presented. It was revealed that the resulting software architecture for each ad hoc mobile node is still capable of supporting existing UDP/TCP/IP-based applications. It is an important element since the ability to execute current applications is attractive to many computer users. The performance of the Associativity Based Routing (ABR) protocol was evaluated in terms of route discovery time, end-to-end delay, packet loss, and amount of control overhead in an indoor environment. The feasibility and practicality of ABR was demonstrated and the impact of using beacons on battery life was examined. In the next chapter, we will discuss in greater depth the communication performance of a practical ad hoc wireless network in an outdoor environment.

Chapter 8

COMMUNICATION PERFORMANCE OF AD HOC NETWORKS

8.1 Introduction

A communication network must be able to transport user traffic toward its targeted destination. The communication performance of a network can affect the satisfaction level of users. The ability to send information at high speeds demands low end-to-end delay. In addition, users would like to transmit a variety of information, such as data, audio, and video.

An ad hoc network is not useful if it cannot offer acceptable communication services. Given the dynamics of the network topology, the underlying protocols must be able to cope with these dynamics efficiently while at the same time yielding good communication performance. In this chapter, we will examine the communication performance of an ad hoc wireless network and how it affects existing applications. Results discussed in this chapter were obtained from several experimental field trails performed in Atlanta, Georgia.

8.1.1 ABR Beaconing

Periodic beaconing has been implemented within ABR to gather longevity information. The beacon structure used is shown in Figure 8.1,which reveals the ABR base header (BH), with the type field defined as $BEACON$, encapsulated by the data link header (in this case, the Ethernet header). The ABR beacon is a control message generated at the ABR layer and broadcast by a node to its immediate neighbors. The transmitted beacons are received and updated by neighboring nodes to reflect longevity information.

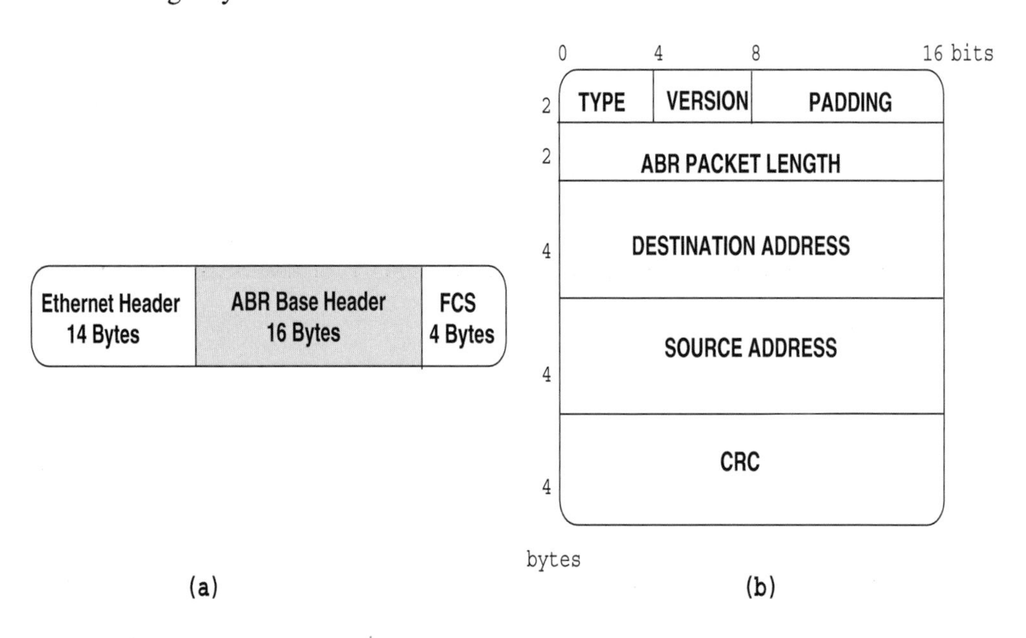

Figure 8.1. The ABR beacon structure.

The beaconing interval is an important issue worthy of investigation since it has strong implications on: (a) the amount of bandwidth and power consumed, (b) the accuracy of longevity information gathered, and (c) the time to detect a link failure and initiate route repair. In the author's implementation, one can readily change the beaconing interval to observe its impact on communication performance. Several other ad hoc routing protocols (such as RDMAR and STAR) proposed to the

IETF (Internet Engineering Task Force) also utilize periodic updates to detect the presence of mobility, but not association stability[1].

8.2 Performance Parameters of Interest

In order to further evaluate the communication performance of the ad hoc wireless network, several experimental field trials were performed in an outdoor environment. Performance results were then recorded. The parameters of interests include:

(a) **Route discovery time:** The time needed for the source node to discover a route to the destination.

(b) **End-to-end delay:** The delay experienced by a packet from the time it was sent by the source till the time it was received at the destination.

(c) **Communication throughput:** The useful data rate in bits per second (bps).

(d) **Packet loss:** The percentage of packet loss during 100 `ping` of data.

(e) **Route reconstruction time:** The time taken upon triggering route reconstruction or repair (when mobility occurs) till completion.

The above performance parameters were evaluated with different packet size, beaconing interval, and route length.

8.3 Route Discovery (RD) Time

Ad hoc routing protocols that are source-initiated on-demand-based initiate a route discovery (RD) whenever a route that is desired by the source is not immediately available in the route cache. The RD process is invoked by the source node sending a BQ control packet that is broadcast in search of valid route(s) to the destination. After some delay, if the network is not partitioned, the BQ control packet will ultimately reach the destination node. The destination node uncovers the route path information contained inside the BQ control packet. A REPLY control packet is then sent back via the reverse path, so that ultimately, the source is informed about the discovered route.

In the field trial, ABR invoked an RD process each time a `ping` to a new destination was performed. We defined the RD time as the time between when an

[1] Association stability information is needed to support long-lived ad hoc routing.

RD control packet was sent and when the route reply packet was received by the source node. The results of 100 RD requests for different beaconing intervals and route length. Evaluating RD with varying data packet size is omitted, however, since there is no direct relationship between the two variables.

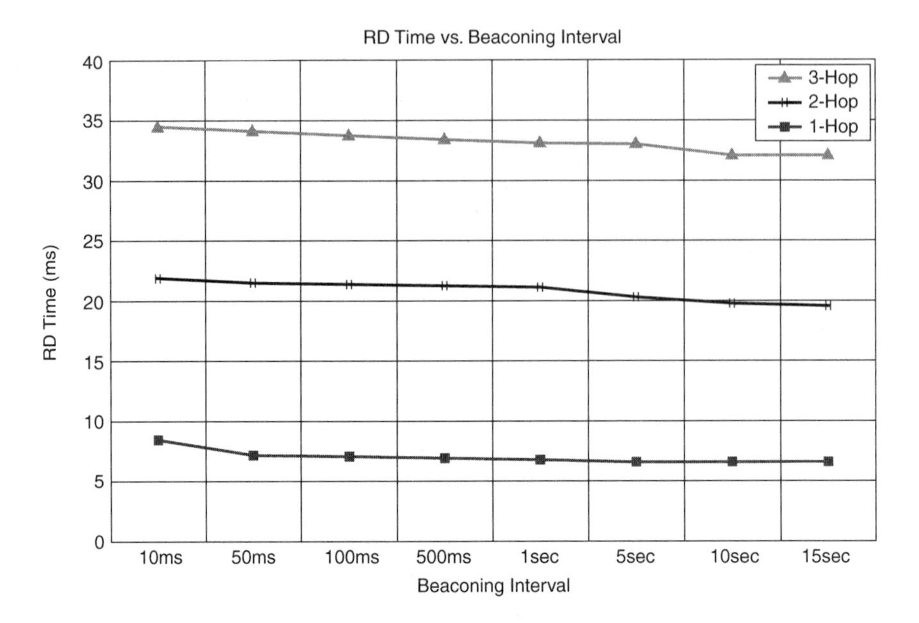

Figure 8.2. Route discovery time vs. beaconing interval for different hop counts (outdoor field trial results).

8.3.1 Impact of Beaconing Interval on RD Time

Next, by changing the beaconing intervals of all the concerned ad hoc mobile computers under test, the route discovery times are recorded. Figure 8.2 reveals that RD time increases slightly when beaconing frequency increases. This relationship is almost linear. The difference between the maximum and minimun RD values over different beaconing intervals for one-, two-, and three-hop routes are 1.92, 2.39, and 2.50 msecs, respectively. Beaconing at high frequencies has a greater impact on the RD time. This additional delay experienced by nodes in the route is due to the increase in channel contention.

8.3.2 Impact of Route Length on RD Time

Figure 8.2 shows that RD time increases as route length increases. With increasing route length, the total propagation delay is increased. This also applies to the

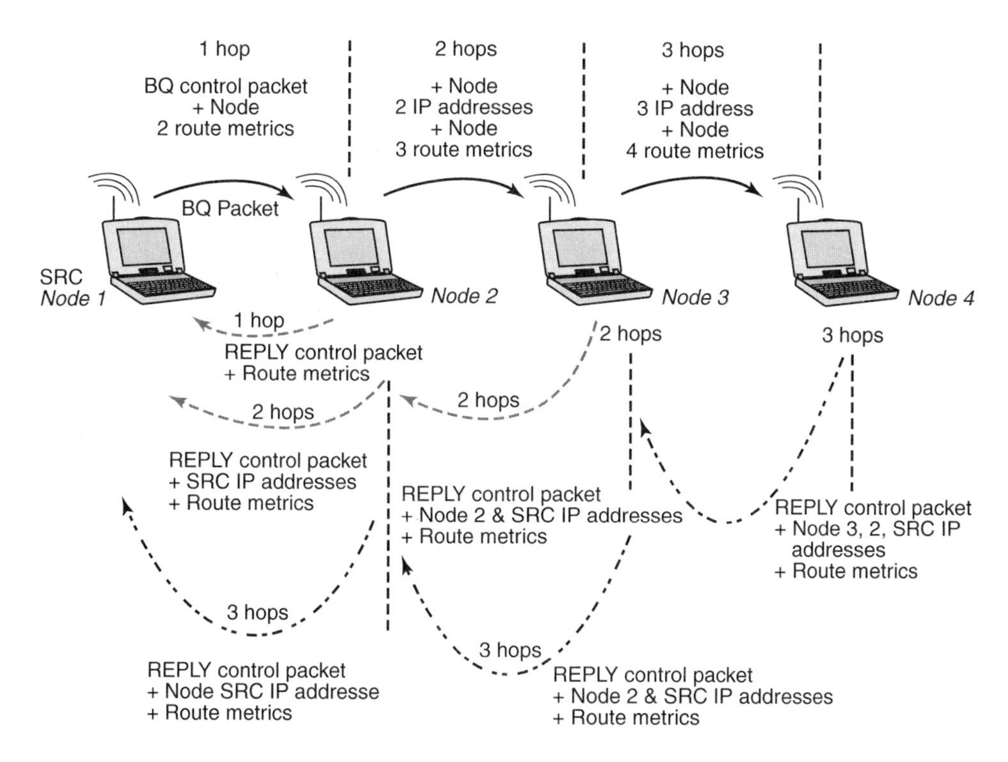

Figure 8.3. Changes in BQ/REPLY control packet sizes during the route discovery process.

total processing and queuing delay incurred at the nodes in the route. Table 8.1 shows that the RD times recorded for one-second beaconing interval increases with increasing route length.

It is clear from Table 8.1 that the RD time has a non-linear relationship with route length. If the RD process utilized fixed-size messages, one would expect the delay time to increase by a fixed amount for every additional hop. However, the RD time increases by 14.3 ms and 12.04 ms for the two- and three-hop routes, respectively. This can be explained by examining in detail how the RD control messages are generated and forwarded. Figure 8.3 shows an example of the ABR RD process for the one-, two-, and three-hop routes. As shown, the source node sends BQ packets to discover routes.

The BQ packet flow shown in Figure 8.3 indicates the basic and additional information that must be included in the BQ packet at every IN toward the destination. The figure also indicates that the BQ packet becomes larger as it is forwarded, since it has to record the IP address of every node it transits. The REPLY packet,

Table 8.1. RD Time at one second beaconing interval for different hop counts

Route Length	Average RD time
One Hop	6.64 ms
Two Hops	20.94 ms
Three Hops	32.98 ms

however, is larger when the route is longer since the IP addresses of all nodes in the route are included in the packet sent by the destination. With the presence of these variable-sized control packets during the RD process, the non-linear relationship between RD time and route length is, therefore, observed.

8.4 End-to-End Delay (EED) Performance

EED generally accounts for all delays along the path from the source to the destination. This includes the transmission delay, propagation delay, processing delay, and queuing delay experienced at every node in the route. To measure the EED, packets are sent from the source to the destination using the `ping` utility, over different route lengths and beaconing intervals. The EED is taken as one-half the RTT. The impact of varying route length and beaconing interval on EED shall be discussed below.

8.4.1 • Impact of Packet Size on EED

Varying the transmission packet size has a direct influence on the EED of ad hoc wireless routes. This is because the larger the packet size, the longer the packet transmission, propagation, and processing times. This is further verified by Figures 8.4, 8.5, and 8.6. The increase in EED as a result of increasing packet size exhibits a relatively linear relationship.

Average EED			
Hop Count	**Min Size**	**1000 bytes**	**Max Size**
One hop	3.25ms	10.40ms	340.00ms
Two hops	6.20ms	19.70ms	26.20ms
Three hops	8.80ms	29.20ms	38.30ms

Table 8.2. Average end-to-end delay at minimum, 1000-byte, and maximum packet size.

Table 8.2 shows the average EED for different packet sizes. The minimum and maximum packet sizes are 64 and 48,856 bytes (for one-hop route experiments) or 1448 bytes (for two- and three-hop experiments). As seen from Table 8.2, EED in the one-hop case increases from 3.25 ms to 340 ms when the packet size is increased from 64 bytes to 48,856 bytes. This corresponds to an average increase of 0.69 ms per 100 bytes. For two-hop scenarios, EED increases from 6.20 ms to 26.20 ms when the packet size is increased from 64 bytes to 1,448 bytes. This reveals an average increase of 1.44 ms per 100 bytes, which is approximately twice the value of the one-hop scenario. Finally, the EED in the three-hop experiments increases from 8.80 ms to 38.30 ms, with an average increase of 2.13 ms per 100 bytes. This value is approximately triple that of the one-hop case.

8.4.2 • Impact of Beaconing Interval on EED

Another important issue is the effect of varying beaconing frequency of ad hoc mobile computers on EED performance. Results presented in Figures 8.4, 8.5, and 8.6 indicate that varying the beaconing frequency does not have a significant impact on EED performance, provided that other factors (such as packet size and route hop count) are kept constant. An exception is the case when beaconing is performed at very high frequencies. In such situations, an increase in EED is observed due to the presence of severe contention over the wireless media.

This phenomenon is a result of the MAC protocol operation. The WaveLAN cards used in the field trial is based on CSMA/CA (Carrier Sense Media Access with Collision Avoidance) protocol. This protocol prevents collisions at the moment they are most likely to occur, that is, when the channel is released. All nodes are forced to wait for a random number of time-slots before sensing the medium again for an opportunity to resume retransmission.

8.4.3 • Impact of Route Length on EED

So far, EED has been examined as a result of different packet size and beaconing interval. Ad hoc networks are multi-hop networks, so it is important to evaluate EED for different route length as well. A comparison of Figures 8.4, 8.5, and 8.6 shows that EED increases as route length increases. Each intermediate node that participates in the route introduces additional delays in transmission, propagation, and processing of a packet. Table 8.2 lists the average EED measured at 64- and 1000-byte packet sizes, and shows that a packet propagating through the network experiences approximately the same amount of delay at every route hop it transits. For a 64-byte packet, each additional route hop introduces about 3 msec of delay. For a 1000-byte packet, each additional hop introduces approximately 10 msec of delay.

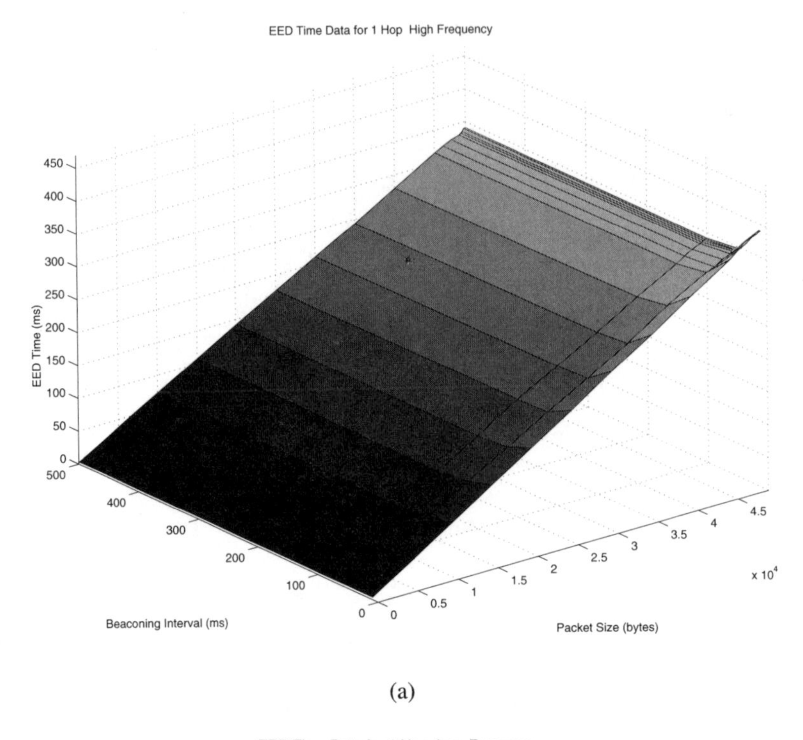

(a)

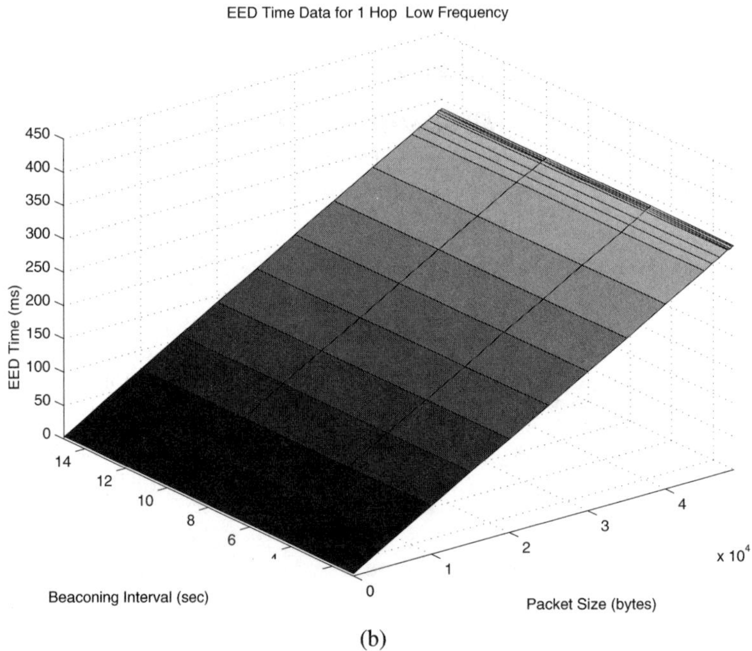

(b)

Figure 8.4. EED vs. packet size and beaconing interval for a one-hop route while beaconing at: (a) High frequencies, and (b) Low frequencies.

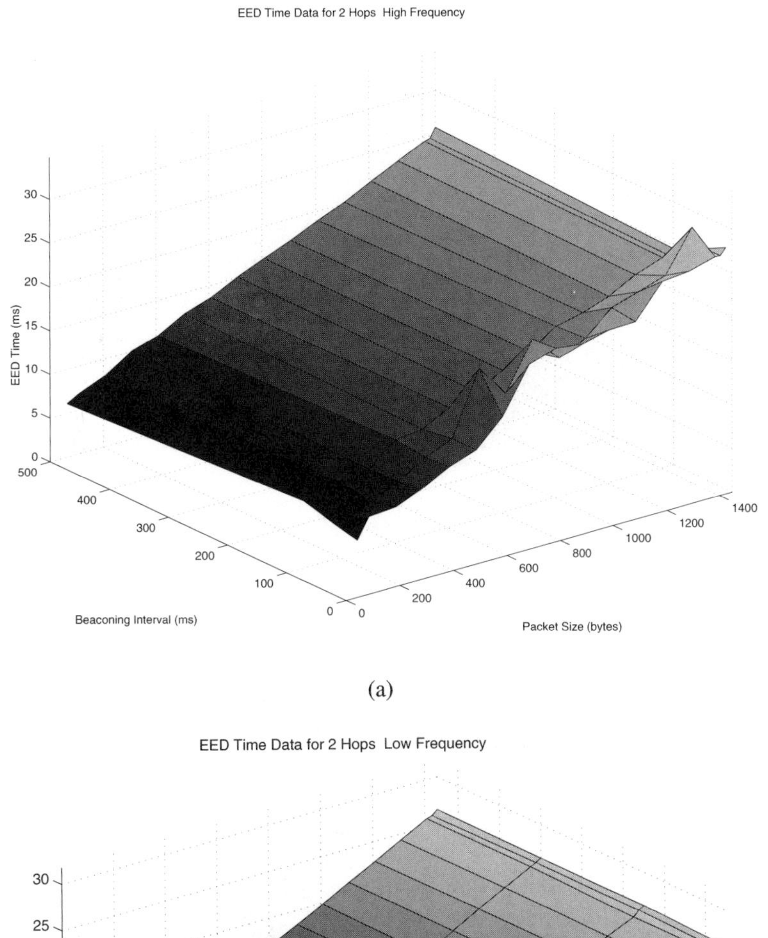

(a)

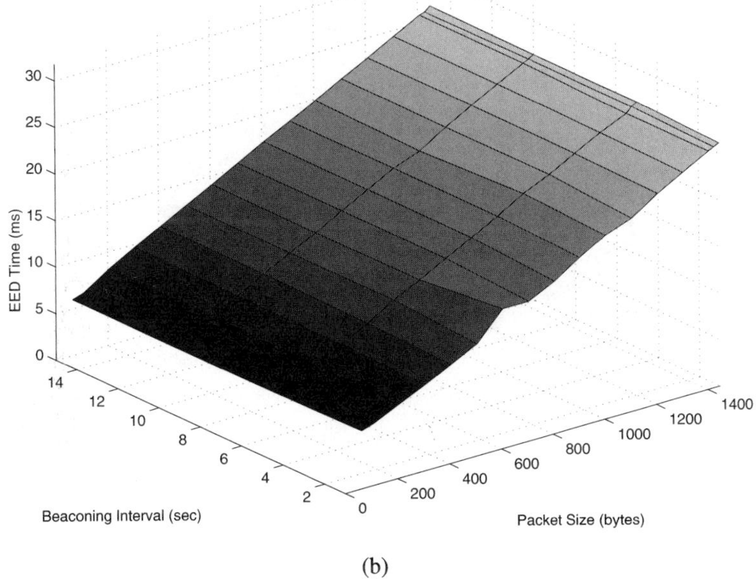

(b)

Figure 8.5. EED vs. packet size and beaconing interval for a two-hop route while beaconing at: (a) High frequencies, and (b) Low frequencies.

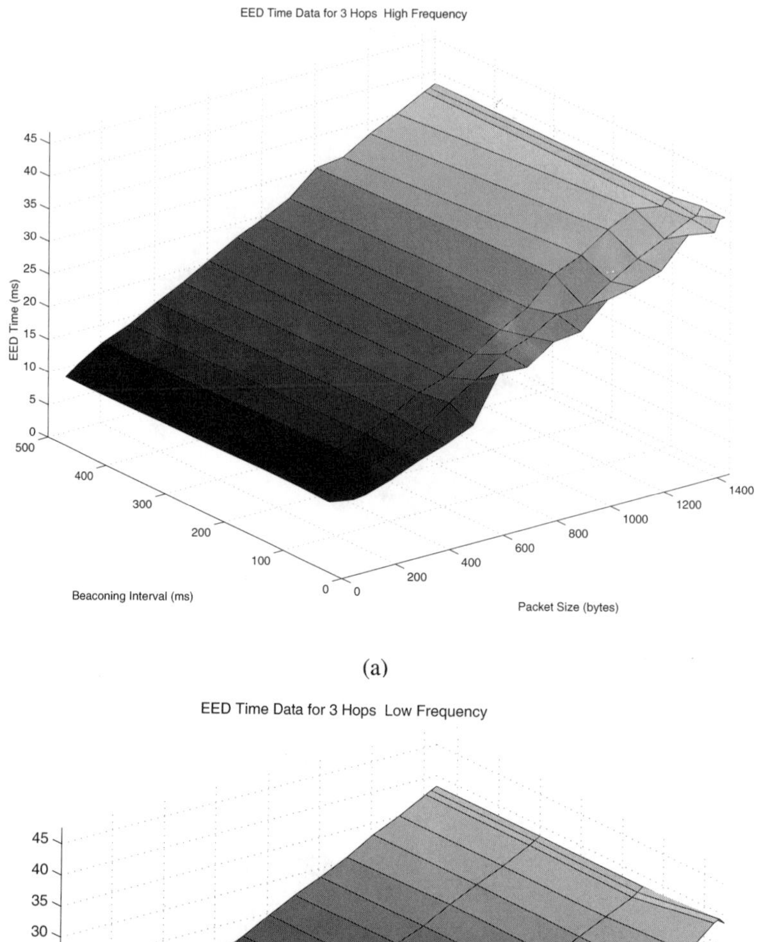

(a)

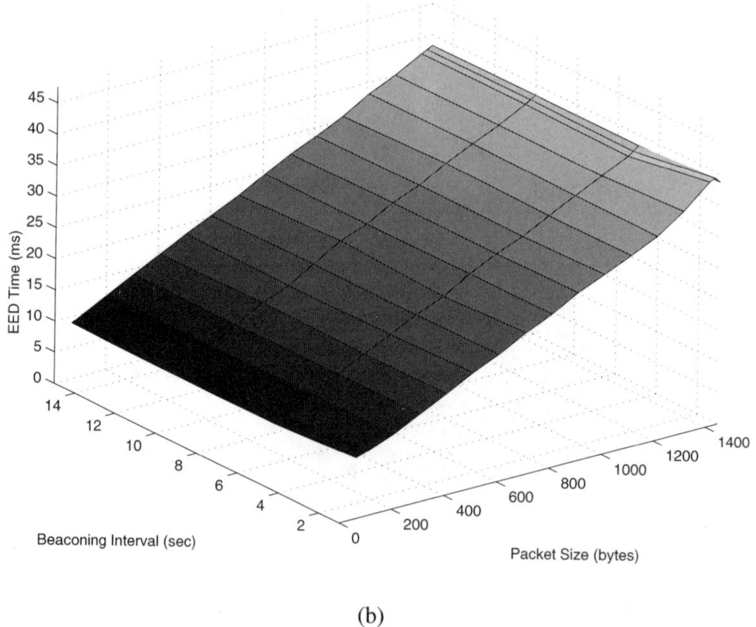

(b)

Figure 8.6. EED vs. packet size and beaconing interval for a three-hop route while beaconing at: (a) High frequencies, and (b) Low frequencies.

The importance of the above observations and deductions extends not only to the understanding that the per-hop delay is relatively constant, but also to the issue of scalability. The delay for an MTU[2]-sized packet transfer over a one-hop link is 11 ms. Hence, if we assume a network with a diameter of ten hops, then the resulting EED is only 110 msec. Therefore, this delay value is reasonable and acceptable for most data applications.

8.5 Communication Throughput Performance

Ad hoc wireless networking will continue to support existing TCP/IP-based applications. Communication throughput can affect the operation and performance of some applications. Hence, it is essential to investigate throughput performance for ad hoc wireless networks. A way to measure throughput is by sending data from the source to the destination using `ping`. The average throughput is, therefore, the ratio between the total data received and the total delay incurred. The impact of varying *packet size*, *beaconing interval*, and *route length* on communication throughput will be analyzed and discussed below.

8.5.1 • Impact of Packet Size on Throughput

The results obtained from the `ping` experiments are plotted in several 3-dimensional graphs, as shown by Figures 8.7, 8.8, and 8.9. Figure 8.7b shows the variation of throughput performance with increasing packet size. Considering the specific case when the beaconing interval is 15 s, communication throughput increases from 157.5 kbps to 775.4 kbps when packet size is increased from 64 to 1008 bytes. This corresponds to an increase of 392% in throughput. When the packet size is increased to 5056 bytes, the communication throughput reaches 1.05 Mbps. After reaching a packet size of 10,056 bytes, there is no significant change in throughput and it saturates at 1.16 Mbps.

A similar impact of packet size on throughput is observed for both two- and three-hop routes. Consider Figure 8.8b, which shows the throughput for the two-hop route experiments. Again, at the 15s interval, when packet size increases from 64 bytes to 1,008 bytes, throughput increases from 85.3 kbps to 406.2 kbps. This corresponds to an increase of 376.2%. As the packet size continues to increase, throughput increases at a slower rate. In the two-hop route scenario, the maximum packet size permissible using the `ping` application is 1448 bytes, and the maximum obtainable throughput is 439.4 kbps.

[2]MTU (Maximum Transfer Unit) refers to the maximum size in octets of the data that can be sent in one packet on the network.

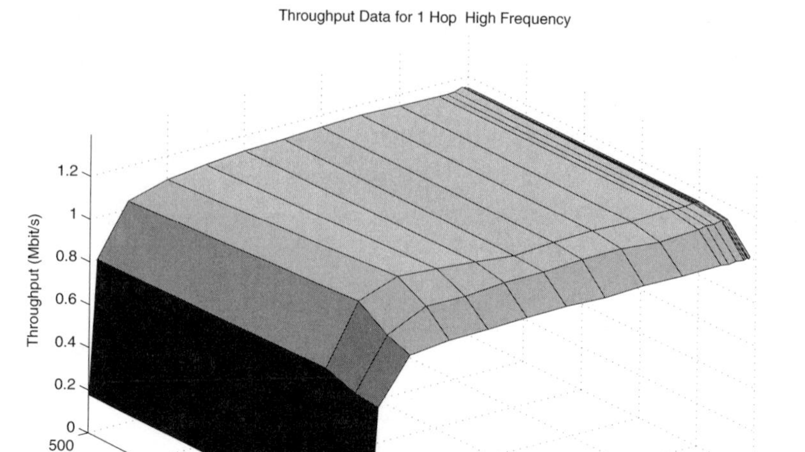

(a)

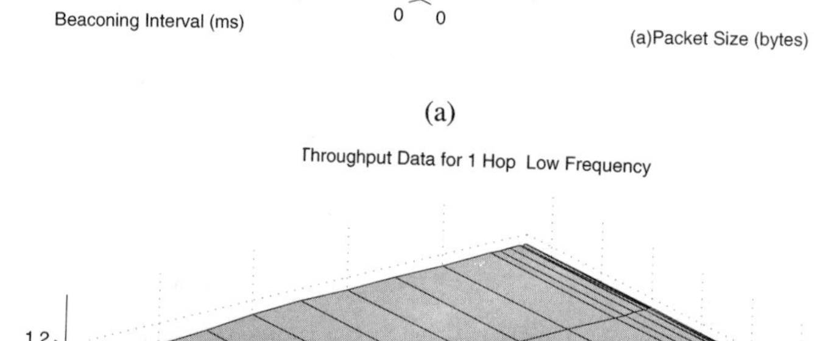

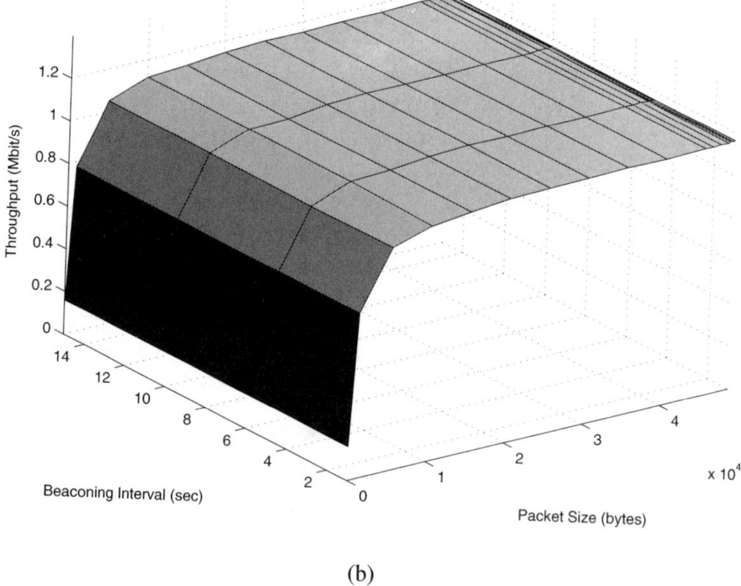

(b)

Figure 8.7. Throughput vs. packet size for a one-hop route while beaconing at: (a) High frequencies, and (b) Low frequencies.

Results on throughput for three-hop routes is shown in Figure 8.9b. At the 15s beaconing interval, as packet size increases from 64 bytes to 1,008 bytes, throughput increases from 57.2 kbps to 282 kbps. This corresponds to an increase of 392%. When the packet size is further increased, throughput gradually increases. For three-hop route experiments, since the maximum packet size permissible is 1,448 bytes, the maximum obtainable throughput is 302.8 kbps.

Based on the above observations, one can deduce that communication throughput increases dramatically when packet size increases beyond 64 bytes. With a further increase in packet size, communication throughput increases at a slower rate until the saturation point is reached.

The use of a large packet size can increase the performance of ad hoc networks in terms of throughput. However, at large packet sizes, there is a higher probability that a packet is corrupted. This behavior is likely to occur in a wireless environment due to its high bit error rate compared to a wired medium. Moreover, contention can be a problem when traffic load is high. However, the latter problem is not observed in the experiment since only one traffic session is executed each time, and each node is positioned with a minimum number of neighbors. Therefore, the number of contentions arising from other traffic sessions and neighbors is minimized. Because of this, the optimum packet size for ad hoc networks cannot be determined easily, since it depends on various parameters, such as link loss probability, contention from on-going traffic, and number of neighboring nodes.

8.5.2 • Impact of Beaconing Interval on Throughput

By varying the beaconing interval of ad hoc mobile computers from 15 secs to 10 ms, severe channel contention is expected over the shared wireless medium. However, the experimental results show that there is no significant impact on throughput performance. It can be seen from Figure 8.7b that no significant change in throughput is noted as the beaconing interval is varied from 15 secs to 1 sec. In the high-frequency beaconing experiments, the results in Figure 8.7a also show similar behavior as before, except at very low beaconing intervals (10 ms–50 ms), where throughput drops by as much as 200 kbps. This is due to the presence of severe contention in the wireless channel.

The results (Figures 8.8 and 8.9) obtained from two- and three-hop configurations reveal the same characteristic. The beaconing interval has no significant impact on throughput except at extremely high beaconing frequencies, which is the result of high contention in accessing the shared medium among beacons and data packets. In practical scenarios, the impact can be slightly more severe when more traffic loads are placed on the network. In general, beaconing intervals of

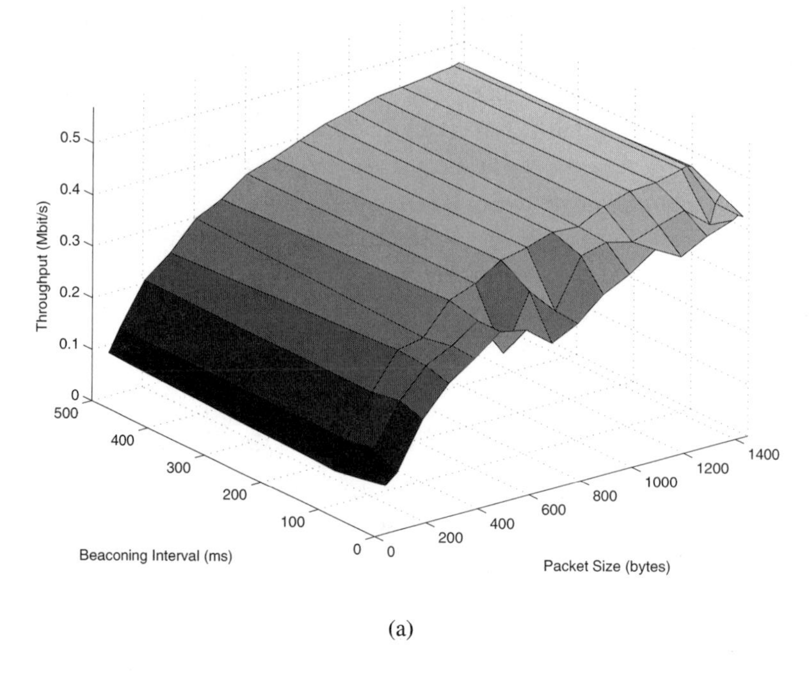

(a)

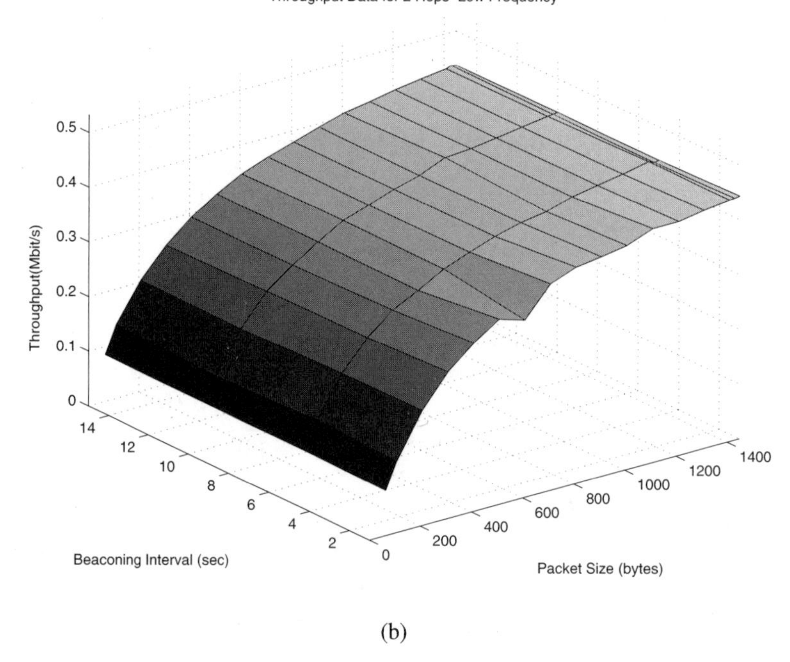

(b)

Figure 8.8. Throughput vs. packet size for a two-hop route while beaconing at: (a) High frequencies, and (b) Low frequencies.

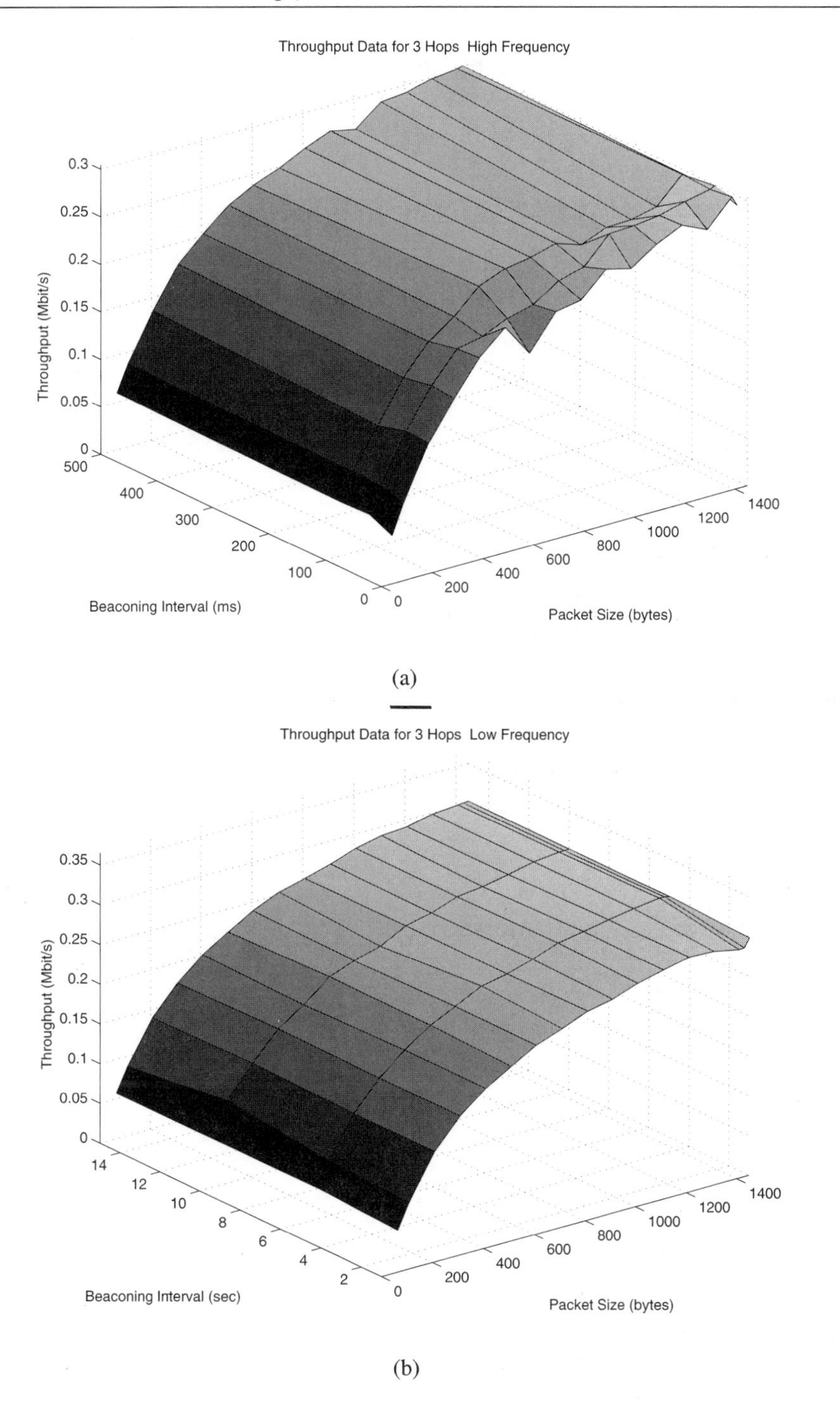

Figure 8.9. Throughput vs. packet size for a three-hop route while beaconing at: (a) High frequencies, and (b) Low frequencies.

one second or longer have negligible impact on throughput performance in most configurations.

8.5.3 • Impact of Route Length on Throughput

In an ad hoc wireless network, data packets can be forwarded along several links (hops) until they arrive at the destination. Moving data packets over multiple wireless links results in a greater delay, hence affecting communication throughput. Figures 8.7, 8.8, and 8.9 illustrate the throughput performance for one-, two-, and three-hop routes, respectively. At a packet size of 1000 bytes and the ad hoc mobile hosts beaconing at 1-second intervals, the average throughput for one-hop routes is 780 kbps. For two-hop routes, throughput drops to 410 kbps. Finally, for three-hop routes, throughput drops to 280 kbps.

The same amount of throughput degradation is observed in all experiments. At two hops, throughput decreases by one-half of the one-hop case. When the route length is three hops, throughput decreases to one-third of the one-hop case. Therefore, from our observation, throughput is expected to decrease to approximately $\frac{1}{N}$ of the one-hop throughput in light-load environments (as in the experiments), where $\mathcal{N}$ is the route hop count. This impact on throughput, however, will be different in a heavily loaded environment or when the number of neighboring nodes increases.

8.6 Packet Loss Performance

In a wireless environment, interference and multipath fading can affect the successful transmission and reception of information. Since an end-to-end ad hoc wireless route may comprise multiple wireless links, packet loss performance is therefore important.

8.6.1 Impact of Packet Size on Packet Loss

For a one-hop route, 100 pings are sent from the source to the destination. The packet size of the data sent is varied for each communication session. After 100 pings, the amount of packet loss is recorded and the results are plotted. As shown in Figure 8.10a, at high beaconing frequencies, packet loss tends to increase at larger packet sizes. At low beaconing frequencies, packet size has a lesser effect on packet loss performance (see Figure 8.10b).

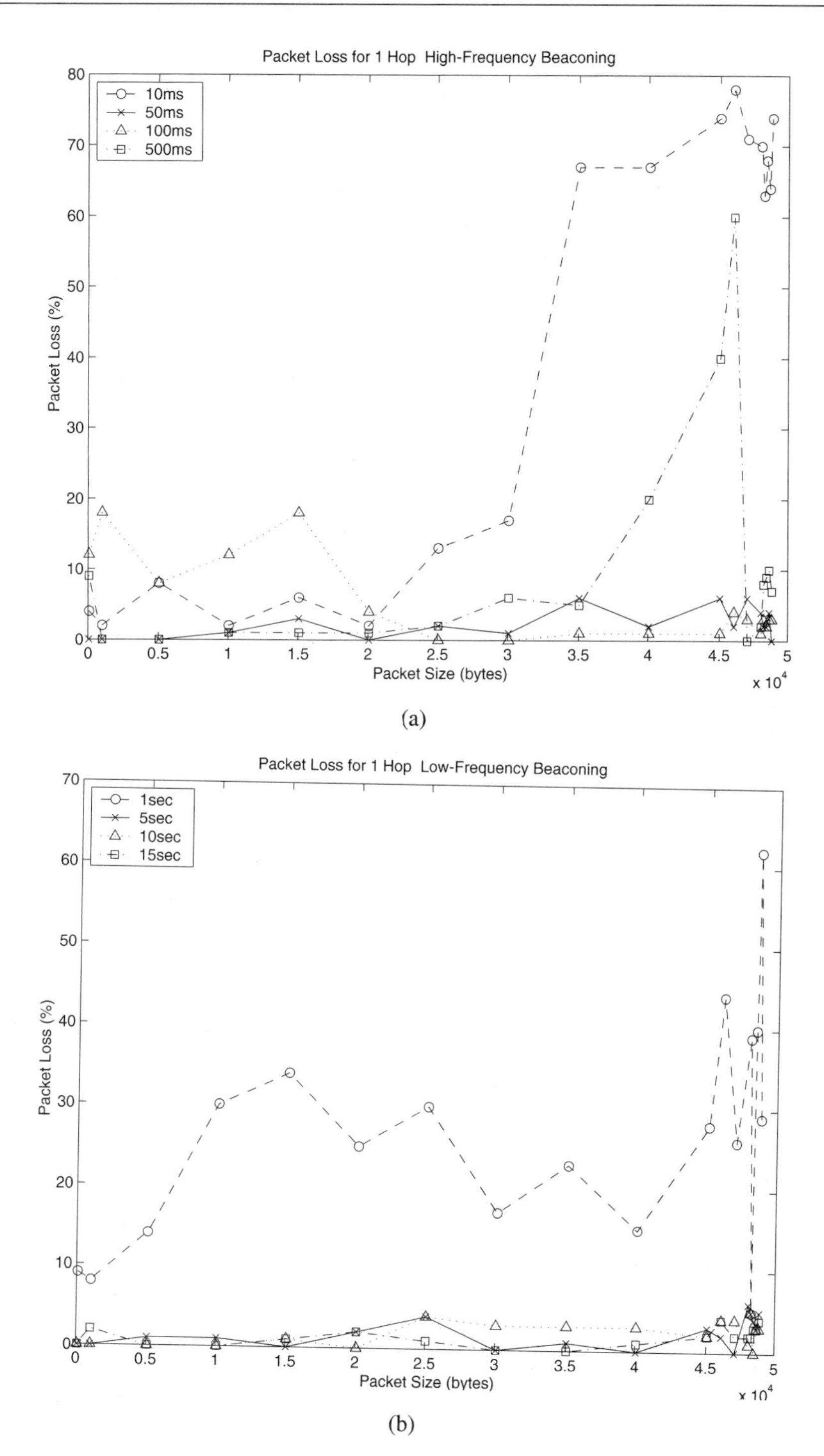

Figure 8.10. Packet loss performance vs. packet size for a one-hop route at: (a) High-frequency beaconing, and (b) Low-frequency beaconing.

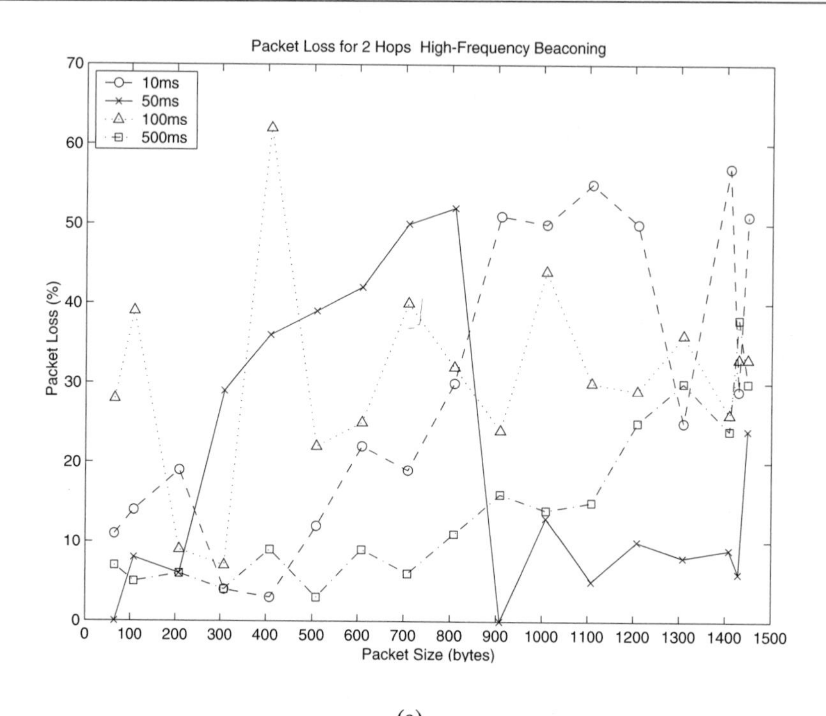

(a)

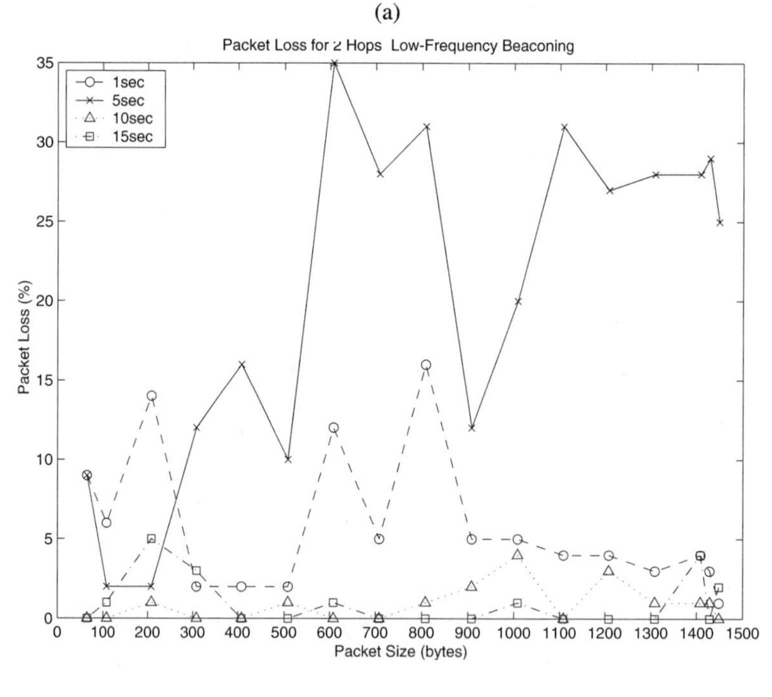

(b)

Figure 8.11. Packet loss performance vs. packet size for a two-hop route at: (a) High-frequency beaconing, and (b) Low-frequency beaconing.

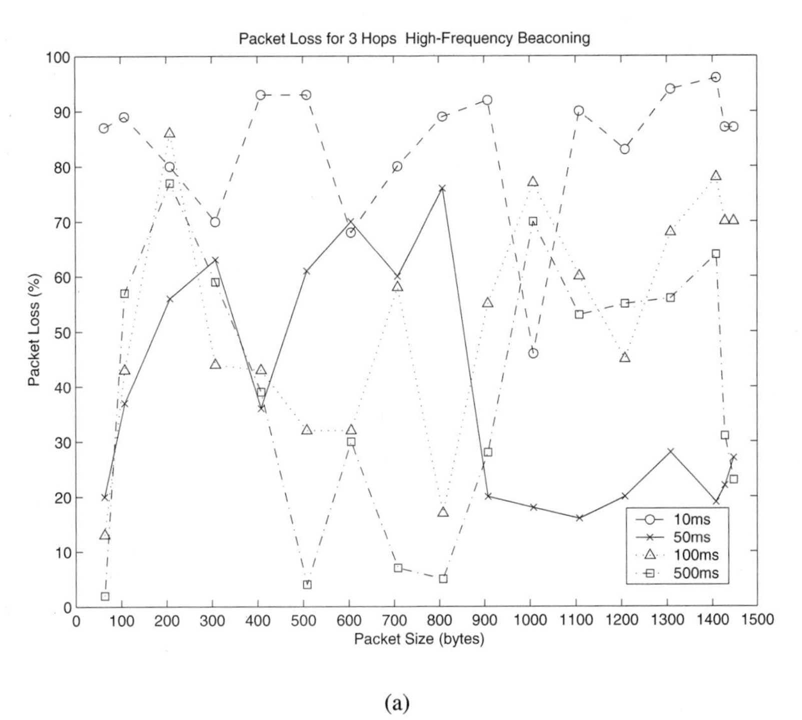

(a)

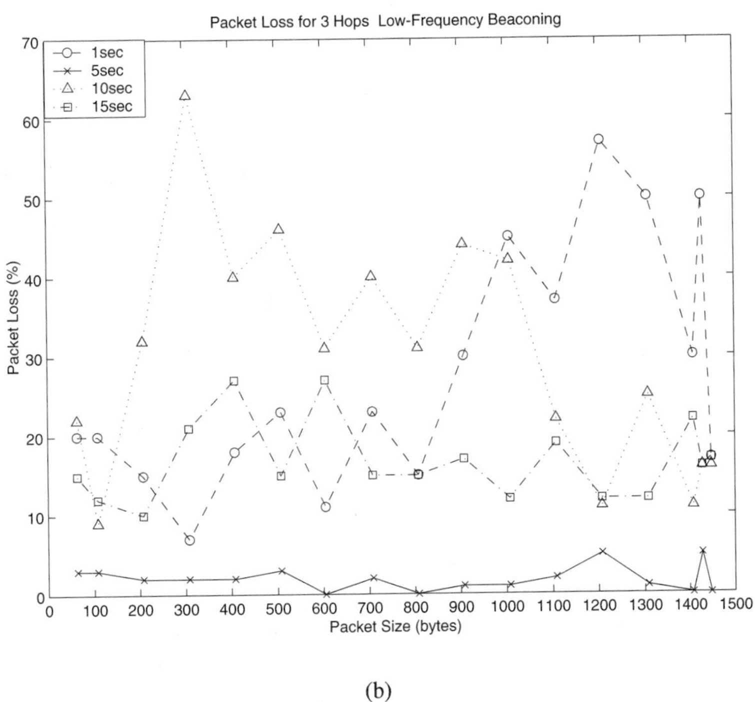

(b)

Figure 8.12. Packet loss performance vs. packet size for a three-hop route at: (a) High-frequency beaconing, and (b) Low-frequency beaconing.

8.6.2 Impact of Beaconing Interval on Packet Loss

Hence, increasing the beaconing frequency or the packet size of data to be transmitted over an ad hoc wireless route does not have a linear impact on the percentage of packet loss.

8.6.3 Impact of Route Length on Packet Loss

The above test is repeated but over two- and three-hop routes. Figures 8.11a and 8.11b present the results obtained at high and low beaconing frequencies over two-hop routes. At high beaconing frequencies, packet loss performance is almost independent of packet size. Collision in the channel can also significantly contribute to packet loss, in addition to environmental factors. At low beaconing frequencies, however, the percentage of packet loss increases when packet size is large.

Finally, for three-hop routes, the results at high and low beaconing frequencies are illustrated in Figures 8.12a and 8.12b. The former indicates that there is hardly any direct relationship between packet loss performance with increasing packet size. The latter figure also exhibits similar characteristics. When an ad hoc route is lengthened to multiple hops, the overall packet loss characteristic is dependent on each wireless link in the route.

8.7 Route Reconfiguration/Repair Time

Another experiment is conducted to evaluate how fast the ABR routing protocol can cope with repairing a route which is invalidated by host mobility. The route reconstruction (RRC) time is the time when a host determines that another host has moved till the time when the route is re-established.

Average RRC Time (ms)	Standard Deviation
13.15	0.05

Table 8.3. Average RCC time for 50 RRCs as a result of intermediate node movements.

As shown in Figure 8.13, with a three-hop route, a `ping` session is initiated and a continuous stream of 64 bytes of data are sent from Laptop $\mathcal{A}$ to Laptop $\mathcal{E}$ via laptops $\mathcal{B}$ and $\mathcal{C}$. The laptops are beaconing at 1 sec interval. To emulate

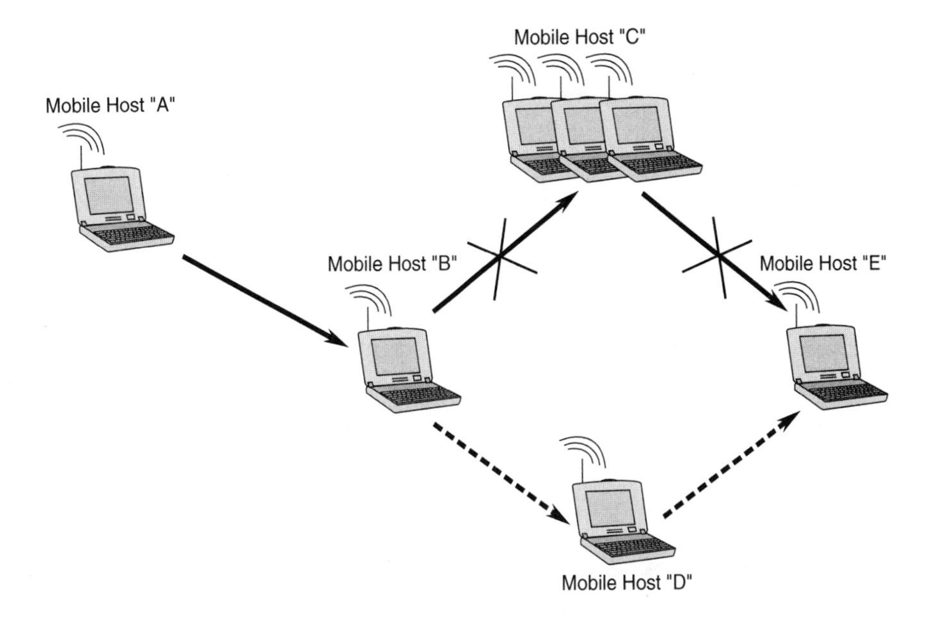

Figure 8.13. Mobility experimental setup.

movement, we remove Laptop $\mathcal{C}$'s WaveLAN card[3]. This triggers a route repair process, which is initiated by the upstream node, i.e., Laptop $\mathcal{B}$. This node will attempt to find an alternate route using the localized route repair mechanism. The newly discovered alternate route is shown by the dotted lines. To repeat the test, Laptop $\mathcal{D}$'s and $\mathcal{C}$'s WaveLAN cards are removed and replaced interchangably, for 50 attempts. This alternation, therefore, provides the route reconstruction times. Throughout the experiment, ABR is able to reconstruct the route without flaw. The results for route reconstruction time are recorded and the average time is reported in Table 8.3. The results revealed that this time is equivalent to a two-hop route search time.

8.8 TCP/IP-Based Applications

To reveal if existing TCP/IP-based applications could be supported by the ad hoc wireless testbed, several field trials running TELNET, HTTP, and FTP applications were performed.

[3]In the field trial, one can physically move the laptop until it is out of radio range of its upstream and downstream neighbors. However, removing the WaveLAN card is a good method to ensure that the radio connection is indeed removed.

8.8.1 Running TELNET over Ad Hoc

TELNET is a remote terminal protocol. It allows a user at one end to establish a TCP connection to a remote login server at the other end. Hence, when user invokes `telnet`, the application program on the user's machine becomes the client and it establishes a TCP connection to the remote server. The remote server is specified by the client, through the use of the server hostname or IP address.

Once a `telnet` connection has been established, keystrokes from the user's keyboard will be accepted by the client and relayed over to the server for processing and/or execution. Outputs from the server is then relayed back to the client over the established TCP connection. As a result, the service is *transparent* because it gives the appearance that the user's keyboard and display are directly attached to the server machine.

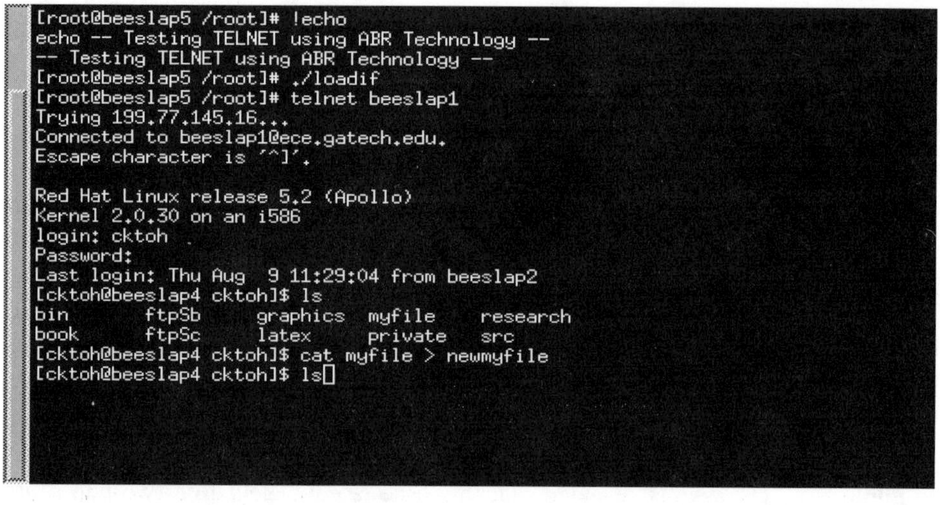

```
[root@beeslap5 /root]# !echo
echo -- Testing TELNET using ABR Technology --
-- Testing TELNET using ABR Technology --
[root@beeslap5 /root]# ./loadif
[root@beeslap5 /root]# telnet beeslap1
Trying 199.77.145.16...
Connected to beeslap1@ece.gatech.edu.
Escape character is '^]'.

Red Hat Linux release 5.2 (Apollo)
Kernel 2.0.30 on an i586
login: cktoh
Password:
Last login: Thu Aug  9 11:29:04 from beeslap2
[cktoh@beeslap4 cktoh]$ ls
bin       ftpSb     graphics  myfile    research
book      ftpSc     latex     private   src
[cktoh@beeslap4 cktoh]$ cat myfile > newmyfile
[cktoh@beeslap4 cktoh]$ ls
```

Figure 8.14. Snapshot of an ad hoc client TELNETing to a remote ad hoc server.

Figure 8.14 shows a window capture of a TELNET session at the client (beeslap5). Five ad hoc mobile hosts are lined up such that a 4-hop route is formed. The TELNET session occurs from beelaps5 to beelaps1 (which is the server). The capture shows that the a TCP/IP/ABR connections is succesfully established over multihop wireless links and the user can login as if it was local and proceed with reading and access files and other tasks. Clearly, this feature is useful since client-server interaction is still possible and supported in an ad hoc mobile environment.

8.8.2 Running FTP over Ad Hoc

File transfer and access in a computer network is viewed to be an important function. In a distributed network environment, there are diskless machines who have little or no file storage ability. Such devices can create, store, and update files to a remotely connected file server, which is another machine but equipped with large storage and processing capability. There are also cases where remote storage is used to archive data. FTP is one of the earliest version of file transfer software.

FTP uses TCP to ensure reliable transfer of file data. Although this sounds simple, in actuality it is complex. Not only will a TCP connection be established prior to data transfer but also the need for authorization checks, naming, and representation among heterogeneous machines. FTP clients need to supply a valid login name and password to the server before being allowed to perform file transfer/access operations.

```
-- Testing FTP over TCP/IP/ABR protocols --
[root@Beeslap2 /root]# !ftp
ftp beeslap4
Connected to beeslap4.
220 beeslap4 FTP server (Version wu-2.4.2-academ[BETA-18](1) Mon Aug 3 19:17:20
EDT 1998) ready.
Name (beeslap4:root): cktoh
331 Password required for cktoh.
Password:
230 User cktoh logged in.
Remote system type is UNIX.
Using binary mode to transfer files.
ftp> bin
200 Type set to I.
ftp> hash
Hash mark printing on (1024 bytes/hash mark).
ftp> put myfile
local: myfile remote: myfile
200 PORT command successful.
150 Opening BINARY mode data connection for myfile.
#
226 Transfer complete.
61 bytes sent in 0.00807 secs (7.4 Kbytes/sec)
ftp> ▯
```

Figure 8.15. Snapshot of a client FTPing to a remote server and accessing files over a multihop wireless connection.

FTP is interactive in that the user can enter specific commands to store and retrieve files. Large file transfers are supported by FTP and a timer is used to close a session if the connection is inactive after some time period. To be able to support FTP over an ad hoc wireless network is attractive and desirable. There is no need for transfer of floppy disk from one host to another. The transfer of bits can be done over-the-air, in a multihop fashion.

To verify that this function can be achieved, an experiment involving 4 laptops are performed where an `ftp` operation is executed at the source node (client) and the destination (server) is actually 3 wireless hops away. This function can be performed successfully over TCP/IP/ABR protocols without any problems. Both large and small files are transported and transfer times are observed and recorded.

8.8.3 Running HTTP over Ad Hoc

Web access is getting more and more popular these days with the arrival of the Internet. The web server allows one to advertise information to others. Web users can download information (video and audio clips, text files, etc.) into their local machines.

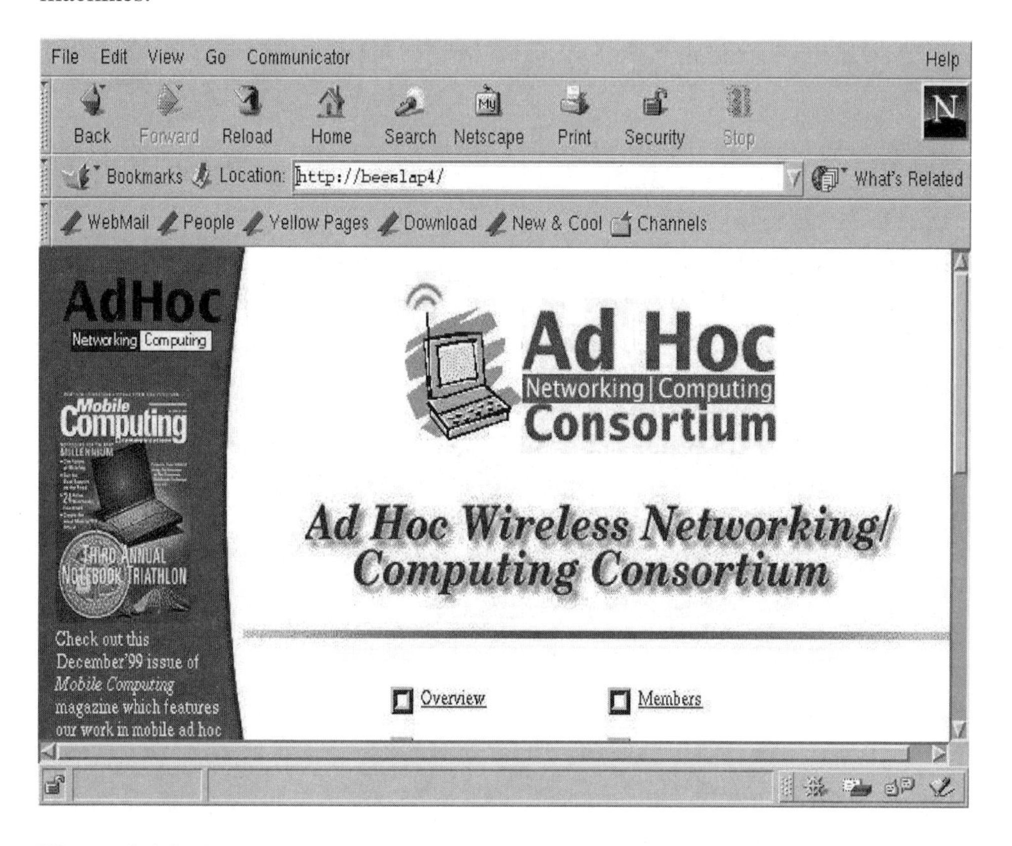

Figure 8.16. Snapshot of a loaded web page at an ad hoc mobile host which is acting as a client.

More recently, we saw the arrival of electronic commerce. E-commerce is definitely more than a transaction between a buyer and a seller, although it is the focus of the majority of businesses. There are four key components to e-commerce programs beyond the transaction, namely: (a) marketing, (b) merchandising, (c) promotion and (d) customer support. By integrating value added information with the rapid delivery modes the Internet offers, a business can quickly satisfy the needs of a client and strengthen the client-vendor relationship.

Online auctioning over the Internet has also gained in popularity. Even without the patter of an auctioneer, online auctions are bidding to become a sizable slice of the Internet commerce pie.

To ensure that web access and services can be supported in ad hoc networks, an ad hoc mobile host is configured to be a web server. The client (beeslap2) then accesses a specific web page stored at the web server (beeslap4), which is several wireless hops away. The client is running a web browser and by specifying the HTTP address of the specific web site, an underlying connection is established from the client to the server. Information about the web page is retrieved and displayed at the client, as shown in Figure 8.16.

8.9 Conclusions

Ad hoc wireless networking is evolving in both the design of ad hoc routing protocols and that of prototype implementation and feasibility studies. In this chapter, experimental field trails of a working ad hoc wireless network are described and a series of experimental results on communication throughput, EED, packet loss, and RD time are presented.

We examined the impact of varying *beaconing interval*, *packet size*, and *route length on* communication performance. We performed mobility experiments to measure route reconstruction time incurred. Successful testing of existing TCP/IP applications such as TELNET, HTTP, and FTP were also reported.

Experimental results obtained from field trials revealed that the beaconing interval has little impact on throughput, EED, and RD time, with the exception of very low beaconing intervals (below 100 milliseconds). Packet size, however, can significantly affect ad hoc end-to-end delay and communication throughput.

Based on the experience gained in the field trials, implementing an ad hoc wireless network using current mobile computer technology, wireless adapters, and ad hoc routing software is considered feasible and the resulting communication performance is acceptable for most existing data applications. One awaits to see how soon such systems would be available for purchase, deployment and usage.

Chapter 9

ENERGY CONSERVATION: POWER LIFE ISSUES

9.1 Introduction

Mobile computing is evolving rapidly with advances in wireless communications and wireless networking protocols. Despite the fact that devices are getting smaller and more efficient, advances in battery technology have not yet reached the stage where a mobile computer can operate for days without recharging. While research is on-going to build long-lasting batteries, sometimes we wonder if there is an *electrochemical* limit. If so, then advanced power conservation techniques are necessary. Hence, in this chapter, we will discuss properties of batteries and power management at the *device*, *protocol*, and *application* layers.

Wireless transmission, reception, retransmission, and beaconing operations all consume power. Many existing routing protocols use periodic transmission of route update messages to maintain the accuracy of routing tables. In wireless networks, beaconing can also be used to signify the presence of neighboring nodes and indicate the *spatial*, *temporal*, *connection*, and *signal stability* of these nodes. Hence, the power consumed as a result of periodic beaconing and its impact on existing applications need to be examined.

9.2 Power Management

Device manufacturers have always been striving for lower power consumption in their products so that these devices are efficient to operate. However, many such power-limiting efforts are concentrated on the individual device level. A mobile computer, however, is composed of many different devices, such as hard and floppy disk drives, LCD displays, CD/DVD ROMs, etc. Each of these devices has its own power requirements, operational characteristics, and usage patterns, which make power management in the overall system complicated. The answer to more comprehensive power management (PM) is advanced power management (APM)[44], which was followed recently by operating system power management (OSPM), and advanced configuration power interface (ACPI)[45].

Most of these techniques are incorporated into existing mobile computers; hence, they are worthy of mention here. In APM, one or more layers of software are present to support PM in computers with power-manageable hardware. APM's objective is to control the power usage of a system based on the system's activity. Power is reduced gradually as more system resources remain unused until the system suspends.

ACPI, on the other hand, defines new ways of power control. It enables an operating system to implement *system-directed PM*. The ACPI hardware interface is a standardized way to integrate PM throughout a portable system's hardware, OS, and application software. The ACPI gives the operating system direct control over the PM and plug-and-play functions of a computer. This shall be elaborated in later sections.

9.2.1 Smart Batteries and Battery Characteristics

To facilitate PM in mobile computers, changes have to be made in batteries as well. One of the most important factors required in systems today is the ability to *read* the battery's remaining power. This single piece of information is useful in many cases: APM or ACPI can put a system in a certain power-saving state when the power level drops below a specific threshold.

Smart batteries can support the tracking of power life. Through a series of

Figure 9.1. The smart battery circuit module of a laptop battery pack (source IBM).

specifications [46][47][48][49][50], the industry has come up with general guidelines providing a comprehensive approach toward *reading*, *selecting*, and *charging* smart batteries. The smart battery hardware of a Li-On battery packet used in an IBM ThinkPad is revealed in Figure 9.1.

Several parameters are used to measure the performance of batteries, such as: (a) self-discharge rate, (b) cycle life, (c) operating temperature range, (d) energy density, and (e) cell balancing. Their definitions are stated below:

(a) **Self-discharge rate:** The rate at which a battery releases its energy without being used. Energy release may occur when a battery just sits on a shelf. Ideally, we want this self-discharge or leakage rate to be as low as possible.

(b) **Cycle life:** The number of times a battery may be charged and discharged. Every time a battery is charged and discharged, it uses one cycle. Most rechargable batteries can be recharged and reused quite a few times before the chemical property starts to fade away.

(c) **Operating temperature range:** The temperature range for a battery to work normally. Beyond this range, either the battery works poorly or its cycle life is reduced.

(d) **Energy density:** The amount of energy a battery can provide in one basic weight or volume unit. The unit of measurement is Watt x hour / Kilogram or Watt x hour / Litre.

(e) **Cell balancing:** Mismatches in the characteristics of battery cells, and consequently the discharge rate and capacity between them, can cause one cell in a pack to reach its peak charge before the others. It is important in this case to quit charging the pack with the effect of leaving the other cells partially charged. Likewise, when a pack is discharged, caution must be taken to terminate use when any cell reaches the preset undervoltage limit, even if other cells have not yet reached that limit. Embedded controller chips can help in alleviating problems like this by adjusting the charging and discharging rates of individual cells.

A high-performance battery is expected to have low self-discharge rate, long cycle life, wide operating temperature range, and high energy density. There are currently three major portable battery technologies to choose from for mobile computers, namely: (a) nickel cadmium (NiCd), (b) nickel metal hydride (NiMH), and (c) lithium-ion (Li-ion).

The popular Li-ion batteries have the key advantage of higher energy density at a higher voltage than NiCad and NiMH. Their energy density is more than two times that of NiCad. To get a higher voltage, Li-ion cells can be stacked from one (3.6V nominal) to eight (28V nominal) cells in series. Figure 9.2 shows the internals of a battery pack used in an IBM ThinkPad. The side view clearly shows that cells are connected in series to yield the desire VA rating.

The typical discharge and storage (self-discharge) characteristics of one Li-ion battery cell are shown in Figures 9.3a and 9.3b, respectively. Both characteristics are highly dependent on the surrounding temperature. In general, cell voltage decreases with a higher discharge capacity ratio (i.e., discharge current) and declines more rapidly at low temperature (a result of higher cell impedance), which is shown in Figure 9.3a. Figure 9.3b shows that higher temperature storage leads to greater self-discharge or capacity loss. Under a proper temperature-controlled environment, for example, 25°C, capacity loss can be kept below 10% per month.

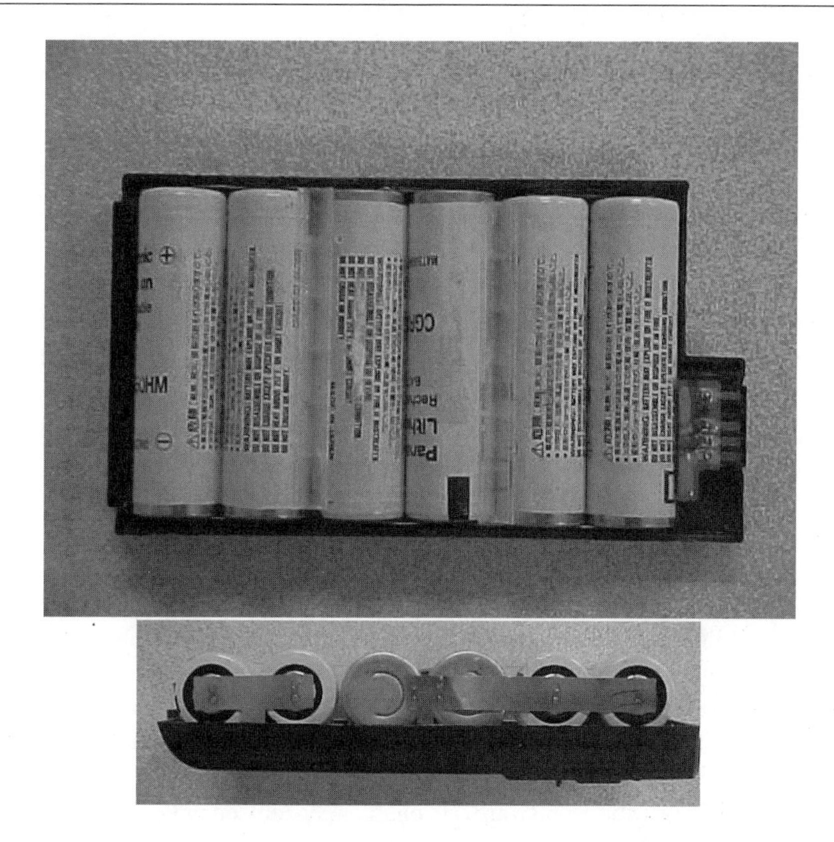

Figure 9.2. The internals of a laptop battery pack showing multiple cells are connected to provide the required VA rating.

9.3 Advances in Device Power Management

9.3.1 Advance Power Management (APM)

Device manufacturers have always been striving for low power consumption, partly for economic reasons and partly because the overall system imposes limitations (such as thermal issues). However, such efforts have been concentrated on the individual device level. A mobile computer, however, comprises several different devices, such as hard disk drives, LCD dislays, CD/DVD ROMs, etc. Each of these devices has its own power requirement, usage characteristic, and profile, which makes PM of the overall system complicated. All these form the motivation to devise a comprehensive device PM framework. An example is APM (advanced power management)[44].

APM, as described in the specifications, consists of one or more layers of software that support PM in computers with power-manageable hardware. APM defines a hardware-independent software interface between hardware-specific PM software and an operating system PM policy driver. The main objective of APM is to control the power usage of a system based on the system's activity. Power is gradually reduced as more system resources remain unused until the system suspends. The five states defined in the APM specifications are summarized in Table 9.1.

Figure 9.4 shows the software architecture of the APM system, where APM-aware applications interact with the APM driver, and with the help of the APM BIOS, applications can control the underlying hardware. Alternatively, if BIOS support is absent for certain devices, specific APM-aware device drivers can be used to access devices directly.

9.3.2 Advance Configuration and Power Interface (ACPI)

With mobile computers incorporating more features, the system BIOS itself is no longer capable of providing adequate management. A new form of operating system (OS)-oriented PM known as OSPM (operating system power management) has evolved from this need. OSPM uses ACPI (advanced configuration power interface)[45]. ACPI defines a new level of power control and its hardware interface provides a standardized way to integrate PM throughout a portable system's hardware, OS, and application software. It enables system designers to implement a

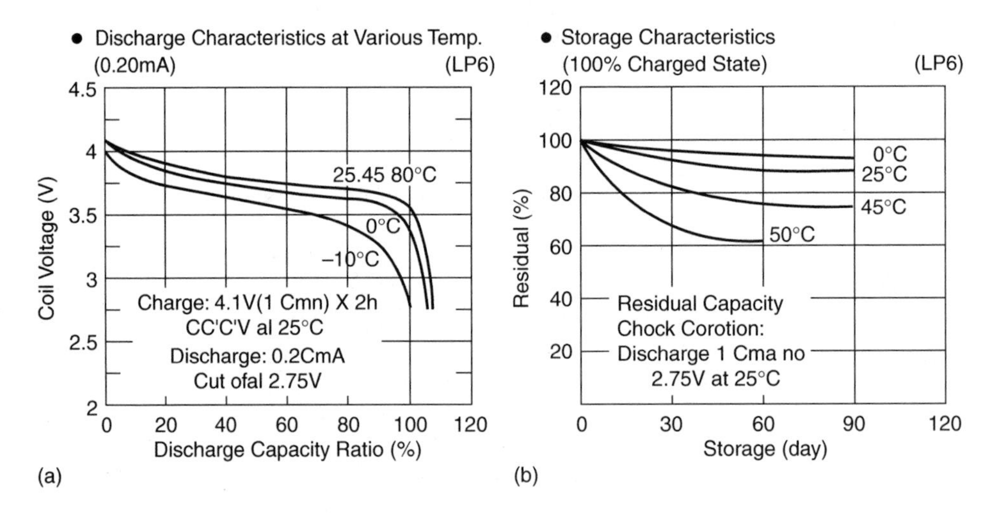

Figure 9.3. (a) Discharge characteristics, and (b) storage characteristics of Li-ion batteries.

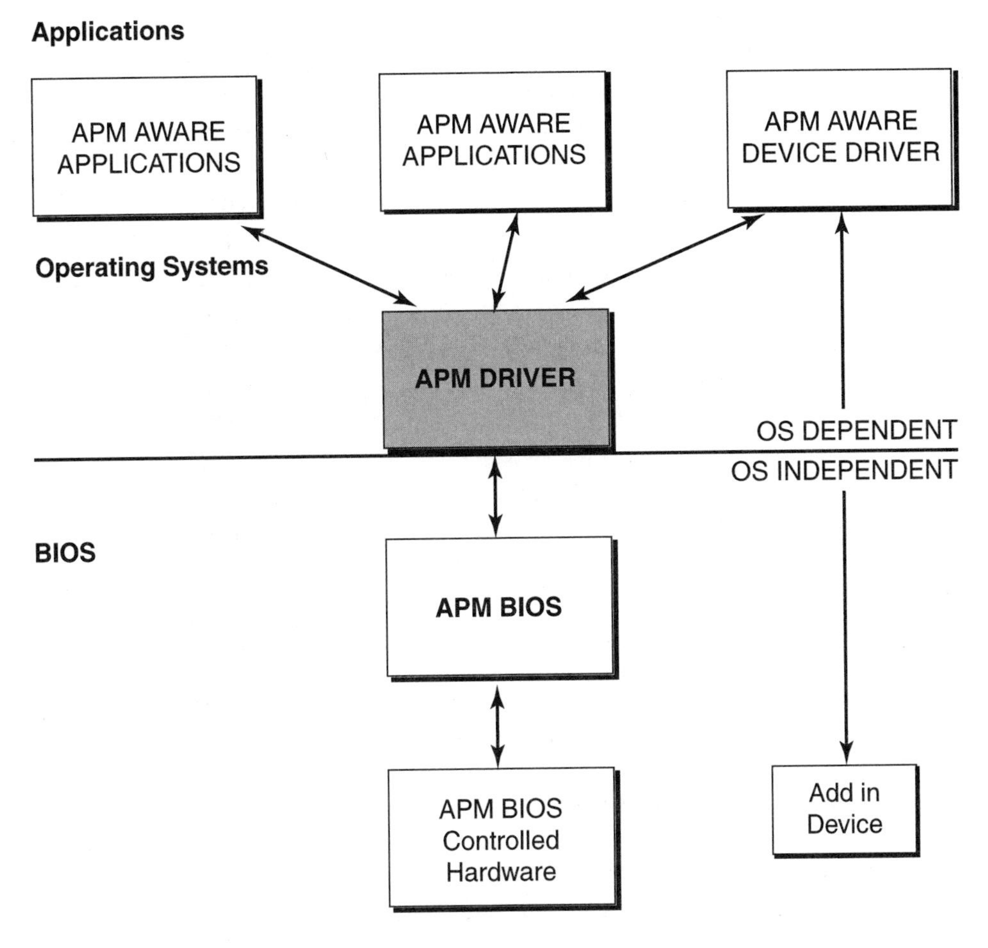

Figure 9.4. APM software architecture.

spectrum of PM features with different hardware designs while using the same OS driver.

ACPI uses the existing collection of PM BIOS code, APM APIs, etc., to form a well-specified PM mechanism. It provides support for an orderly transition from existing (legacy) hardware to ACPI hardware, and it allows both mechanisms to exist in a single machine and be used as needed. The functional areas covered by ACPI include:

- **System power management:** ACPI defines mechanisms for putting the computer as a whole in and out of the system's sleeping states. It also provides a general mechanism for any device to wake the computer.

Table 9.1. A Summary of APM states and their implications.

System States	Characteristics
Full On	Devices are on, system is working but no power management.
APM Enabled	System is power managed and devices are power managed when needed.
APM Standby	System and most devices are in a low power state. Should eventually return to APM Enabled state.
APM Suspend	System is not working and is in a low power state with maximum power savings. Most devices are not powered and CPU clock is stopped. System may enter hibernation state to save operational parameters. Wakeup events can return the system to APM Enabled State.
Off	System is not working and power supply is off.

- **Device power management:** ACPI tables describe motherboard devices, their power states, the power planes the devices are connected to, and controls for putting devices into different power states. This, therefore, enables the OS to put devices into low-power states based on application usage.

- **Processor power management:** When the OS is idle but not sleeping, it can use commands specified by ACPI to put processors in low-power states. In these low-power states, the CPU does not run any instructions, and wakes when an interrupt occurs. The OS determines how much time is being spent in its idle loop by reading the ACPI PM timer. Depending on this idle time estimate, the OS will put the CPU into different low-power states.

- **Battery management:** The OS determines the low-battery and battery warning points, and also calculates the battery remaining capacity and battery remaining life. Battery management is concerned with power conversion, charge/discharge control, safety (over-temperature, short-circuit, over-voltage, and current protection), and cell balancing. Smart battery systems today incorporate a communications link between the battery pack and the host or

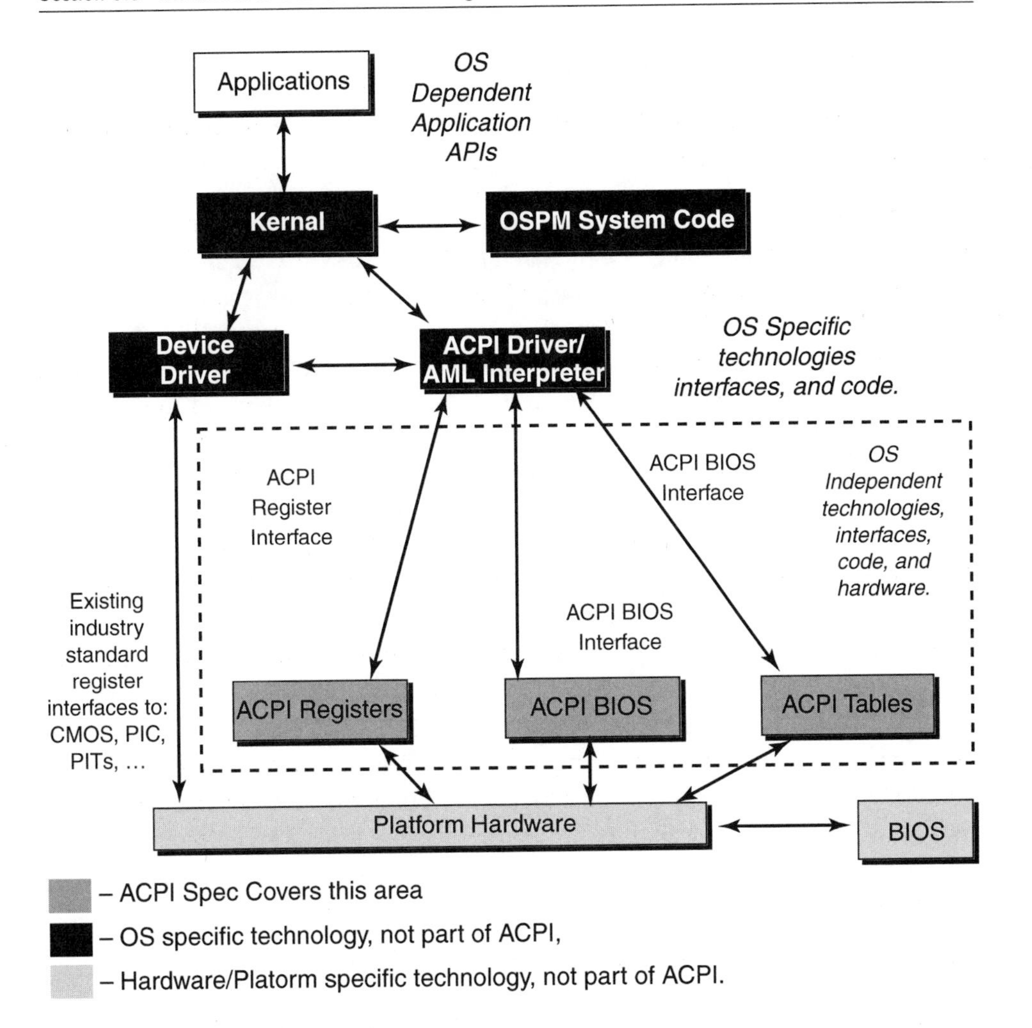

Figure 9.5. ACPI architecture and interfaces.

charger, and allow the host to manage the battery.

- **Thermal management:** ACPI addresses system thermal management by providing a simple and scalable model that allows OEMs to define thermal zones, thermal indicators, and methods for cooling thermal zones. ACPI moves hardware cooling policies from firmware into the OS. By allowing the OS to monitor the system temperature, new cooling decisions can be made based on application load on the CPU in addition to the thermal heuristics of the system. This also enables the OS to shut down the computer in case of high-temperature emergencies.

- **System events:** ACPI provides a general event mechanism encompassing thermal events, PM events, docking, device insertion, and removal.

Figure 9.5 shows the software and hardware components relevant to ACPI and how they relate to one another. Compared to APM, ACPI is OS/kernel-centric and its specification describes the interfaces between components, the contents of the ACPI tables, and the related semantics of other ACPI components. With this structure, power management functionality is moved into the OS, making it available on every machine on which the OS is installed. Although the level of functionality (power savings, etc.) varies from machine to machine, users and applications will see the same power interfaces and semantics. The ACPI BIOS refers to the portion of the firmware that boots the machine and implements interfaces for sleep, wake, and restart operations. The ACPI tables describe the interfaces to the hardware. Note the presence of the ACPI driver/AML (ACPI Machine Language) interpreter, which is capable of interacting with non-ACPI device drivers.

The Intel Power Monitor[51] is a software based power analysis tool that monitors system activity on computers. It enables independent software vendors to develop power efficient mobile applications. It can be used to verify if a particular software is using battery energy unnecessarily. The identification and control of such power hungry software can prolong the lifetime of the computer. As shown in Figures 9.6 and 9.7, system, CPU, and disk power usage can be monitored and recorded.

9.4 Advances in Protocol Power Management

In addition to bandwidth constraint, power limitation in mobile devices is a serious factor and has motivated research into *power-aware* and *power-efficient* mobile protocols. Algorithms designed to conserve power currently address data-link, network, and transport layer protocols. We will discuss some of the current work in this area.

9.4.1 Power Conservation at the Data-Link Layer

Power conservation techniques [52][53][54][17] can be performed at both the link and access sub-layers. Firstly, at the MAC layer, unnecessary collisions should be eliminated as much as possible since retransmission incurs power consumption. In cellular-type wireless last-hop networks, reservation and polling based channel access techniques are proposed to overcome unnecessary collisions. Secondly, in

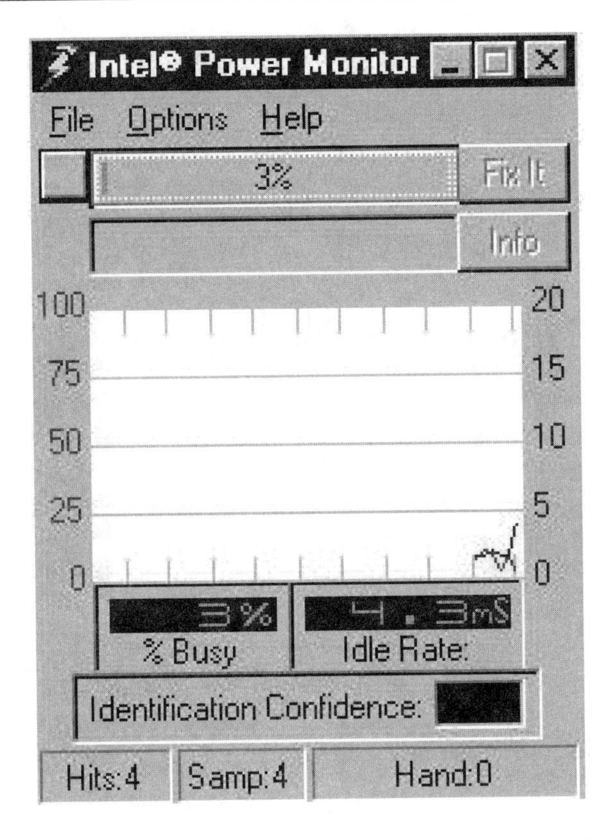

Figure 9.6. The Intel power monitor (source Intel).

another scenario, instead of having the receiver powered on at all times in a wireless broadcast network, we could instead broadcast a schedule depicting the transmission times for each mobile device, hence allowing mobile devices to switch to standby mode to save power. Thirdly, in a single-transceiver system, switching from transmit to receive mode is required to support both uplink and downlink communications. By providing a means for allocating contiguous slots for transmission and reception, power consumed as a result of transceiver mode switching can be reduced. In terms of channel reservation, power may be conserved by supporting the request of multiple slots with a single reservation packet. From another perspective, considerable power savings can be obtained by intelligently turning off radios when they cannot transmit or receive packets.

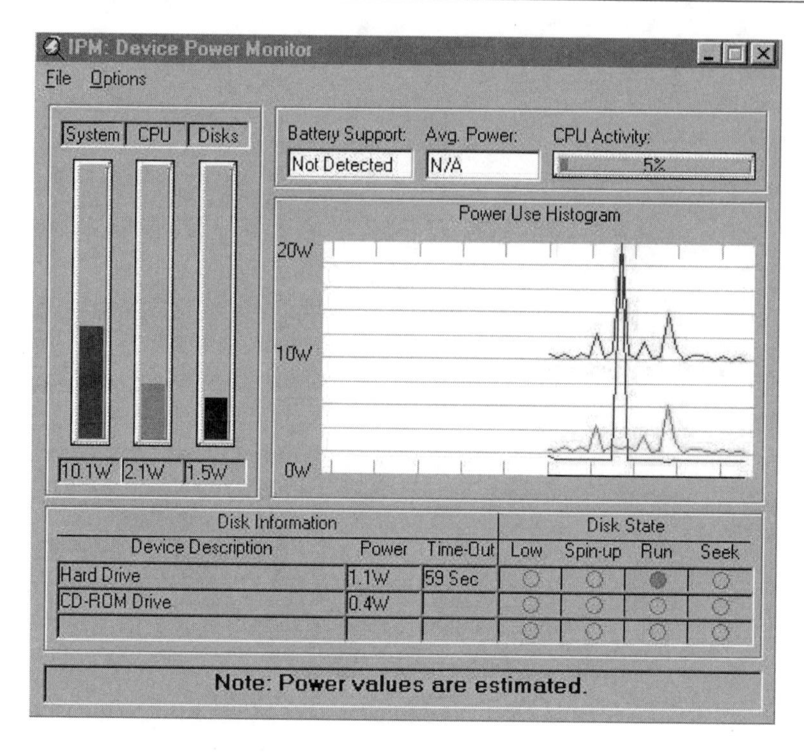

Figure 9.7. Intel Power Monitor showing system, CPU, and disk power levels, along with information about disk states (source Intel).

9.4.2 Power Conservation at the Network Layer

An important function of the network layer is to support routing. There are several proposals for power-aware and power-efficient routing protocols [55][56]. In wireless last-hop networks, mobile IP is used to support the routing and tunneling of packets as a mobile host migrates from one base station to another. Most of the work, therefore, is done via signaling over the wired network. However, in ad hoc wireless networks, there are no base stations and each node acts as a router and packet forwarder. Hence, the computation and communication (packet processing, transmission, reception, etc.) load can be quite high.

In [55], new power-aware metrics are used for determining routes in wireless ad hoc networks. It was argued that routing protocols that derive routes based on minimizing hop count or delay will result in some nodes depleting their energy reserves faster, causing them to be powered down at an earlier stage. In addition, routing packets through lightly loaded (i.e., referring to route relaying load[57][40]) nodes is considered energy conserving since there is less likelihood of contention. Some of the metrics proposed for power-aware routing are: (a) minimize energy consumed per packet, (b) maximize time to network partition, (c) minimize variance in node power levels, (d) minimize cost per packet, and (e) minimize maximum node cost.

One can also reduce power by avoiding periodic route updates, as commonly found in pro-active/table-driven routing protocols. Such protocols result in intensive route updates when the frequency of link changes increases. Improvements in route discovery technqiues can also help to reduce excessive power consumption. Transmitting packets with a more compact header (other than source routing) is another power efficient technique.

9.4.3 Power Conservation at the Transport Layer

Supporting end-to-end communications is a key function of the transport layer. An example of a reliable transport protocol is TCP. In TCP, ARQ (automatic repeat request) is used to provide error control. ARQ employs error detecting codes to detect errors and request retransmission. Repeated retransmissions consume power and should be avoided as much as possible while still maintaining a certain level of communication performance. Several proposals for power efficient error control schemes have been proposed [58]. In addition, several enhanced versions of TCP for a wireless last-hop environment have evolved. Most of these are split-connection based protocols where the TCP connection between the source on a wired network and the destination on a wireless network is transparently split into two transport segments. The protocol used to handle the wireless link is enhanced to improve communication performance by handling packet losses locally. An example is the indirect-TCP scheme [59].

Summarizing, a variety of techniques can be used to cope with power scarcity. Some of these techniques are listed in Table 9.2.

9.5 Power Conservation by Mobile Applications

Intel Corporation has investigated the power consumption of applications running on a single computer. The power measurement shown in Figure 9.8 was made on a 166MHz Pentium processor with an MMX technology-based platform with 16MB EDO memory and 256KB pipelined burst level two cache. The results include all components in the notebook's interior with the exclusion of the LCD panel power. It shows the distribution in power consumption. Although CPU power consumption is high, collectively, the power consumed by other devices are in fact higher. APM was enabled during the power measurements. Figure 9.9 shows a breakdown of

Table 9.2. Power Conservation Techniques at Various Protocol Layers

PROTOCOL LAYER POWER CONSERVATION TECHNIQUES
Data-Link Layer - Avoid unnecessary retransmissions. - Avoid collisions in channel access whenever possible. - Put receiver in standby mode whenever possible. - Use/allocate contiguous slots for transmission and reception whenever possible. - Turn radio off (sleep) when not transmitting or receiving.
Network Layer - Consider route relaying load. - Consider battery life in route selection. - Reduce frequency of sending control message. - Optimize size of control headers. - Efficient route reconfiguration techniques.
Transport Layer - Avoid repeated retransmissions. - Handle packet loss in a localized manner. - Use power-efficient error control schemes.

power consumed by different applications. The results identified MPEG2 movie playback as the worst-case application. Note also that word processing consumes quite a bit of power, too. It is speculated that 3-D games are likely to be the next worst-case applications. Hence, there is a need to conserve power, even at the application level.

Mobile applications today provide a variety of functions, such as email, word processing, video conferencing, etc. They all have to perform under the constraints of limited bandwidth and power. While many research projects are related to context- and location-aware mobile applications, very few projects are devoted to power-aware applications. Recently, there have been proposals for network resource-aware applications and the introduction of an adaptive mobile QoS framework. Basically, adaptive mobile applications can alter their QoS requirements so that mobile users can continue to execute their programs without severe interruptions. Also, moving power-intensive computations from a mobile host to the base station can reduce power consumed at the mobile host. However, this is

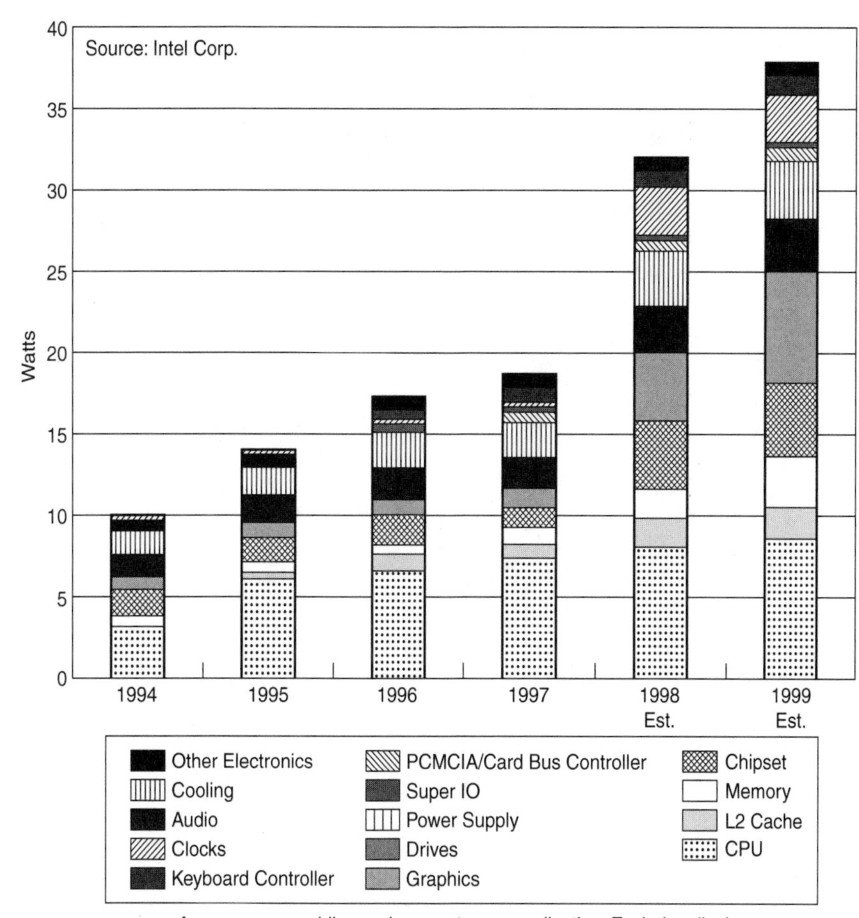

Figure 9.8. Power consumption within a computer (source: Intel).

only applicable to a cellular wireless last-hop network.

Another way of managing low-power availability for mobile clients is the use of proxies. Proxies can be designed to make applications adapt to power and/or bandwidth constraints. Proxies can intelligently cache frequently-used information, suppress video transmission and allow audio, and employ a variety of methods to conserve power consumed by the mobile host.

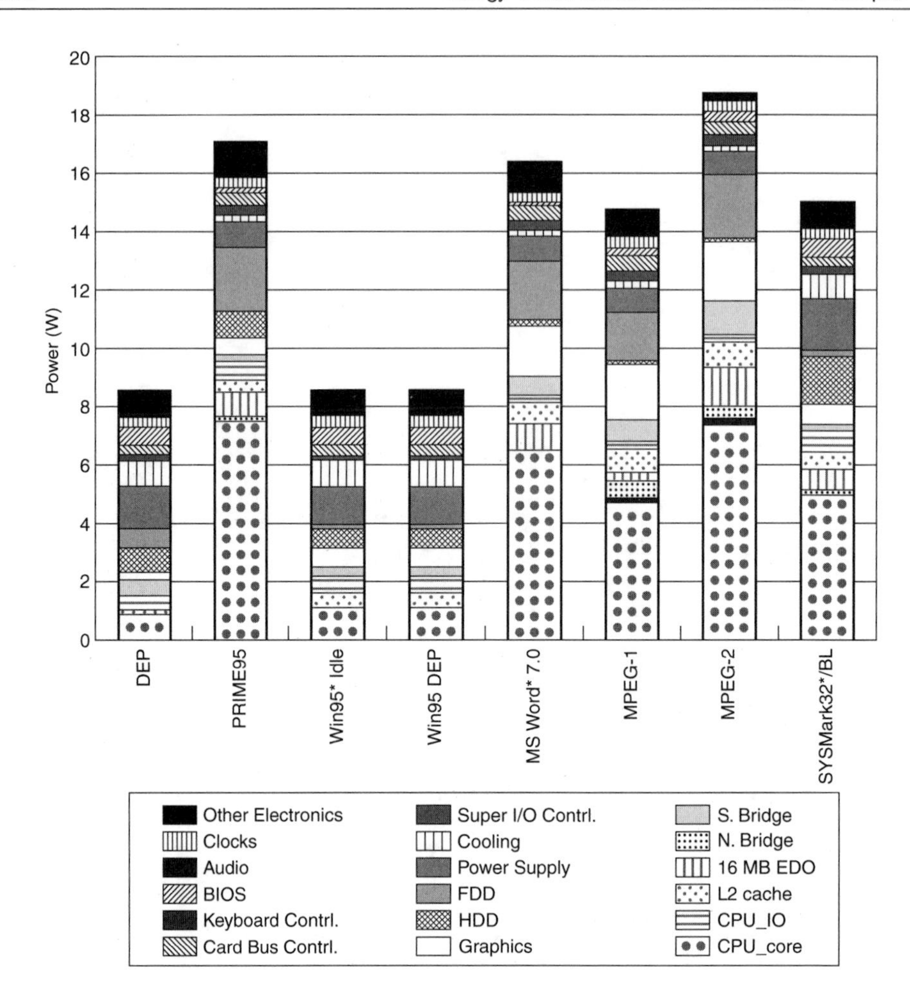

Figure 9.9. Breakdown of power consumption by different applications (source: Intel).

9.6 Periodic Beaconing On Battery Life

In this section, a narration of a series of experiments conducted to evaluate the effect of periodic beaconing on the battery life of an ad hoc mobile computer will be made.

It is imperative to select an appropriate beaconing interval so as not to upset the overall power degradation characteristic of the system while at the same time causing no noticeable side-effects for existing applications. It was discovered that the actions taken by the system in preparation for power shutdown actually draw

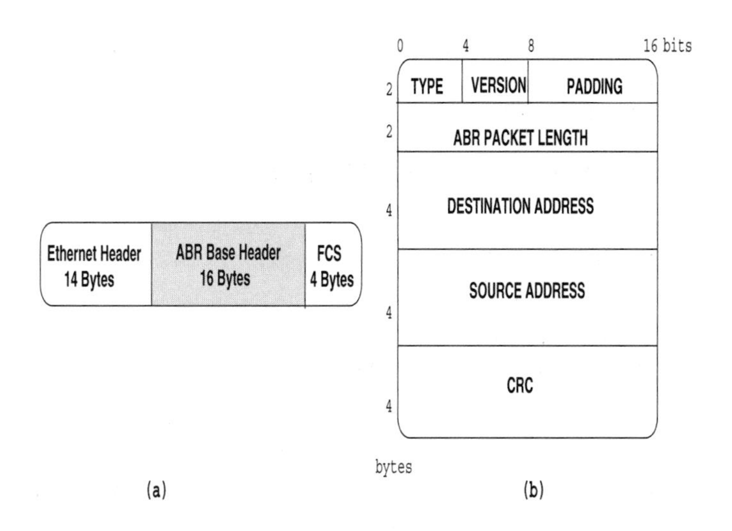

Figure 9.10. Details of: (a) ABR beacon, and (b) ABR base header.

more power than usual. The impact of neighboring nodes' beacons on power life needs to be examined too.

The ABR beacon structure is shown in Figure 9.10. It reveals the ABR base header (BH), with the type field defined as $BEACON$, encapsulated by the data link header (in our case, Ethernet).

The ABR beacon is a network control message generated at the ABR layer and broadcast[1] to the network. It contains the identity of the beacon generator and its intended recipients. The broadcast beacons are received at the destination nodes and are used to derive connection stability information. This information is then used by ABR to derive long-lived routes. Note that beacons can contain the remaining power life information of a node. When these beacons are propagated to their immediate neighbors, neighboring mobile nodes will know when the node is likely to power down.

9.7 Standalone Beaconing

A laptop equipped with a wireless WAVELAN adapter and enhanced with ABR communication software is powered up and left on its own. Beacons are periodically transmitted at preset intervals. The scenarios of interests include:

[1]This does not imply flooding the network. The beacon is sent via radio and all nodes within the radio cell range can therefore receive this beacon.

- Standalone beaconing at high frequencies
- Standalone beaconing at low frequencies
- Beaconing in the presence of neighboring nodes at low and high frequencies

The parameter of interest is the percentage of battery life remaining over time for each beaconing period. In the following sections, the results and observations for each scenario are discussed below.

9.7.1 At High-Frequency (HF) Beaconing

In this experiment, a single mobile computer was set up to beacon at one-millisecond intervals. At the beginning of the experiment, the mobile computer battery was fully charged. The computer was booted with the Linux-ABR kernel (which has the ad hoc routing code incorporated) and the Ethernet interface was configured. The beaconing process was activated soon after. Since there were no other mobile computers in the neighborhood, the power usage was independent of the receive power. The computer had the `Xwindow`, `xload`, and `netstat` applications running. The APM daemon in the OS was disabled and the BIOS APM was carefully prevented from operating.

Figure 9.11 shows a capture of the `netstat` program displaying the interface statistics of the system. WaveLAN cards were configured on interface *eth0*. The `netstat` program updates its readings every second, so the increase by 10 in the number of transmitted packets (TX-OK) every time denotes the transmission of beacons every 100 msecs. The value of RX-OK denotes the number of correctly received beacons in the same one-second interval. In this case, since no other laptops were operating in the same area, the value was always zero. Over regular time intervals (i.e., 20 mins), the percentage of battery life remaining was noted for beaconing intervals of 10, 50, 100, and 500 ms and the results are shown in Figure 9.12. At the 10 ms beaconing interval, the battery life is degrading at the fastest rate, while at 50, 100, and 500 ms, the battery life has relatively familar power degradation characteristics. In addition, it is noticeable that at the 10 ms beaconing interval, the computer has a shorter lifetime (by about 40 mins) compared to at other intervals. At 10 ms, the CPU load is high (as indicated by `xload`) and the computer seems to be slow in forking new applications and responding to commands.

An important observation was made during the last 20 mins of operation by the computer. As shown in Figure 9.12, there is a sudden and substantial increase in power consumption before the computer eventually shuts down. This phenomenon can be attributed to the various operations undertaken by the OS and BIOS in preparation for power down, such as saving all active files, system, and application states.

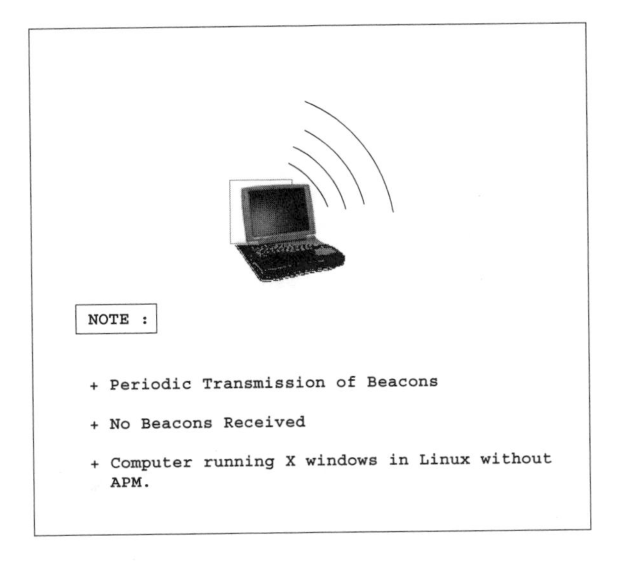

(a)

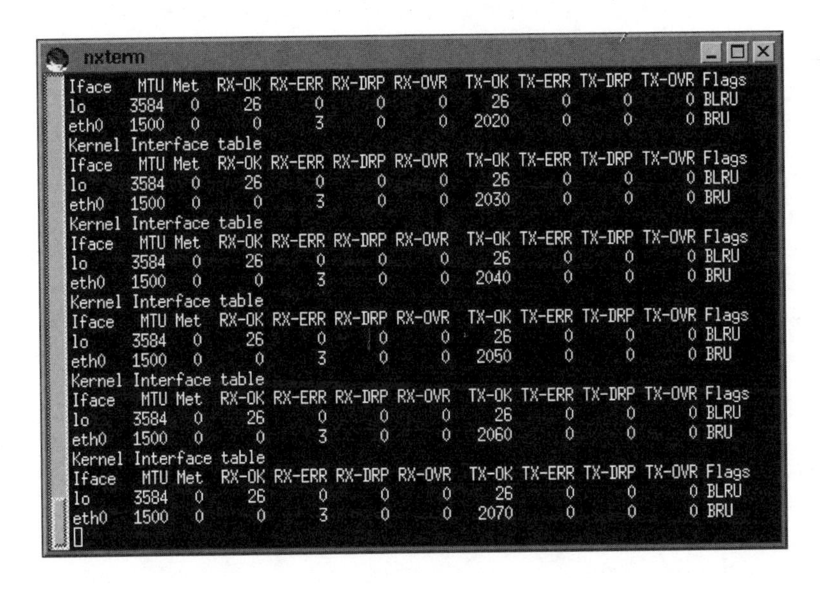

(b)

Figure 9.11. (a) System setup, and (b) Window capture showing periodic standalone ABR beaconing at 100 msec intervals.

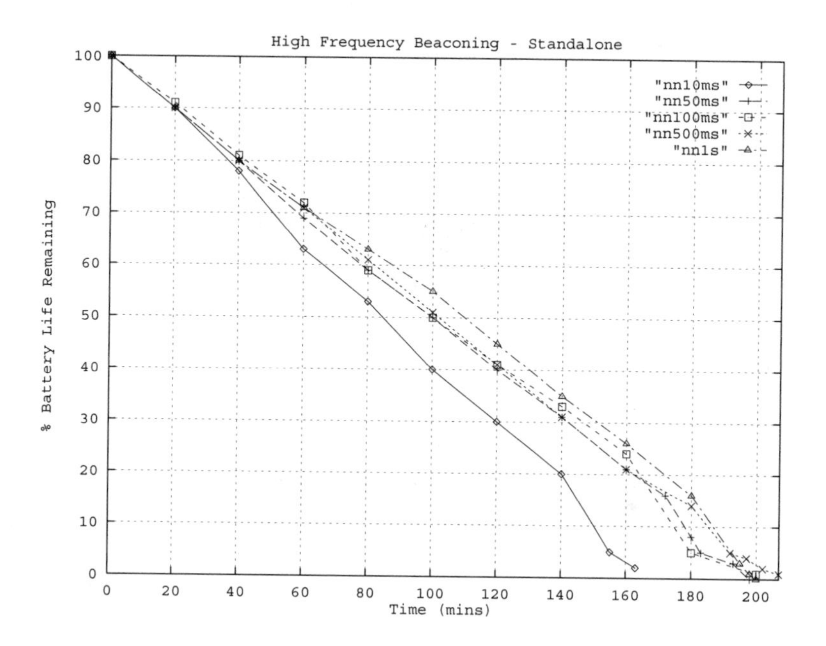

Figure 9.12. The effects of HF beaconing on battery life.

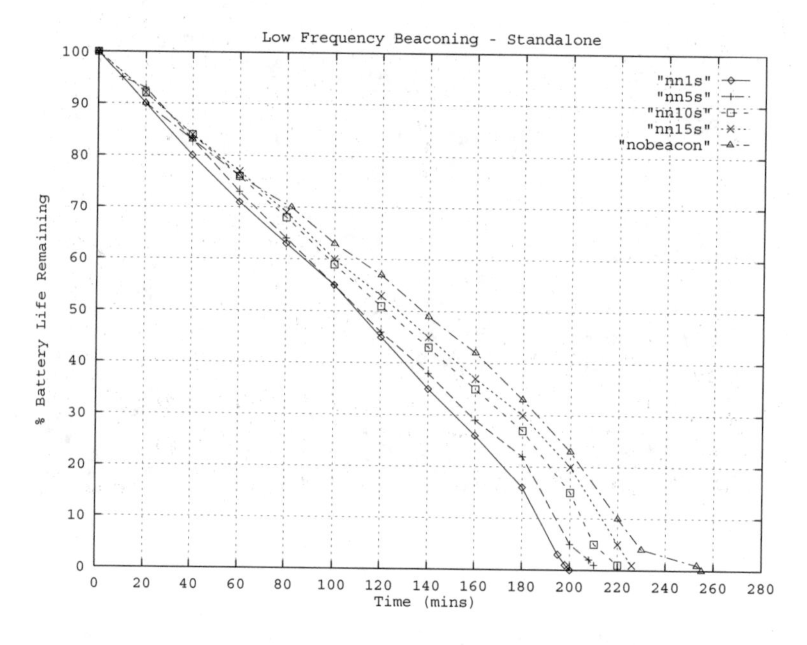

Figure 9.13. The effects of LF beaconing on battery life.

9.7.2 At Low-Frequency (LF) Beaconing

In another scenario, the beaconing interval was increased to investigate the effects of slow beaconing on power consumption. The remaining power life at beaconing intervals of 1, 5, 10, and 15 seconds was observed. The system setup was essentially the same as the earlier case and readings were taken at regular time intervals.

The results obtained are shown in Figure 9.13. For up to the first 40 mins, the difference in power degradation for different beaconing intervals is not significant. Beyond this time, however, the curves began to spread. At 180 mins, the difference in remaining battery life between 1s and 15s beaconing intervals is about 15%. Although there are noticeable differences in remaining battery life for a computer using different beaconing intervals, this does not affect on-going applications and the user may not notice any difference.

Figure 9.13 also reveals an important observation. For 1s and 5s beaconing intervals, the remaining power life suddenly plunges at 180 mins. This is similar to the phenomenon observed for high beaconing intervals and is a result of the actions taken by the OS and BIOS in preparation for power down. However, for 10s and 15s beaconing intervals, this phenomenon happens 20 mins later, at 200 mins. This shows that with a longer beaconing interval, the system will operate longer before it needs to prepare for power down. For completeness of this discussion, it can also be observed from Figure 9.13 that for the case with no beacons, this phenomenon occurs at 230 mins.

9.7.3 Comparing HF and LF Standalone Beaconing

By referring to Figures 9.12 and 9.13, it can be concluded that LF beaconing results in a longer operating time of the mobile computer. Comparing the results for 10ms and 15s beaconing intervals, the difference in lifetime is about 60 mins. This significant difference further advocates that careful selection of an appropriate beaconing interval is important for overall *usefulness*, *efficiency*, and *availability* of the mobile computer.

For both cases, the interesting observation where a greater amount of power is drawn during the last 20 mins of operation, where the computer system invokes procedures and actions in preparation for power down. It is observed that such procedures actually draw more power than pure beaconing alone. This is a result of additional I/O operations performed during the power down process.

9.8 HF Beaconing with Neighboring Nodes

The scenarios discussed so far are related to a single mobile computer beaconing at periodic time intervals. However, in ad hoc mobile networks, ad hoc nodes rely on the presence of neighbors to forward packets to their destinations. Consequently, one needs to know to what degree the presence of neighboring ad hoc nodes will impact the power life of mobile computers. Figures 9.11a and 9.14a reveal the difference between the earlier and current setup.

9.8.1 With One Neighor

A common ad hoc configuration is to have pairs of nodes. In such a case, a node will not only transmit beacons but also receive neighboring beacons. At HF beaconing, the same intervals are used as in standalone beaconing, that is, 10, 50, 100, and 500 ms. The results obtained are shown in Figure 9.15. At a 10 ms beaconing interval, the battery life is degrading at the fastest rate, while at 50, 100, and 500 ms, the battery life has relatively similar power degradation characteristics. In addition, the lifetime for a mobile computer beaconing at 10 ms is about 40 mins shorter than when beaconing at 50, 100, and 500 ms. At 20 mins after the initiation of the experiment, we started to observe a deviation of remaining power life for the 10ms beaconing interval, while the rest had indistinguishable remaining battery life. The phenomenon that occurred during the last 20 mins of operation, where there was a sudden increase in power consumption in preparation for power down, was also observed in this experiment.

9.8.2 With Two Neighbors

Another scenario is to have several ad hoc mobile devices in proximity. The results obtained for two neighbors are shown in Figure 9.16. At the 10 ms interval, the battery power life degrades faster than at 50, 100, and 500 ms and the time to prepare for power down is also faster, occurring at 120 mins. For 50, 100, and 500 ms and 1s beaconing intervals, the preparation for power shutdown occurs at 140, 160, 160, and 160 mins, respectively. Little difference is observed for 100ms, 500ms and 1s beaconing intervals, signifying that such beaconing intervals have little impact on the overall degradation of battery life.

9.9 Comparison of HF Beaconing with and without Neighbors

With the presence of neighbors, the computer's overall power degradation is a function of both transmit and receive power since neighboring beacons are also re-

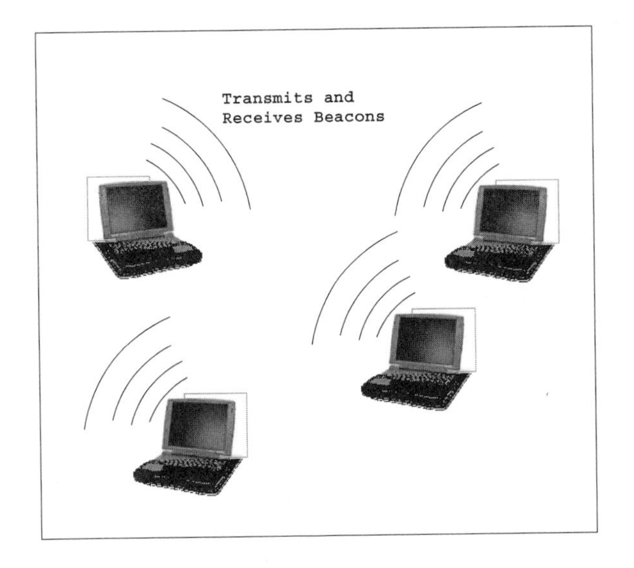

(a)

```
nxterm
Iface   MTU Met  RX-OK RX-ERR RX-DRP RX-OVR  TX-OK TX-ERR TX-DRP TX-OVR Flags
lo     3584   0      0      0      0      0      0      0      0      0 BLRU
eth0   1500   0   9952      7      0      0   9590      6      0      0 BRU
Kernel Interface table
Iface   MTU Met  RX-OK RX-ERR RX-DRP RX-OVR  TX-OK TX-ERR TX-DRP TX-OVR Flags
lo     3584   0      0      0      0      0      0      0      0      0 BLRU
eth0   1500   0   9972      7      0      0   9600      6      0      0 BRU
Kernel Interface table
Iface   MTU Met  RX-OK RX-ERR RX-DRP RX-OVR  TX-OK TX-ERR TX-DRP TX-OVR Flags
lo     3584   0      0      0      0      0      0      0      0      0 BLRU
eth0   1500   0   9992      7      0      0   9610      6      0      0 BRU
Kernel Interface table
Iface   MTU Met  RX-OK RX-ERR RX-DRP RX-OVR  TX-OK TX-ERR TX-DRP TX-OVR Flags
lo     3584   0      0      0      0      0      0      0      0      0 BLRU
eth0   1500   0  10012      7      0      0   9620      6      0      0 BRU
Kernel Interface table
Iface   MTU Met  RX-OK RX-ERR RX-DRP RX-OVR  TX-OK TX-ERR TX-DRP TX-OVR Flags
lo     3584   0      0      0      0      0      0      0      0      0 BLRU
eth0   1500   0  10032      7      0      0   9630      6      0      0 BRU
Kernel Interface table
Iface   MTU Met  RX-OK RX-ERR RX-DRP RX-OVR  TX-OK TX-ERR TX-DRP TX-OVR Flags
lo     3584   0      0      0      0      0      0      0      0      0 BLRU
eth0   1500   0  10052      7      0      0   9641      6      0      0 BRU
```

(b)

Figure 9.14. (a) Beaconing in the presence of neighboring nodes , and (b) Window capture showing both transmited and received beacons.

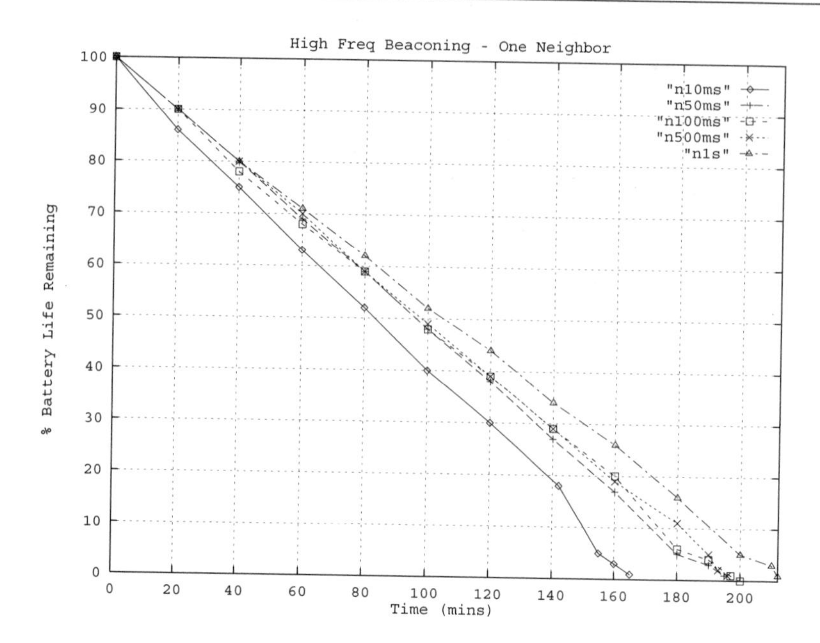

Figure 9.15. The effects of HF beaconing on battery life when a neighbor is present.

ceived. Compared to the results obtained for standalone beaconing, the remaining power life now decreases at a faster rate and this is true for all beaconing intervals under test.

9.9.1 With One Neighbor

As shown in Figure 9.17a, there are noticeable increases in power consumption for the 10ms beaconing interval during the first 40 mins of the experiment. Thereafter, the differences between the results are minute. Figure 9.17b shows the case for the 50ms beaconing interval. Here, the impact of an additional neighbor can be observed between 80 and 180 mins of experimental time. For 100ms and 500ms beaconing intervals, the difference in remaining power life with one and without neighbors is noticeable over a wide range of experimental time (i.e., from 20 to 180 mins). The worst-case percentage difference is observed to be about 7% at the 50ms beaconing interval.

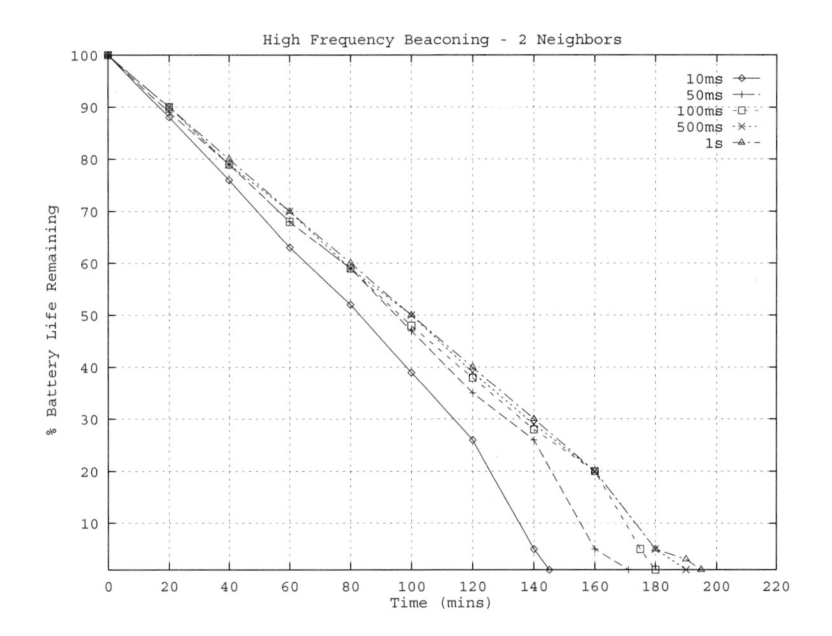

Figure 9.16. The effects of HF beaconing on battery life in the presence of two neighbors.

9.9.2 With Two Neighbors

With the presence of two neighbors, more beacons are received by the monitored ad hoc mobile host. At 10, 50, 100, and 500 ms beaconing intervals, little difference exists for the power degradation characteristics of the one- and two-neighbor scenarios. A noticeable occurrence is that the power shutdown time for the two-neighbors case at the 10ms beaconing interval is now occurring much earlier, at 120 mins. At 50, 100, and 500 ms beaconing intervals, this time is lengthened and occurs at 140, 160, and 160 mins respectively. Comparing the battery life at 10, 50, 100, and 500 ms with the presence of two neighbors, and increasing the beaconing interval by 5, 10, and 50 times, improves the battery lifetime by 20.97%, 25.87%, and 32.87%, respectively.

9.10 LF Beaconing with Neighboring Nodes

9.10.1 With One Neighbor

With the presence of a neighboring host beaconing at the same LF intervals conducted in earlier experiments, results for remaining battery life are recorded and

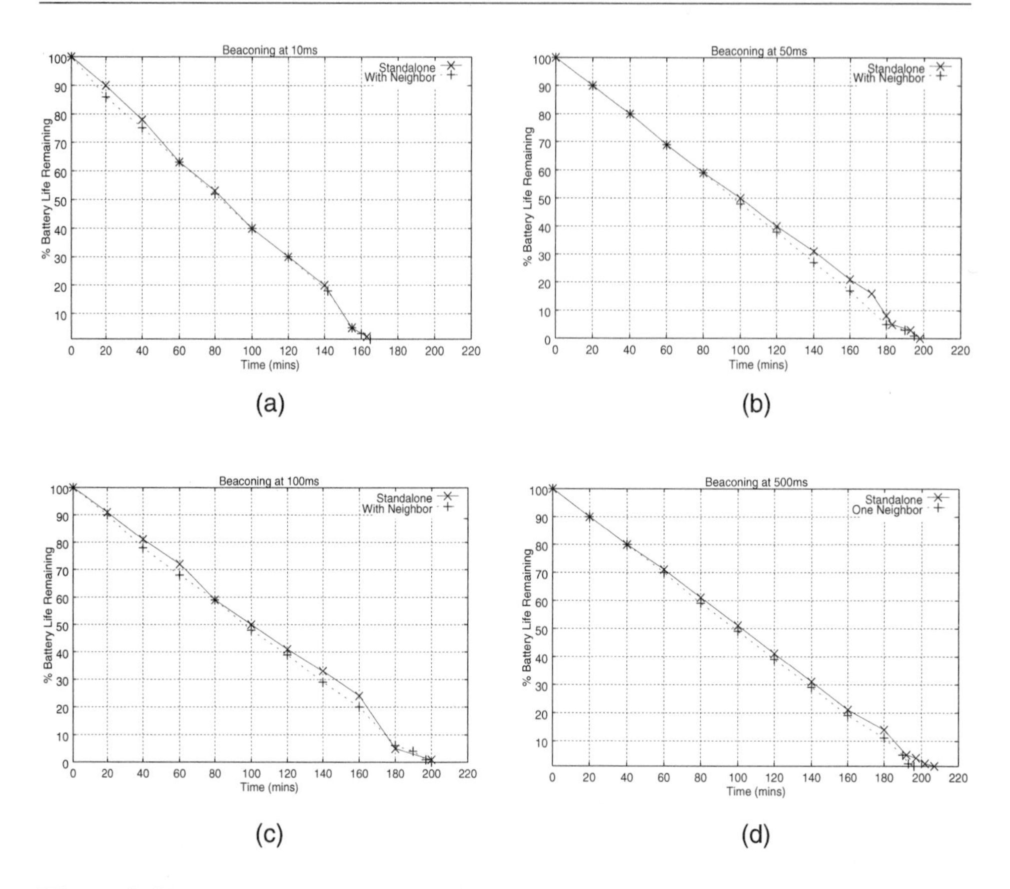

Figure 9.17. Power degradation characteristics with one and no neighbors at HF beaconing.

shown in Figure 9.19. Again, the results show that there is little difference in the remaining battery life at 1, 5, 10, and 15 secs during the first few minutes of the experiment. Compared to the HF case with the presence of a neighbor, the computer system has a longer lifetime, about 50 mins more if we compare the 10ms and the 1s beaconing intervals. Similar to previous experiments, the same power down phenomenon was observed during the last 20 mins of operation.

9.10.2 With Two Neighbors

Figure 9.20 shows the effects of LF beaconing on battery life for the case of two neighbors. Compared to other beaconing intervals of 5, 10, and 15 secs, the remaining battery life is degrading fastest at 1 sec. Minute differences are observed for

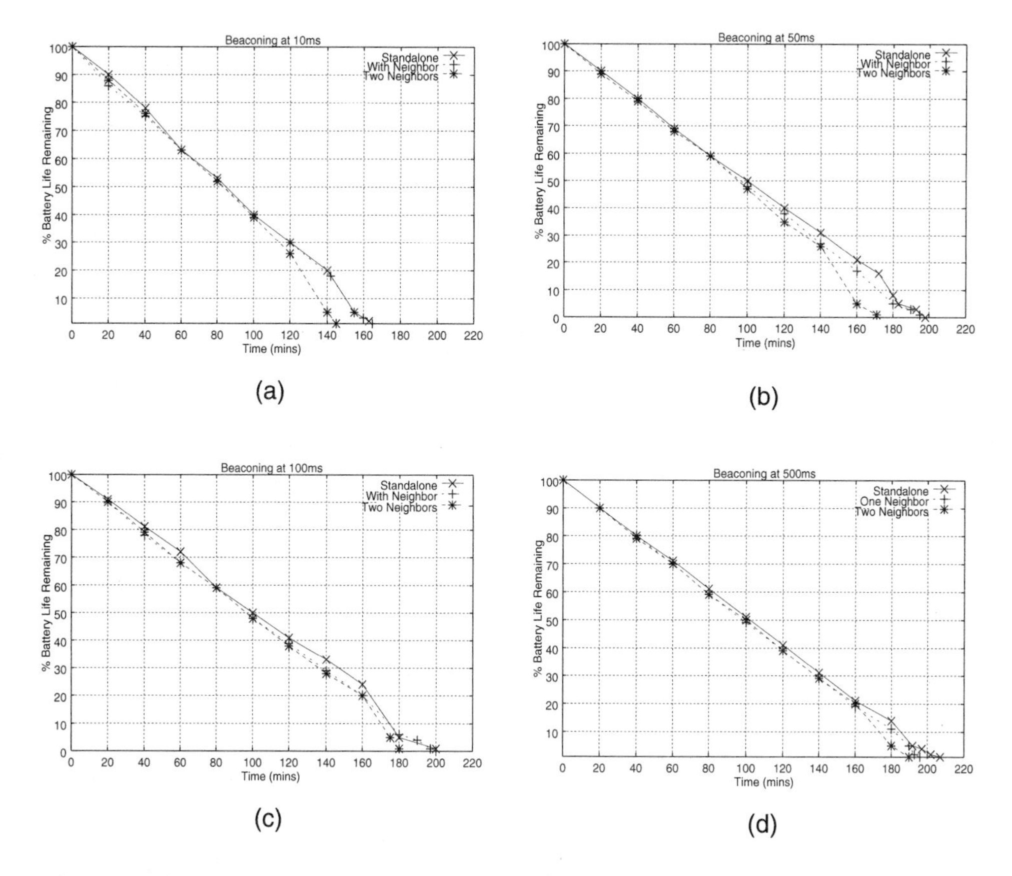

Figure 9.18. Power degradation characteristics with one, two and no neighbors at HF beaconing.

the results at 5, 10, and 15 secs. Comparing the results obtained here with those for one neighbor (Figure 9.19), the overall lifetime of the mobile node has decreased slightly. For 1 and 5 secs, the mobile host now turns off at around 190 mins instead of 218 mins for the earlier case.

9.11 Comparison of LF Beaconing with and without Neighbors

We will now compare the remaining battery life for the same beaconing interval with and without a neighbor. The differences are illustrated in Figure 9.21.

Our first observation is that the differences in remaining battery life are minute except for the case of 10 and 15 secs beaconing intervals. *At large beaconing intervals, the receive power contribution tends to overwhelm transmit power con-*

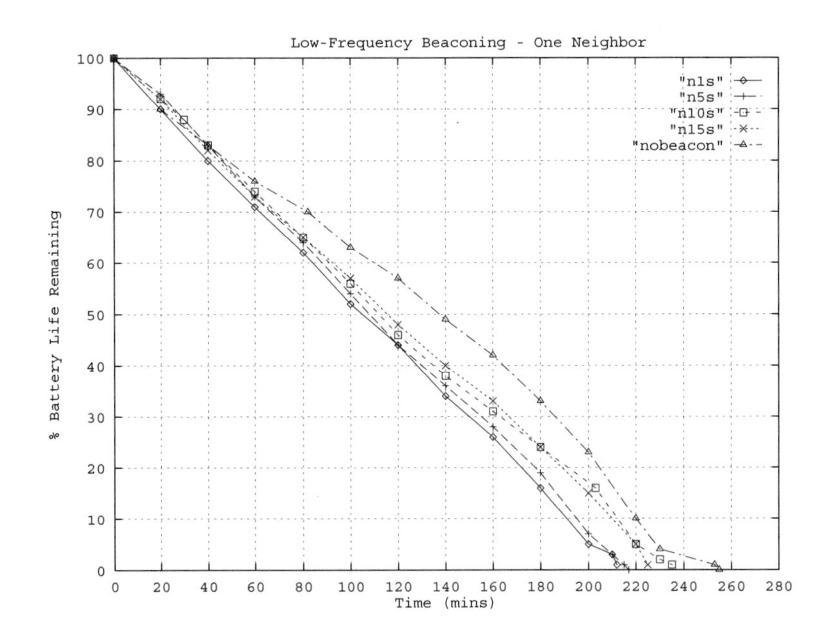

Figure 9.19. The Effect of LF beaconing on battery life when a neighbor is present.

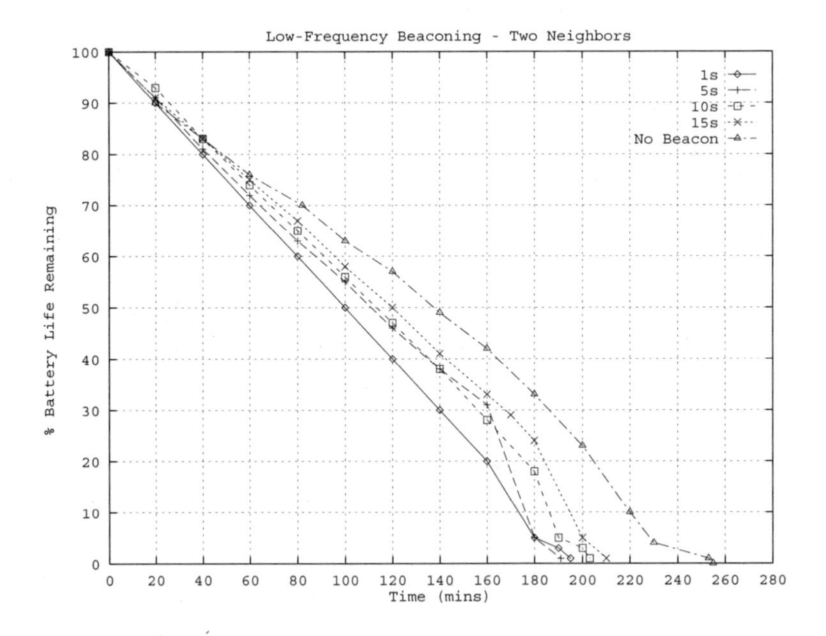

Figure 9.20. The Effect of LF beaconing on battery life when two neighbors are present.

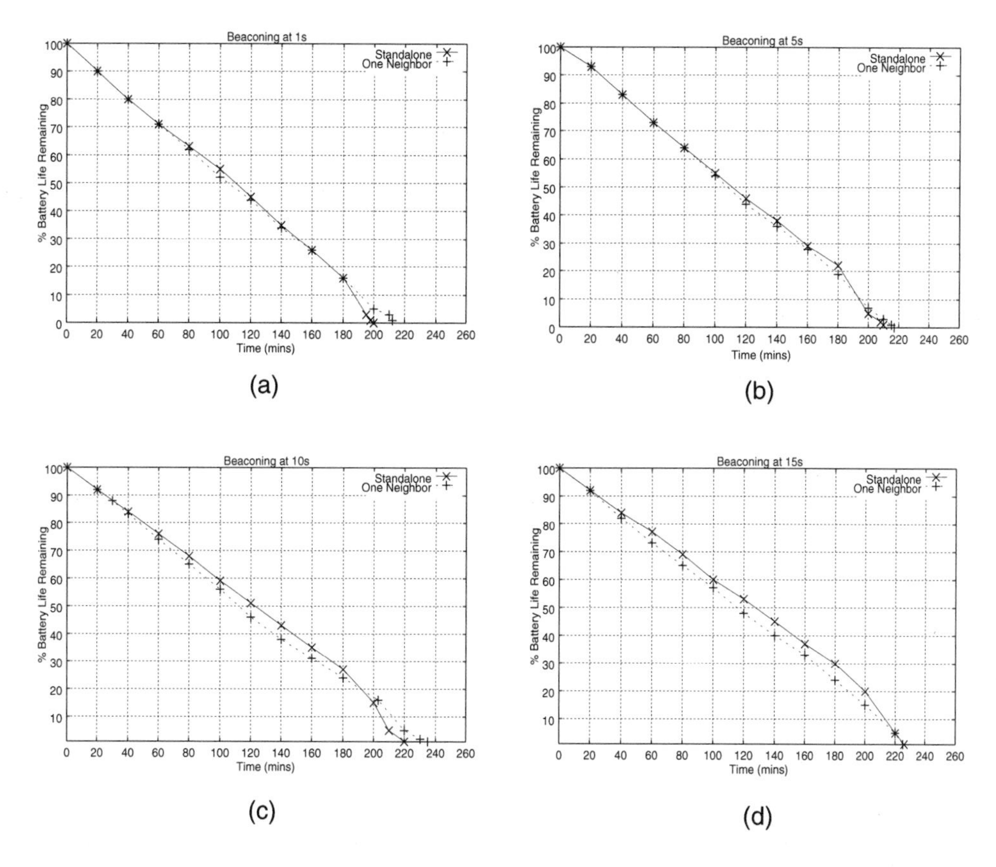

Figure 9.21. Comparing the difference in power degradation with one and no neighbors at LF beaconing.

sumption. Given the fact that transmit power is greater than receiver power for WaveLAN radios, this explains why the difference in remaining power life is only noticeable at very high beaconing frequencies. The second observation made is the phenomenon where the computer system enters power down at about 180 mins, which is similar to the case when there are no neighbors present. This shows that receiving additional beacons from a neighbor does not significantly change the power down time of the mobile node.

The results for the case of two neighbors are shown in Figure 9.22. The battery life degradation characteristic for two neighbors is inferior to the case for one neighbor, but this difference is relatively minute at 1, 5, 10, and 15 secs beaconing intervals. At the 1 sec interval, the computer shutdown preparation time occurs at 160 mins, which is 20 mins earlier than the case with one neighbor. The differ-

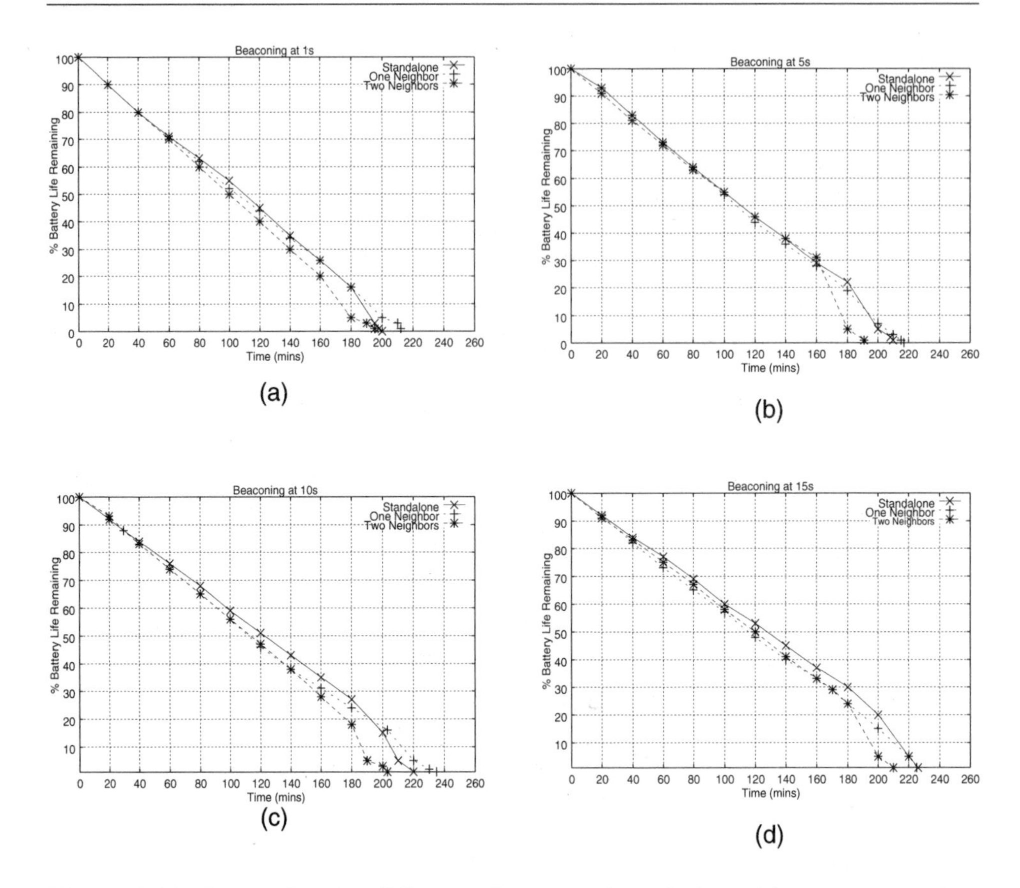

Figure 9.22. Comparing the difference in power degradation with one, two, and no neighbors at LF beaconing.

ence in battery lifetime for one and two neighbors at the 1s beaconing interval is almost indistinguishable, while those at 5, 10, and 15 secs are at 26, 32, 15 mins, respectively.

9.12 Deductions

Based on the above discussions, we arrived at the following deductions:

- Beaconing at extremely HF (i.e., at 10ms interval) can significantly shorten the battery life of a mobile computer and also affect its speed of application execution. This is, therefore, not advocated in practical wireless ad hoc mobile systems.

- Minute differences are observed if a standalone ad hoc mobile host beacons at 50, 100, and 500 ms, and at 1s intervals.

- Beaconing at 1, 5, 10, and 15 secs does not significantly lengthen the lifetime of an ad hoc mobile node significantly (only 20 mins more as seen from our experimental results).

- The power degradation characteristic of a standalone ad hoc mobile computer does not differ greatly from one that has a periodic beaconing interval of 5, 10, or 15 secs.

- For both LF and HF beaconing, the interesting phenomenon where more power is drawn when the OS and BIOS take action in preparation for system power down is observed.

- With the presence of a neighbor, the receive power consumption as a result of receiving neighboring beacons does not contribute much to the overall power degradation when the beaconing interval is small. This is attributed to the fact that transmitting beacons consume more power than receiving beacons. In addition, not all transmitted beacons will be received due to the presence of collisions in the channel.

- Minute differences are observed at LF beaconing in terms of the power degradation characteristics for one- and two-neighbor scenarios.

- The power degradation characteristics for one and two neighbors are similar at HF beaconing with the exception of power shutdown time. The case with two neighbors has an earlier power shutdown time.

Despite these time-consuming experiments and discoveries made so far, there are several issues that have yet to be investigated in greater depth. They include:

- Performing the LF and HF beaconing experiments with APM enabled and comparing the results obtained with those without APM.

- Performing the LF and HF beaconing experiments in the presence of more neighbors to reflect a community network scenario.

- Performing the LF and HF beaconing experiments using a different radio (with different underlying MAC protocols and different transmit and receive power characteristics).

- Differentiating the proportion of power consumed by periodic beaconing and that consumed by the mobile computer (keyboard, screen, disk drives, etc.).
- Presenting the impact of beaconing on the communication performance (i.e., throughput, packet loss, delay, etc.) of ad hoc mobile networks.

9.13 Conclusions

Beaconing is a technique used by many routing protocols for ad hoc mobile networks. Beacons can be used to either update routing information or just to denote the presence of a node in the network. They can also be used to convey *remaining power life* information and also *available service* information. Consequently, beaconing is an important element in the implementation of ad hoc wireless networks and should not be a limiting factor. One particularly challenging task in mobile computing is the drive for low power consumption. Although a lot of work has been done in terms of limiting the power consumption of hardware devices in a system, the same has not been pursued for communication protocols and applications.

In this chapter, the motivation for power management and a brief overview of current power management techniques and smart batteries used in current laptops are presented. Power management needs to be performed at various layers, including device, data link, network, transport, and application layers.

The findings derived from a series of experiments associated with beaconing and its impact on the battery life of an ad hoc mobile computer are discussed. It is imperative to select an appropriate beaconing interval for mobile devices so as not to upset the overall power degradation characteristic while at the same time causing no noticeable side-effects on existing applications running on the device. It was also observed that the actions taken by the computer system in preparation for power shutdown actually draw more power than usual to save the appropriate operating states.

Chapter 10

AD HOC WIRELESS MULTICAST ROUTING

10.1 Multicasting in Wired Networks

In wired networks, changes in network topology are rare and link capacities are considered abundant. This is, however, not the case for mobile ad hoc networks. In a mobile ad hoc network, mobile nodes establish wireless connections among themselves on-the-fly, without any centralized coordinators (e.g., base stations, mobile repeaters, etc.). These connections can be direct connections (one radio hop) if the two communicating nodes are within radio range of each other. Otherwise, multi-hop connections are used in which packets are forwarded by one or more intermediate nodes until the destination node is reached. Because every mobile node can move, changes in network topology are frequent. Furthermore, the bandwidth of wireless links is an order of magnitude lower than that of wired links. As a result, new routing protocols for both unicast and multicast communications are required. These protocols must be highly adaptive to be able to cope with highly dynamic network conditions. In addition, they must be lightweight in terms of control overhead and power consumption. The design of such protocols is very challenging given the unique characteristics of mobile ad hoc networks.

The multicast concept has been around for several years, although deployments are not as ubiquitous as expected. The most notable is the IP multicast architecture and its wide-area implementation, the MBone[60]. Several multicast routing techniques and protocols have been invented and successfully utilized. All these architectures and protocols are discussed below, including their shortcomings when applied to an ad hoc mobile environment.

10.1.1 IP Multicast Architecture

In 1989, Deering [61] proposed the IP multicast architecture to enable point-to-multipoint communications in TCP/IP networks and the Internet. In this architecture, a *multicast group* is identified by a single IP address[1], which serves as a multicast group address. To receive multicast packets, a receiver needs to join and become a member of a particular multicast group. By contrast, a multicast sender can transmit multicast packets to any multicast group without being a member and without knowing the receivers.

Deliveries of multicast packets from multicast senders to all intended multicast members are handled by the network, with help from multicast routing protocols. Because of this, *all routers in the network are required to support multicast routing*. Overall, this is a simple and flexible architecture, although the fact that multicast senders have no control over the multicast delivery process can make administrative and security policies cumbersome. Despite these weaknesses, this architecture is widely accepted and is the basis for the design of many protocols discussed in this chapter.

10.1.2 Multicast Tunnels and the MBone

The requirement that every router in the network must be multicast-enabled has prohibited large-scale multicast deployment in the Internet. To circumvent this problem, multicast traffic is encapsulated in unicast packets and transmitted from one multicast network to another multicast network, where it is decapsulated and processed as multicast packets. The connection through which the encapsulated

[1]Presently, class D IP addresses, which range from 224.0.0.0 to 239.255.255.255, are reserved for multicast use.

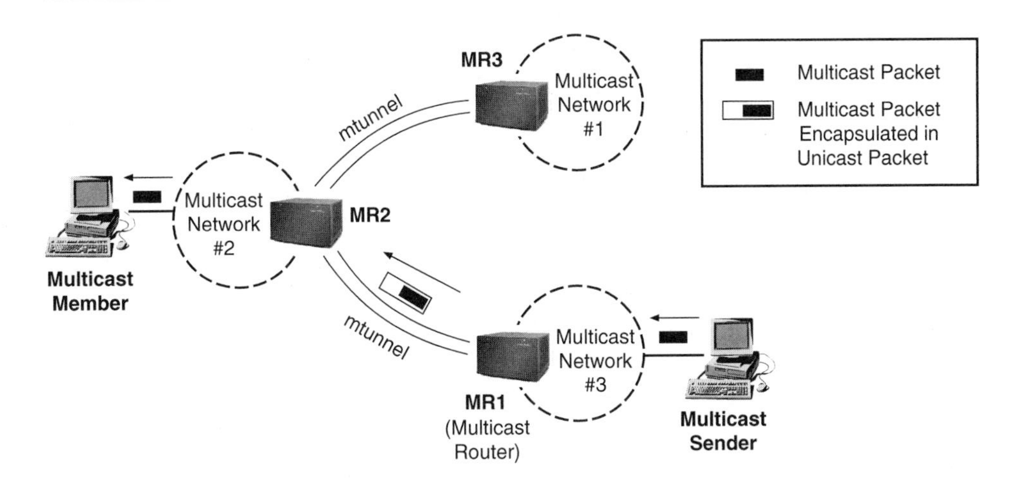

Figure 10.1. The concept of multicast tunnels in the MBone.

packets are transmitted is called the *multicast tunnel*, or *mtunnel*, and is shown in Figure 10.1. In this figure, a multicast packet is sent from the multicast sender in Multicast Network #3. Multicast router (MR1) encapsulates this packet inside a unicast packet and sends it through an *mtunnel*, which has been established between MR1 and MR2. The packet is then decapsulated by MR2 and the resulting multicast packet is sent to the multicast member in Multicast Network #2. By doing this, all routers in transit networks are not required to be multicast-enabled.

Using *mtunnels*, the semi-permanent multicast backbone (MBone) [60] was established on top of the Internet in 1992. Today, several thousands of multicast networks are connected to the MBone. DVMRP (Distance Vector Multicast Routing Protocol)[62] is used as a multicast routing protocol on the backbone, while any protocols can be used in local networks as long as mechanisms are provided to interoperate with DVMRP.

10.1.3 Multicast Routing Algorithms

In the IP multicast architecture described above, actual deliveries of multicast traffic through interconnected networks are an important function of the multicast routing protocols. Currently, several multicast routing protocols are available for wired networks. DVMRP [62] is widely used in the core routers of the MBone. PIM-DM [63] and MOSPF[64] are also used in many local multicast networks. These protocols attempt to provide multicast services in an efficient way by creating and maintaining *source-based shortest path trees* spanning all multicast group members. Both DVMRP and PIM-DM (Protocol Independent Multicast Dense Mode) employ the "broadcast-and-prune" technique [65] and the reverse path forwarding mechanism [66] to derive source-based shortest-path multicast trees, while group memberships and link status information are distributed using the link state mechanism in MOSPF (Multicast Open Shortest Path First). However, these mechanisms incur a lot of control overhead and cannot adapt fast enough to the movements of nodes in mobile ad hoc networks.

Another multicast scheme uses a *single shared tree per multicast group*. Examples are CBT (Core Based Trees) [67] and PIM-SM (Protocol Independent Multicast Sparse Mode [68]. In this scheme, a single shared tree rooted at a selected node called the *core node* or *rendezvous point* (RP) is maintained. Multicast packets are first sent to the core node, where they are forwarded thereafter along the shared tree to all multicast members. The shared tree approach has fewer control overhead, but the path is not necessarily optimal, i.e., the path from a multicast source to a receiver is not necessarily the shortest. Furthermore, in a dynamic network, throughput can deteriorate dramatically unless the core node and shared tree can adapt fast enough to mobility.

10.2 Multicast Routing in Mobile Ad Hoc Networks

Based on the above, existing multicast routing protocols cannot be directly applied to mobile ad hoc networks. Recently, a number of unicast routing protocols for mobile ad hoc networks had been proposed [69]. Most of them aim at providing unicast communications in a highly dynamic environment with minimal control overhead. The same criteria is needed for multicast routing protocols. Furthermore, they have to cope with network mobility, which can be a result of migrations by source nodes, multicast member nodes, or intermediate nodes, in addition to multicast group dynamics (e.g., *join* and *leave* operations).

Multicast routing protocols for mobile ad hoc networks can be classified based on their multicast delivery structures. The multicast delivery structure defines the structure that ultimately forms the path through which each multicast packet can transit to reach all intended multicast group members. All proposed multicast routing protocols can be classified into the following categories:

- **Flooding**

  In this method, no specific structure is enforced. The multicast packet is globally flooded to all nodes in the network. Mechanisms are required to prevent a *broadcast storm* [70], which can be caused by forwarding loops and excessive collisions in the shared wireless medium. This method is robust and well-suited to networks with high mobility. However, bandwidth is severely wasted as a result of unnecessary forwarding of duplicate data.

- **Source-Based Multicast Tree (SBT)**

  In this scheme, a multicast tree is established and maintained for each multicast source node in each multicast group. Thus, in an environment with $\mathcal{G}$ multicast groups where each group has $\mathcal{S}$ multicast source nodes, there will be $(\mathcal{G} * \mathcal{S})$ multicast trees established and maintained. The advantage is that each multicast packet is forwarded along the most efficient path from the source node to each and every multicast group member.

  A multicast tree can be established based on various objectives. The Minimum Steiner Tree (MST) minimizes total cost. This is desirable in many situations, but the exact algorithm is infeasible (NP-complete). Shortest Path Trees (SPTs) minimize the cost between the source node and each multicast member individually. This is simpler and more widely used.

  Overall, the SBT scheme suffers from scalability problems because a lot of overhead is incurred in establishing and maintaining several multicast trees as the number of multicast groups and multicast source nodes increases. Also, it may require prior knowledge of topology information. In a mobile ad hoc network, frequent topological changes become another significant factor in increasing the overall overhead since many source-based trees will be affected and will need to be repaired. Examples of SBT protocols are DVMRP, MOSPF, and PIM-DM.

- **Core-Based Multicast Tree (CBT)**

  CBT is a more scalable approach than the SBT approach. Instead of building multiple trees for each multicast group, a single shared tree is used to connect all multicast group members. Multicast packets are distributed along this shared tree to all members of the multicast group. To establish the shared tree, a special node is designated as the *core node*, which is responsible for creating and maintaining the shared tree. Hence, a core selection algorithm is needed.

  The established shared tree can be either unidirectional or bidirectional. In a unidirectional shared tree, multicast packets must be unicast to the core node, which is the root of the tree. From the core node, the multicast packets will be distributed along the shared tree until they reach all the multicast group members. However, in a bidirectional shared tree, multicast packets can enter the shared tree at any point and they will be distributed along all branches of the shared tree. Obviously, the bidirectional scheme is more efficient in term of both communication performance and forwarding overhead[2].

One disadvantage of the CBT approach is that traffic is concentrated on the shared links, which results in a high tendency for congestion at the shared links. In addition, the multicast packets tend to be forwarded along less optimal paths since they are forced to transit along the shared tree. Moreover, the core node, which is the most critical component in this scheme, becomes the single point of failure. A robust and efficient CBT protocol usually provides mechanisms to dynamically adapt the shared tree into a more efficient configuration and to recover from the core node problem. These extra mechanisms incur additional overhead. Examples of CBT protocols are CBT, PIM-SM, AmRoute [71], AMRIS [72], and AODV [26].

- **Multicast Mesh**

  In the previous two schemes, the multicast delivery structure is tree based because it exhibits the most cost-efficient property by connecting only necessary nodes into an acyclic graph. However, in mobile ad hoc networks, the rate of link changes may be so high that frequent tree reconfigurations are not desirable. Multicast mesh provides another alternative by establishing a *mesh* for each multicast group.

  A *mesh* is different from a *tree* since each node in the mesh can have multiple parents. Care must be taken when forwarding multicast data in multicast mesh to avoid forwarding loops. Using a single mesh structure spanning all multicast group members, multiple paths exist and they are immediately available for use when the primary path is broken. Therefore, a multicast mesh provides multiple redundant paths, avoiding frequent mesh reconfigurations. This minimizes the disruption of on-going multicast sessions and reduces protocol overhead. The multicast mesh approach, however, results in the unnecessary forwarding of multicast packets along all redundant paths in the mesh, causing additional data forwarding overhead. An example protocol is CAMP (Core Assisted Mesh Protocol) [73].

- **Group-Based Multicast Forwarding**

  In this scheme, a group of nodes acts as multicast forwarding nodes for each multicast group. Procedures to establish and maintain this forwarding group are necessary to ensure that all multicast group members are reachable. In

[2]Forwarding overhead is defined as the number of transmissions needed to forward the multicast packet towards the destination.

this scheme, a group of nodes is maintained instead of the links that constitute the tree or mesh. This simplifies the processing required at each node since it does not need to maintain any relationship on multicast packet delivery (e.g., tree links, mesh links) with its neighbors. Multicast packets are forwarded only by forwarding nodes. All received multicast packets that are not duplicated are rebroadcast by the forwarding nodes to their neighbors. Hence, fewer states are kept at each intermediate node and redundant paths are available. Examples of this scheme are ODMRP (On Demand Multicast Routing Protocol) [74] and LBM (Location Based Multicast) [75].

10.3 Existing Ad Hoc Multicast Routing Protocols

10.3.1 Classification

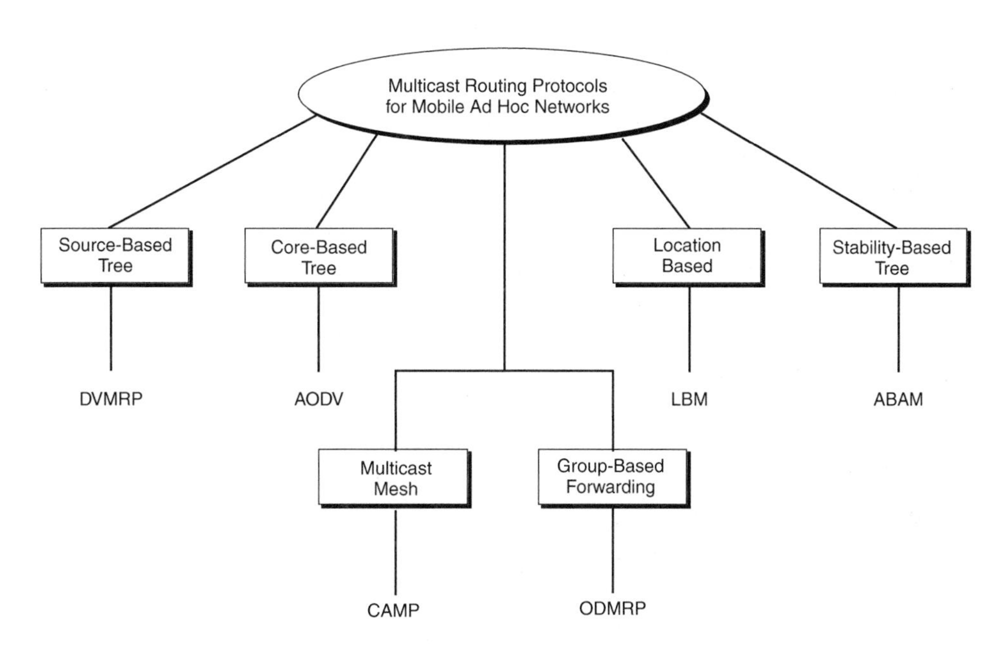

Figure 10.2. A classification of existing ad hoc multicast routing schemes.

10.3.2 DVMRP with Wireless Extension

Distance Vector Multicast Routing Protocol (DVMRP) [62] is a multicast routing protocol initially designed for wired networks. The extensions were presented in [76] to allow DVMRP to function more efficiently in a mobile ad hoc environment.

These extensions are:

(a) leaf-node detection,

(b) dynamic grafting/pruning, and

(c) the use of packet duplication check.

DVMRP maintains source-based multicast delivery trees. The source-based tree is created by first flooding the whole network with the multicast traffic. The reverse path forwarding (RPF) algorithm is used to control the flooding process by having each node rebroadcast only packets received on the shortest path from the source node.

Upon receiving this flooded multicast traffic, the leaf node (i.e., a node that has no downstream nodes) which is not interested in that particular multicast group will send prune messages upstream. The intermediate node receiving this prune message will mark that particular link as "pruned-off" and will not forward any subsequent multicast packets of the corresponding session onto that link.

The non-member intermediate node will also wait to see if all of its downstream links are pruned off. If this is the case, it will send the prune message further upstream. Once this prune process is completed, the optimal source-based tree is established. This *broadcast and prune* process is illustrated in Figure 10.3.

The join/leave operation is controlled by whether or not the prune message is propagated upstream. Periodically, the timer on the pruned branch will expire and the multicast traffic will be flooded again. If there are still no members on the branch, the prune message will be sent; otherwise no prune message is sent and the branch will not be pruned off. This mechanism incurs latency when new members wish to join the group since a new member has to wait until the next flooding period. To eliminate this latency, a new member can explicitly send graft messages upstream to connect the previously pruned-off branch to the multicast tree.

One problem arises when using DVMRP in a mobile ad hoc network. According to DVMRP, the prune message will be initiated at the non-member leaf node and propagated upstream. However, in a mobile ad hoc network, it is difficult for a node to determine whether it is the leaf node or not. Two schemes are possible to resolve this problem. One is to use acknowledgment messages and the other is to exchange routing tables among neighbors. With these methods, a node can easily determine if it is the leaf node of the multicast tree or not and be able to initiate the prune message properly.

The original DVMRP incorporates the distance-vector based mechanism to provide shortest path information to the multicast source nodes. This information is

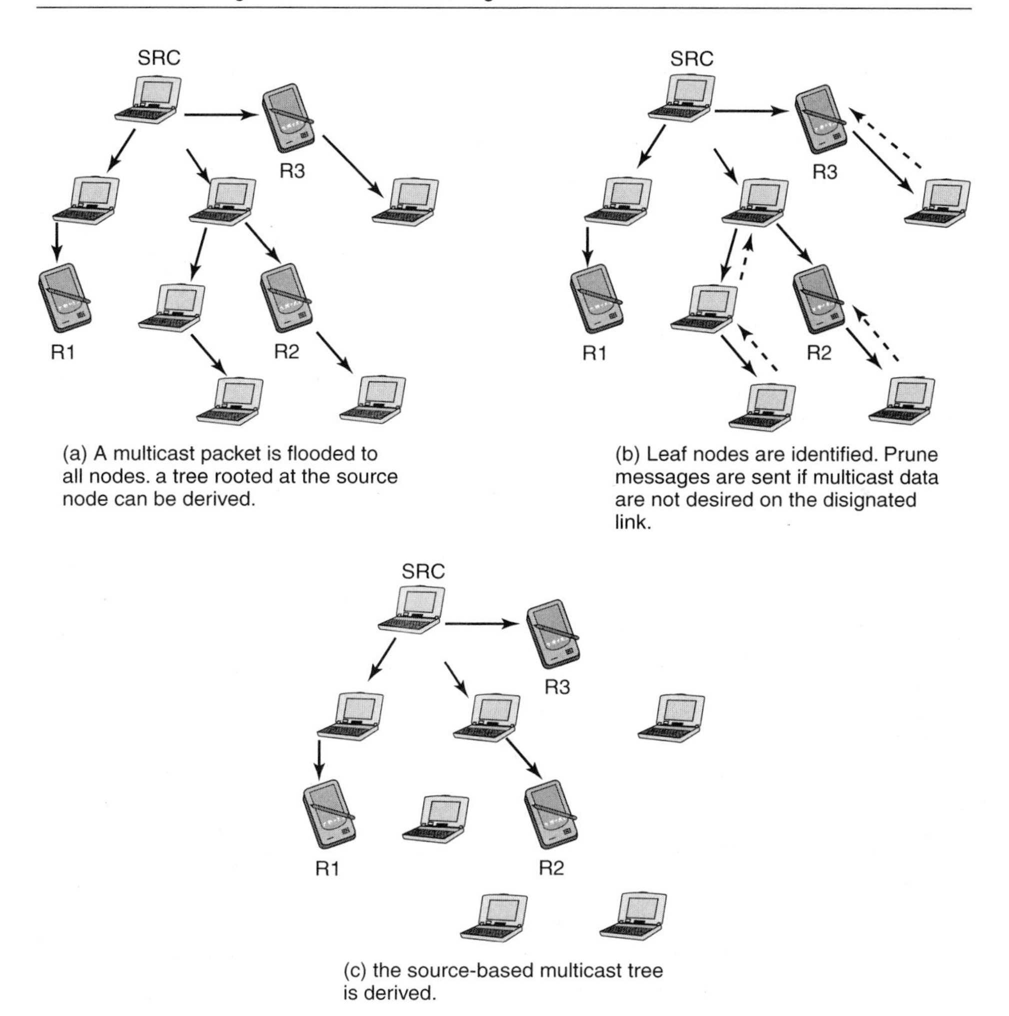

Figure 10.3. The broadcast-and-prune method used in DVMRP.

used in reverse path forwarding (RFP) algorithm. However, in a very dynamic mobile network, routes to the source node can change faster than the routing protocol can keep track. In this case, the RPF algorithm fails to operate efficiently. In [76], instead of RPF, checking for packet duplication is used and ensuring each node rebroadcast only non-duplicate packets.

Dynamic grafting/pruning is another extension which helps in speeding up multicast tree reconfiguration when DVMRP is used in mobile ad hoc network. In the original DVMRP, only dynamic grafting is provided to allow a new member to quickly join the multicast group. In dynamic grafting/pruning, when a node detects that the shortest path back to the multicast source node has changed (for example, when the multicast traffic is received from a different upstream node for duration longer than the threshold duration), it will send a prune message to the current upstream node and a graft message to the new upstream node. With this method, DVMRP multicast tree adapts faster to network mobility. Nonetheless, there exists controversy as to the applicability of DVMRP to mobile ad hoc networks.

10.3.3 AODV Multicast

Ad hoc On-Demand Distance Vector (AODV) [26] initially supports unicast and subsequently covers multicast routing. It is an on-demand protocol based on the destination sequence number concept introduced in DSDV [22]. Sequence numbers are used to determine the *freshness* of routing information so that stale information can be readily detected and discarded. These sequence numbers are updated only by authorized nodes. In unicast mode, only the destination node is allowed to update the sequence number of its own routing information (except possibly when the route is broken). In multicast mode, a node is selected to generate and update the multicast group sequence number. This node is called the *multicast group leader.*

AODV multicast creates and maintains a single shared tree per multicast group. Each multicast group has a group sequence number. This group sequence number is periodically incremented and broadcast throughout the network by the multicast group leader. The first node requesting a route to a particular multicast group automatically becomes the leader of that multicast group. Thereafter, the multicast tree is dynamically built as subsequent nodes join the group. A new member node will select the best on-tree node to graft to. This selection is based on the freshness of the group sequence number and the hop distance to another group member.

To join a multicast group, a joining node first broadcasts a join request packet with the destination address set to the desired multicast group address. This join request will create a route back to the joining node as it traverses through each intermediate node. Once the request packet reaches an on-tree node, a reply packet is sent back to the joining node through the established backward route. It is possible for a node to receive several reply packets if more than one on-tree node receives a join request packet. The reply packet will create a multicast route on its way toward the joining node. However, this multicast route is in a passive state.

A node chooses the best reply packet based on the metric described earlier and sends an activation packet to the node from which the best reply packet is received[3]. On receiving the activation packet, each intermediate node executes the same algo-

[3]This approach is similar to ABAM (Associativity Based Ad Hoc Multicast) except that there is no consideration for link and path stability.

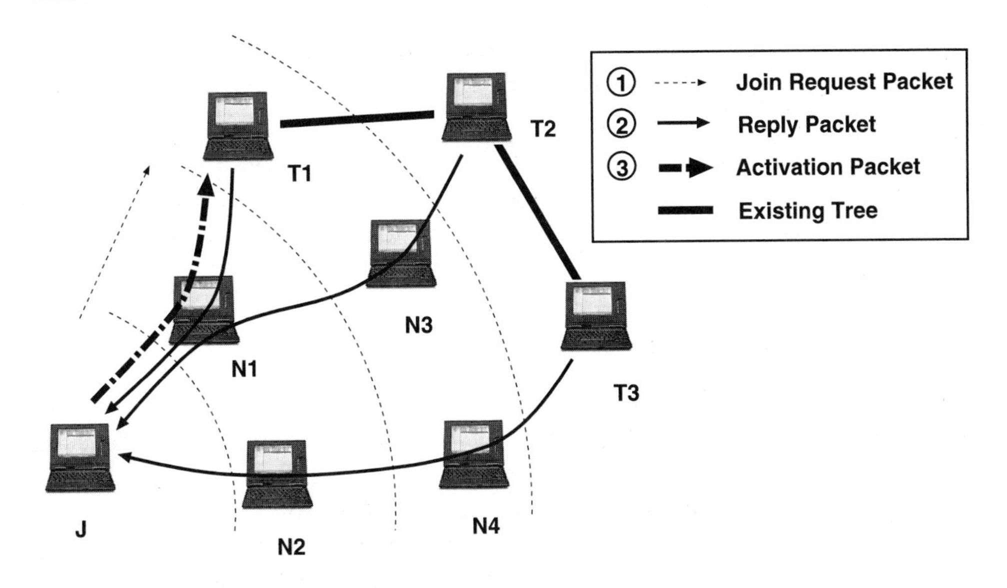

Figure 10.4. AODV multicast join operation.

rithm to choose the best next hop based on the reply packet received earlier. This activation packet will activate the multicast route along the path it traverses. Only one multicast route will be made active. The other passive multicast routes will be removed when their timers expire. Once the activation packet reaches the on-tree node, the multicast communication is enabled. The join process is shown in Figure 10.4.

A leaf node can quit a multicast group by sending a prune message to its next hop on the multicast tree. The next-hop node, on receiving the prune message, will remove the corresponding entry from its multicast route table. If the next-hop node becomes a leaf node as the result of this removal and it is not interested in multicast traffic from this group, it can further prune itself from the tree using the same method.

When a multicast tree link is broken, the downstream node (node further away from the multicast group leader) will initiate the join process as described above. Note that the join request also contains the distance from the multicast group leader so that only the node closest to the group leader can reply. If no replies are received, a network partition is assumed and the joining node will designate itself as a new multicast group leader if it is a member of the group; otherwise, the decision will be passed down to one of its downstream nodes. In the latter case, it may be able to prune itself off if it has only one downstream node.

The periodic network-wide broadcast of the group sequence number also helps to merge partitioned multicast trees together. When one multicast group leader receives a sequence number update for the same multicast group from another multicast group leader, it knows that there is another subtree of the same group. If it has a lower node ID, it will try to initiate the merge[4] process by unicasting the special join request to the newly discovered multicast group leader. Once the join process is completed, the newly discovered group leader will become the group leader for the whole multicast tree.

10.3.4 Multicast Mesh: CAMP

Core Assisted Mesh Protocol (CAMP) [73] establishes a multicast mesh for each *multicast group*. One or more core nodes are delegated to assist in join operations so that flooding is no longer necessary. A multicast mesh can be established by having both sender and receiver nodes join the multicast group.

To join a multicast group, a node first checks whether any of its neighbors are already members of the multicast mesh. If true, it only needs to announce its membership request to its neighboring nodes. However, if there are no mesh members among its neighbors, the join message will be sent to one of the core nodes. The path to the core node is derived from the unicast routing protocol. An acknowledgement message will be sent back if the join process succeeds, in which case, the path that the join message traverses becomes part of the multicast mesh. In the worst case, if the core node is unreachable, an expanded ring search mechanism will be used to find the route to at least one of the mesh member node. Subsequently, the join message is sent to that mesh member. As a result, the core node does not become the single point of failure since CAMP can fall back to using the expanded ring search. Moreover, multiple core nodes can be designated for each multicast group. All the core nodes of the same multicast group use core explicit join messages to establish mesh connectivities among themselves.

When a node wants to leave an existing multicast group, it can simply broadcast a quit notification messages to its neighbors. A node can leave a group only when it is not interested in the multicast group and there is no neighbor depending on it to forward multicast traffic. An anchor table is present in each node which records this dependency information. In the anchor table, a list of anchor nodes is recorded for each multicast group of which the node is a member. The anchor node is the neighbor node that is the next hop in the reverse shortest path to at least one source node. In other words, it is the upstream node in at least one multicast session. This information can be obtained from the unicast routing table. By exchanging this

[4]Investigations are need to understand why spatially separated multicast groups cannot be left isolated and what are the potential benefits of merging.

anchor table among neighbors, each node can determine if it is the anchor node of any of its neighbors.

When a link is broken, it is expected that the multicast mesh has alternate paths ready for use; otherwise, the disconnected receiver has to join back to the multicast mesh. In addition, multicast receivers have to periodically verify that the shortest paths from all source nodes to themselves are included in the multicast mesh. This can be done by examining the message cache, which records all of the recently received data packets. If data packets were not received from the neighbor on the shortest path to the source node, a special control message (*heartbeat* or *push join*) is sent to incorporate the new shortest path into the multicast mesh. A heartbeat is sent when the next-hop node is already a multicast mesh member; otherwise, a push join is sent. A push join has the same effect as a join process, that is, it transforms all nodes along the path to be multicast mesh members. Information about the valid shortest path is obtained from the underlying unicast routing protocol. Hence, CAMP has a dependency on the underlying unicast routing protocol. The use of push join messages is shown in Figure 10.5.

10.3.5 Group-Based: ODMRP

ODMRP[74] is a flooding-based multicast routing protocol for mobile ad hoc networks. Unlike the pure flooding scheme, data is not flooded throughout the network in ODMRP. Instead, data is flooded only throughout the *forwarding group*, which is continually maintained by periodic flooding of control messages. The *forwarding group*, which was first introduced in FGMP (Forwarding Group Multicast Protocol) [77], is a set of ad hoc nodes specially chosen to forward multicast traffic for a particular multicast group. The formation and maintenance of this forwarding group ensures that all these forwarding group nodes provide at least one path from the multicast sender to all receivers. To establish and maintain such forwarding group, ODMRP depends on the following operations:

(a) multicast sender advertisement, and

(b) JOIN-TABLE broadcast by multicast receivers.

When a multicast sender has data to send, it starts the periodic broadcast of JOIN-REQUEST messages. These JOIN-REQUEST messages are flooded throughout the mobile ad hoc network. Each node, upon receiving the JOIN-REQUEST message, will update its unicast routing table with the address of the node from which the JOIN-REQUEST message is received. With this routing table, the unicast path back to the multicast sender is known.

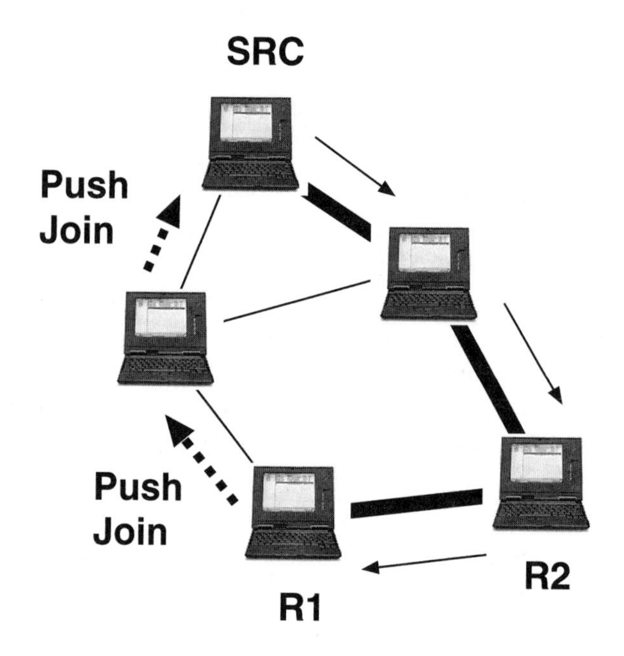

(a) When R1 detects that packets from SRC are not received on the shortest path, push join messages are sent.

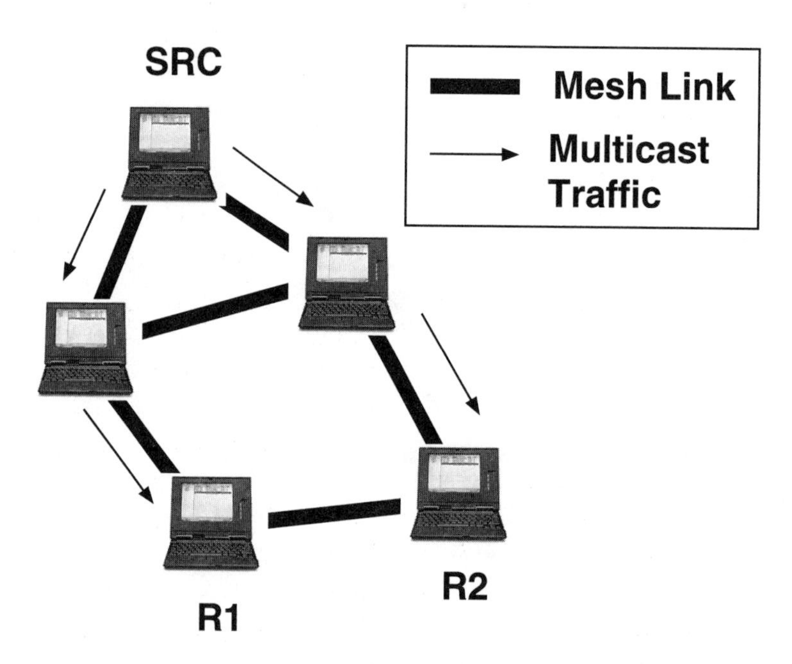

(b) The new shortest path is now included in the mesh.

Figure 10.5. The use of push join messages in CAMP to maintain the multicast mesh.

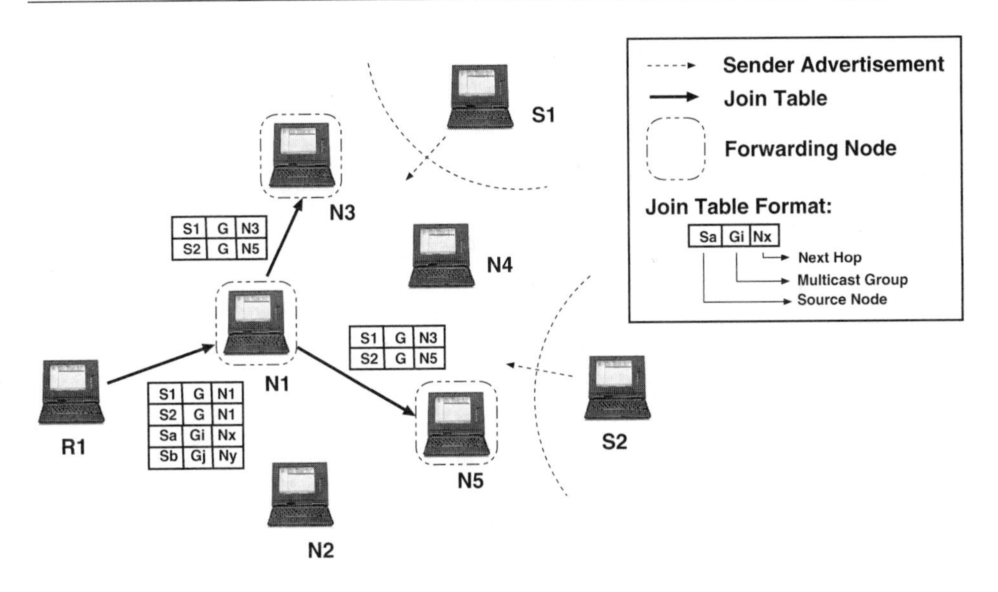

Figure 10.6. ODMRP forwarding group formation using periodic sender advertisements and join table broadcasts.

When a multicast receiver receives the JOIN-REQUEST message, it will update its member table with the address of the multicast sender and periodically broadcast JOIN-TABLE messages. The JOIN-TABLE message contains the list of all multicast senders known to that receiver and also the next-hop nodes towards those multicast senders. These next-hop information are readily available from the unicast routing table. Only the node listed as the next hop in the JOIN-TABLE message will process the JOIN-TABLE message. These nodes will become forwarding group nodes and create the new JOIN-TABLE with the next-hop information from its own message cache. The newly created JOIN-TABLE will be broadcast further. Eventually, JOIN-TABLE information will be propagated back to all multicast senders and all nodes along the way from each receiver to each sender will be included in the forwarding group, as illustrated by Figure 10.6.

By having all these periodic messages, the forwarding group will be continually refreshed. A timeout mechanism is then used to remove stale forwarding group nodes. In addition, multiple paths are generally available through forwarding group so that when a link is broken, data packets are still forwarded along these alternate paths.

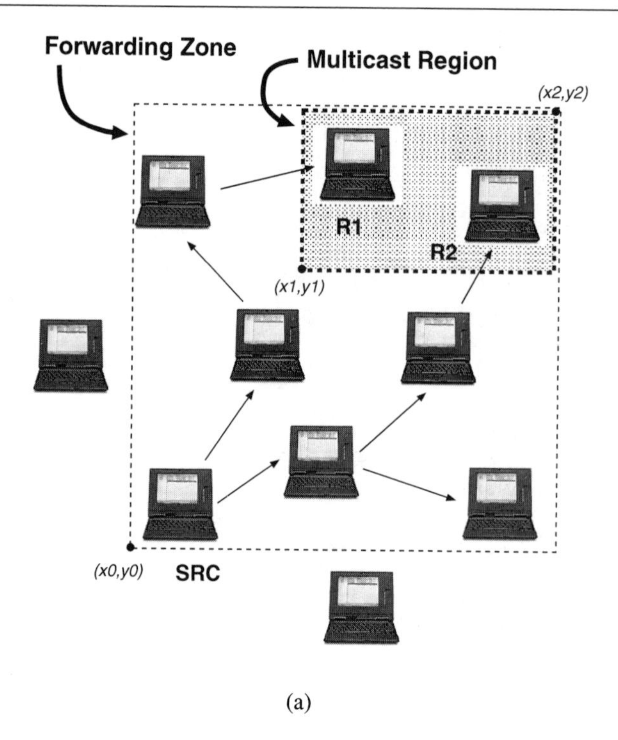

(a)

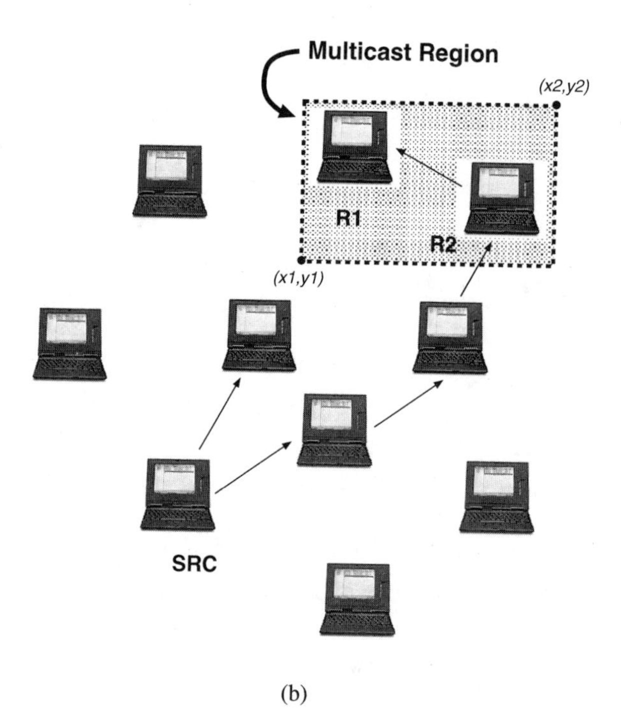

(b)

Figure 10.7. The LBM forwarding scheme.

10.3.6 Location-Based Multicast: LBM

Location-Based Multicast (LBM) [75] uses location information to distribute multicast traffic to the destination multicast group, which is also defined based on location. Unlike the IP multicast model, a multicast group in LBM is now defined as all nodes residing in a specific geographical region called *a multicast region.* The multicast region (the rectangular region defined by four positional coordinates) is specified in each multicast packet by the source node. If the multicast source node itself is not in the multicast region, some other nodes outside the multicast region must help to forward the data. These forwarding nodes, therefore, should reside in the *forwarding region.* All nodes that belong to the forwarding region will forward the data they have received. It is important to ensure that the forwarding region encompasses the multicast region and that network connectivity exists from the multicast source node to the multicast region via nodes in the forwarding region.

LBM assumes that every node knows its own location. This information may be obtained from GPS devices. This assumes that GPS receivers are installed in all nodes in a mobile ad hoc network. A simple solution is to use the multicast flooding scheme. Using this method, a source node appends a multicast region specification to multicast packets before flooding the packets to all nodes in the network. Each node will compare its current location with the multicast region specified in the packet and will accept the packet for further processing only if the locations coincide. Although simple, the flooding scheme results in high control and forwarding overhead.

LBM proposes a method that uses the location of intermediate nodes to confine the flooding scope. This is illustrated in Figures 10.7a and 10.7b. In this method, the source node computes the forwarding zone that encompasses the multicast region and the additional region in between the source node and the multicast region. For simplicity, LBM uses a rectangular shape to define the region, as shown in Figure 10.7a. The forwarding zone specification is appended to each multicast packet in addition to the multicast region specification. Only the node inside the forwarding zone will rebroadcast the packet.

In the second scheme, the distance between each intermediate node and the center of the multicast region is used as the routing criteria. A node will forward the multicast packets only when it is closer to the multicast region than the previous node or it resides in the multicast region.

Shortcomings of LBM

The LBM scheme exhibits some inherent shortcomings. LBM assumes that every node knows its own location. LBM expects that multicast members are nodes inside the multicast region and that all nodes can obtain accurate positional information. Nodes in the multicast region must not be partitioned in terms of connectivity.

Positional information of boundary nodes of both multicast and forwarding region have to be known in advance prior to multicast data delivery. Positional inaccuracy can therefore cause problems.

It is possible that multicast group members **may not** receive multicast packets even though they are properly located in the multicast region. This situation may happen when the network is partitioned within the forwarding zone. A simple workaround is to introduce a *threshold* parameter. The forwarding zone can be expanded by this threshold with the hope to incorporate enough nodes to provide connectivity to all parts of the multicast region.

It is obvious that these solutions only help to reduce the occurrence of the problem, but not to completely eliminate it because the protocol **does not** take into account signal quality and connectivity. For example, two nodes that are near each other do not imply an ability to communicate.

Moreover, by increasing the threshold value, the overhead in multicast data forwarding is also increased. The inaccuracy in positional information obtained from GPS devices should be compensated by increasing the threshold value even further.

10.4 ABAM: Associativity-Based Ad Hoc Multicast

The previous protocols, although attractive in their own regards, have failed to take into account link and route stability. This section discusses how *associativity* can be applied to ad hoc multicasting.

ABAM (Associativity-Based Ad Hoc Multicast) is an on-demand source-based multicast routing protocol for mobile ad hoc networks. It builds a source-based multicast tree focused primarily on the association stability[5] among nodes and their neighbors. A stable multicast tree, therefore, requires fewer tree reconfigurations.

In ABAM, ad hoc multicasting comprises four components, namely:

(a) multicast tree formation (per multicast session),

(b) handling host membership dynamics,

(c) handling node mobility, and

(d) multicast tree deletion/expiration.

[5]Association stability results when the number of beacons received continuously from another node exceeds some predetermined value.

The first component establishes the communication path prior to multicast data transfer. The second component provides the necessary procedures to cope with host membership dynamics. Note that this form of dynamics does not necessarily involve host's migration, but rather, it is based on a user's demand to remain in the multicast session or not. The third component is the mobility management core for ad hoc mobile networks employing a multicast tree approach. This "core" allows the tree to be adaptive, that is, the tree responds to node movement so that the ongoing multicast session can continue as usual. Finally, the last component releases network resources when the established multicast route is no longer necessary and desired by the user.

10.4.1 Multicast Tree Formation

To establish a point-to-multipoint ad hoc mobile multicast tree, a three-phase tree setup approach can be used. In the first phase, the sender (source) sends a BQ-M packet to all related receivers, via a wave-like broadcast. The BQ-M packet contains the address of the intended multicast group. When a valid receiver receives the BQ-M packet, it executes the **route selection algorithm** (RSA) to derive the best route based on a set of routing QoS and stability requirements. The second phase involves the receiver sending a BQ-REPLY packet back to the source. The source, therefore, receives multiple reply packets from various receivers belonging to the same multicast group.

Based on the replied information, the source node may execute a **tree selection algorithm** (TSA) to derive the multicast tree. In the TSA, the desired outcome is a shared-based tree that results in the sharing of wireless links so as to achieve a high link utilization efficiency. Another important aspect of exploiting the shared-based tree concept is to localize mobility to a few "associatively stable" nodes. Once the multicast tree is derived, the third phase is for the source node to send an MC-SETUP packet to all its intended recipients. The MC-SETUP packet contains the route path of the derived multicast tree. Hence, all nodes in this route path have to update their multicast routing tables and be aware if they are going to act as a branching node and/or forwarding node[6] for the multicast session. When the multicast receivers receive the MC-SETUP packet, the route is considered established. The three phases are illustrated by Figure 10.8.

Figure 10.9 reveals the format of an ABAM multicast setup (MC-SETUP) message. The source node specifies the *forwarding*, *branching*, and *receiving* nodes within the MC-SETUP packet. When the source sends out this message, nodes

[6] A *forwarding* node basically has one downstream neighbor to forward the multicast packet to. However, a *branching node* has two or more downstream neighbors.

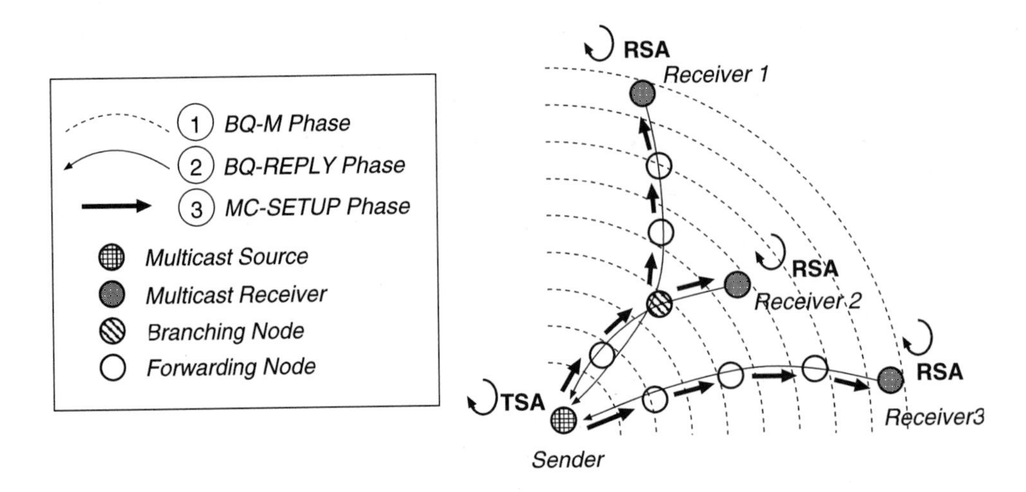

Figure 10.8. The three phases involved in ABAM multicast tree discovery, selection, and formation.

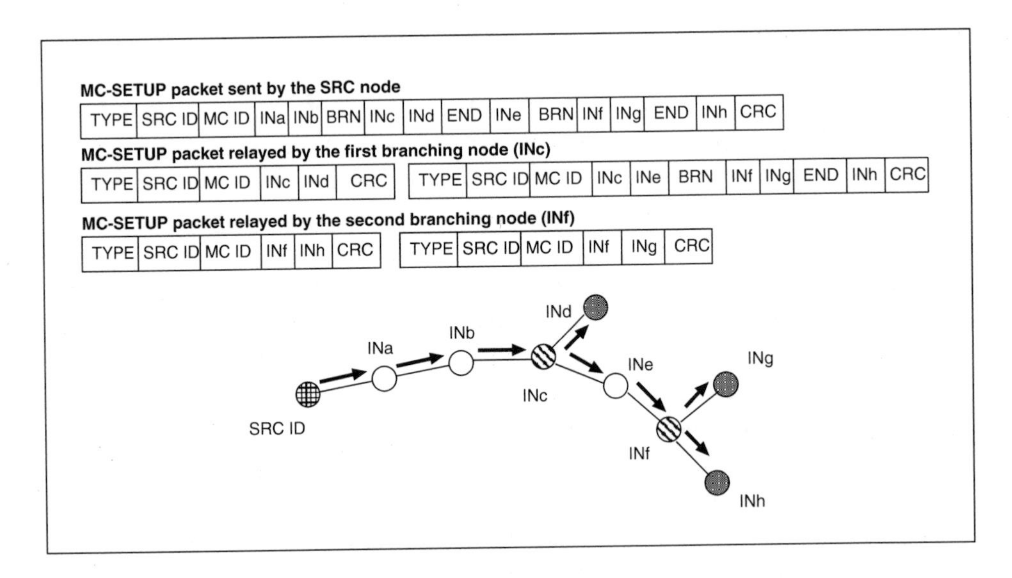

Figure 10.9. Changes in the ABAM MC-SETUP packet as it transits toward the receivers.

Table 10.1. Route Selection Algorithm in ABAM.

ABAM Route Selection Algorithm (RSA)
1. Multicast receivers know the possible routes (R_i) to the source node when it receives the tree search packet.
⇓
2. The receiver discards routes with exceeding route relaying load (i.e., R_x where $R_x < R_i$).
⇓
3. The receiver selects a route/s with highest degree of association stability (i.e., R_y where $R_y < R_x$)
⇓
4. If multiple routes have high acceptable association stability (i.e., $R_y > 1$), select path based on other criteria, such as minimum hop count (i.e., R_{final} is obtained).

that are specifically identified in the packet receive and forward the packet, otherwise the packet is discarded. Hence, return path messaging occurs over a directed broadcast. The multicast session is identified through a multicast ID, which must be unique to avoid conflicts. One way of deriving this ID is to concatenate the sender address with the multicast group address. Note that at the branching node, the MC-SETUP message is split into two or more smaller MC-SETUP packets. Each fragmented MC-SETUP message is concerned with the respective branches originating from the branching node. In this manner, all the multicast routing table entries of the nodes in the tree are updated.

Tree Selection Algorithm (TSA)

In ABAM, a heuristic TSA that is based on common link sharing is used. This algorithm is formally stated in Table10.2. Basically, once the BQ-M process is initiated, the source node will eventually receive information about all source-to-receiver routes. It then computes which of these paths have common links. There-

Table 10.2. ABAM Tree Selection Algorithm

TSAs Based on Common Link Sharing

Let $\mathcal{G}$ be the multicast group.
Let $\mathcal{G}_{\mathcal{I}}$ be the nodes in multicast group $\mathcal{G}$, where I=1, 2, 3,... n.
Let $\mathcal{R}$ be the receivers in multicast group $\mathcal{G}$.
Let $\mathcal{G}_{S}^{\mathcal{R}_{K}}$ be the route path from source node S to receiver R_k, where $K = 1, 2, ..., N$ and N = number of receivers in multicast group $\mathcal{G}$.

Since S will receive MC-REPLYs from concerned receivers, S will therefore be aware of the possible routes $\{ \mathcal{G}_{S}^{R_1}, \mathcal{G}_{S}^{R_2}, \mathcal{G}_{S}^{R_3}, ..., \mathcal{G}_{S}^{R_N} \}$.
Begin
Derive the shared-based ad hoc multicast tree.
Shared portion of the tree = $\{ \mathcal{G}_{S}^{R_1} \cap \mathcal{G}_{S}^{R_2} \cap \mathcal{G}_{S}^{R_3}, ... \cap \mathcal{G}_{S}^{R_N} \}$
End

after, the shared-link multicast tree is derived. The worst case for utilizing this methodology is when there are no common links. In this case, the multicast tree will compose of separate unicast paths from the source to the receivers.

Although link sharing is important, more emphasis should be given to long-lived routes since a tree that exhibits a high degree of link sharing does not necessarily imply that the tree will be associatively stable.

Based on the above algorithm, the computation complexity associated with the TSA is therefore related to the number of receivers per multicast group as well as the route length.

10.4.2 Handling Host Membership Dynamics (Join/Leave)

In ad hoc multicasting, distinctions must be made for multicast group dynamics and dynamics due to node mobility. The former is concerned with either a new node wishing to join an existing multicast group or an existing member node wishing to leave the multicast group. This is depicted in Figure 10.10. Nodes joining and leaving an existing multicast group will require special changes to the routing functions of the nodes involved in the multicast tree. As for the latter dynamics, mobility can be a result of the sender, receiver, or tree nodes[7], and each of these movements will require a route reconfiguration process to ensure that the multicast tree is still valid and that multicast packets are correctly delivered to intended receivers. The procedures to cope with these two dynamics are discussed below.

• Nodes Joining the Multicast Tree

When a multicast tree has been established, nodes that are not part of the multicast tree can initiate a request to join the tree. As shown in Figure 10.10a, an ad hoc mobile node $\mathcal{J}$ initiates a request to join the existing multicast group $\mathcal{M}$ by sending an $L\text{-}JOIN$ request using a wave-like broadcast.

All nodes that are part of $\mathcal{G}_S^{\mathcal{R}_\mathcal{K}}$, which receives the $L\text{-}JOIN$ message, return a $JOIN\text{-}REPLY$ to node $\mathcal{J}$. Because multiple paths may exist from $\mathcal{J}$ to nodes in $\mathcal{G}_S^{\mathcal{R}_\mathcal{K}}$, the latter must select a route path, again based on the RSA. Since $\mathcal{J}$ receives multiple $JOIN\text{-}REPLY$s from nodes in $\mathcal{G}_S^{\mathcal{R}_\mathcal{K}}$, $\mathcal{J}$ will select to join a node that yields a stable route/branch and also fulfills QoS requirements (delays, etc.). $\mathcal{J}$ then sends an $L\text{-}JOIN\text{-}CONF$ message to the selected node over the desired route. Next, nodes in the route update their routing tables. In addition, the branching node is identified and notified. When $\mathcal{J}$ receives the first multicast packet from the multicast group, it knows that the join operation was successful.

Figure 10.11 shows the format of reply packets sent by nodes on the existing multicast tree that received a JOIN-REQUEST message from a new incoming multicast member. Although it is inefficient to send a broadcast JOIN-REQUEST message and wait for all possible nodes on the multicast tree to reply, this need not be the case. An expanding ring search can be imposed so that the "live" of the join search is restricted to the specified number of hops. A shorter hop route implies that the "joining" node is nearby, and if several replies are received for a given "live-count," the best joining route can be selected using the RSA.

• Nodes Leaving the Multicast Tree

Members wishing to quit a multicast group are equivalent to nodes leaving (but note that there is no migration here) a multicast tree. The multicast leave process is relatively simple. A node $\mathcal{L}$ wishing to leave an existing multicast tree sends an L_LEAVE message to its upstream node as shown in Figure 10.10b. The L_LEAVE message propagates upstream along the multicast tree. All nodes that receive the L_LEAVE message then delete their multicast route entries for the concerned multicast group. The route erasure process is performed in a hop-by-

[7] This includes branching and forwarding nodes.

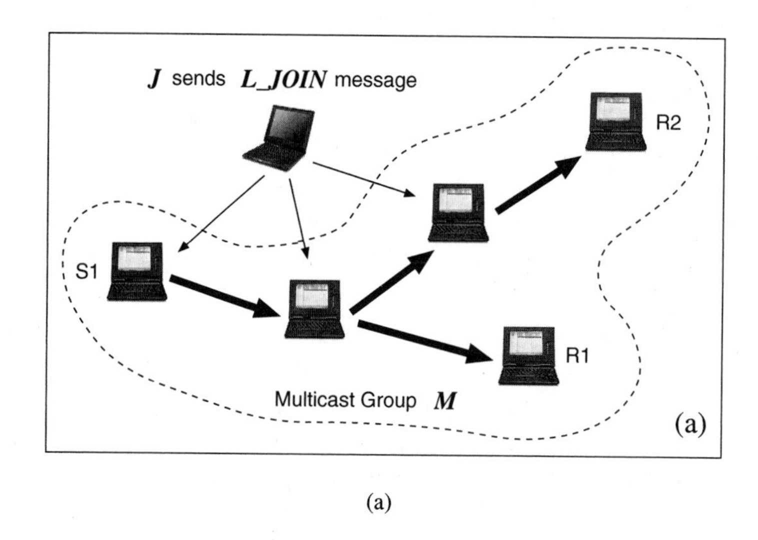

(a)

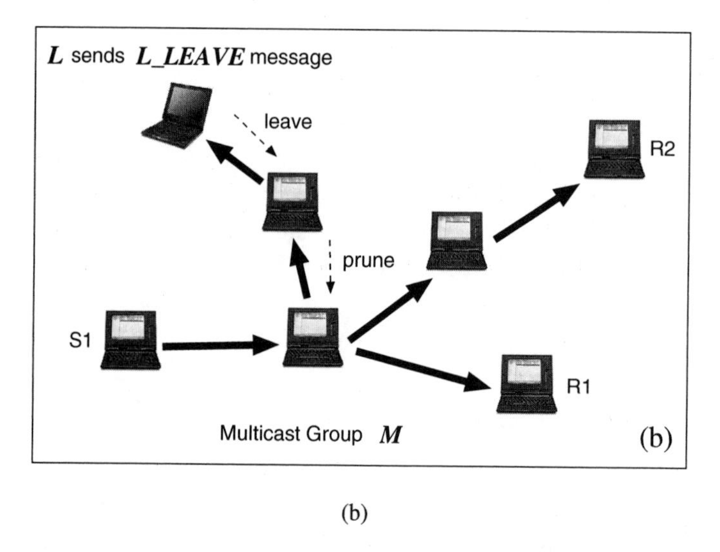

(b)

Figure 10.10. Nodes joining and leaving an ad hoc mobile multicast tree.

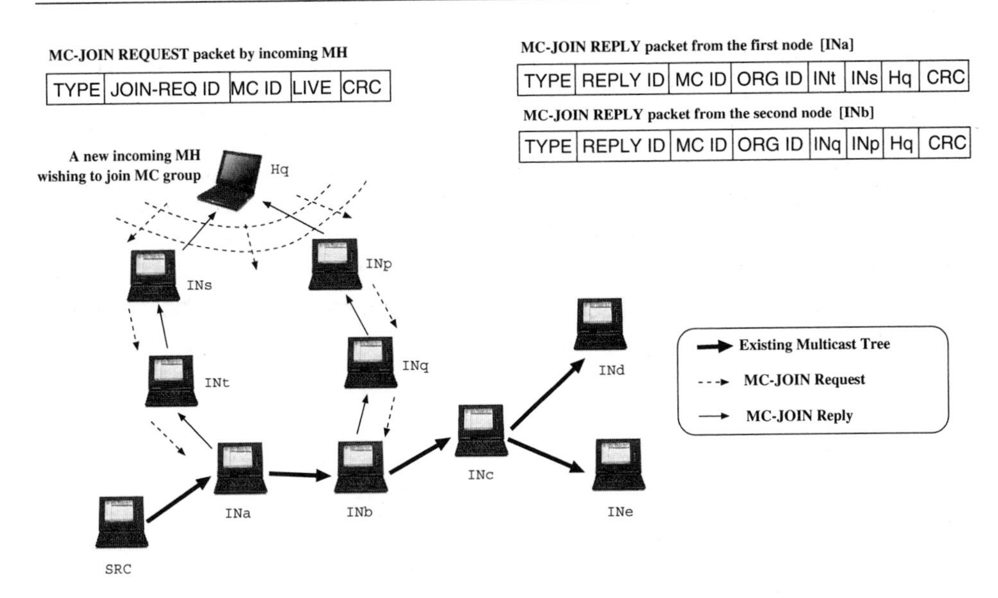

Figure 10.11. The format of reply messages to an ad hoc mobile node wishing to join an existing multicast tree.

hop fashion until the branching node or receiver node is reached.

10.4.3 Dynamics Associated with Node Mobility

Route reconfigurations in ad hoc multicast are non-trivial since distinctions must be made between movements by nodes that are part of the tree and those that are merely leaves of the tree. Because any nodes of the tree can move, the key here is to localize the problem and attempt to resolve the route reconfiguration without disturbing other parts of the tree. There will be cases of concurrent node movements, and hence, several route reconfigurations may be executed. An approach such that ultimately, only one route reconfiguration will be necessary.

• Movement by Multicast Leaf/Receiver

There are two possible approaches to resolving the dynamics associated with receiver migration, namely:

(a) route reconstruction (RRC) initiated by tree nodes, and

(b) receiver-initiated move-join.

In the first approach, if a leaf of the multicast tree moves, RRC procedures will be invoked by the immediate upstream neighbor (N_f in Figure 10.12) in an attempt to discover a new stable partial route to the migrated leaf node. A localized broadcast query (LQ) process is initiated to perform a limited radius search for a partial route to the migrated node. If the migrated node is reachable, it returns an LQ-REPLY message back to node N_f. If N_f cannot find a route to the receiver, it backtracks to the next upstream node and tries to locate a partial route to the destination. Each time it backtracks, the search radius is incremented so that the hop count metric is taken into consideration.

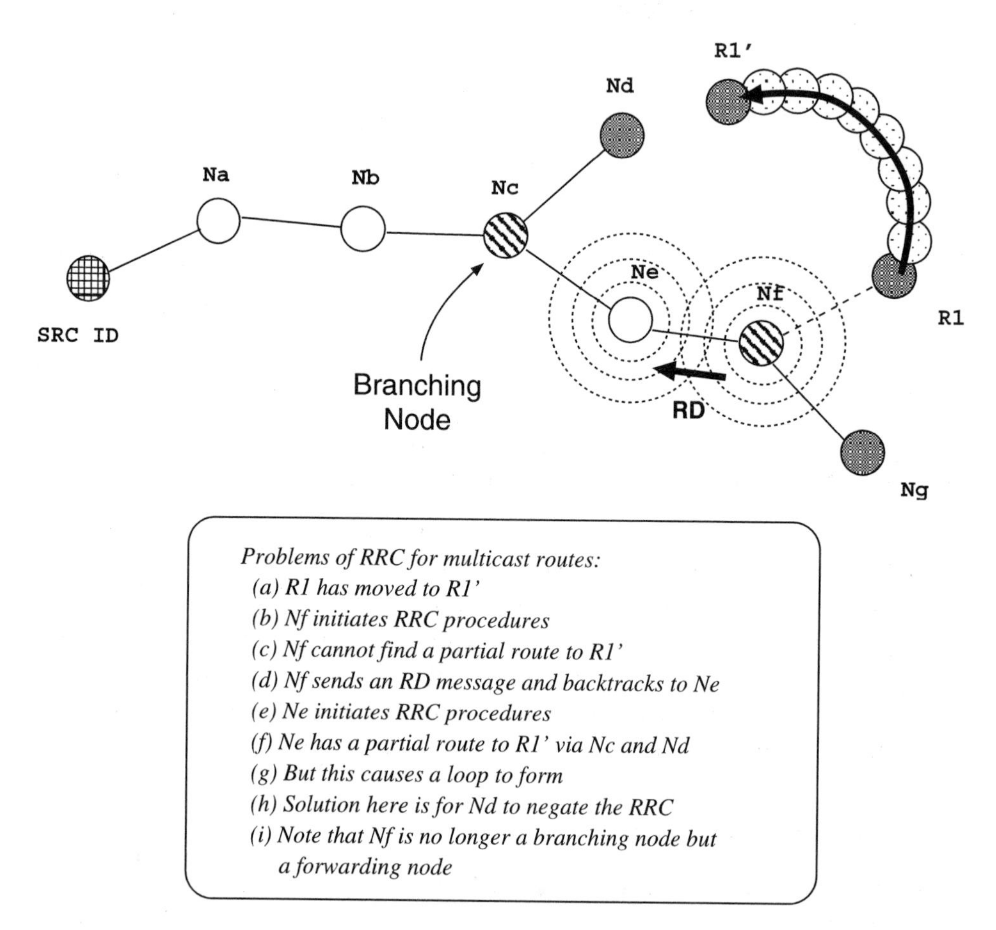

Figure 10.12. ABAM RRC process to handle mobility of multicast receivers.

Recall for unicast routes, the ABR protocol aborts after backtracking to one-half the route hop count to avoid excessive iteration of localized broadcast query search at different points in the route and to speed up the route recovery time. For a multicast route, the "half-route-length" concept cannot be used since the route is now a multicast tree. However, to avoid backtracking to the source, the RRC process in ABAM terminates at the branching node. The disconnected receiver then waits for a timeout period to elapse and requests to join back to the multicast tree.

Another approach is to use the receiver-initiated move-join method whereby the moved node initiates a join request to the existing multicast tree. The upstream node, which senses the departure of the downstream node, then sends a route delete (RD) message upstream, hop-by-hop, until it reaches the first encountered branching node. Thereafter, that branch is pruned from the multicast tree. If the branching node still has two or more other leaves, its role as a branching node remains. Otherwise, it returns to function as a forwarding node. This is further illustrated in Figure 10.13.

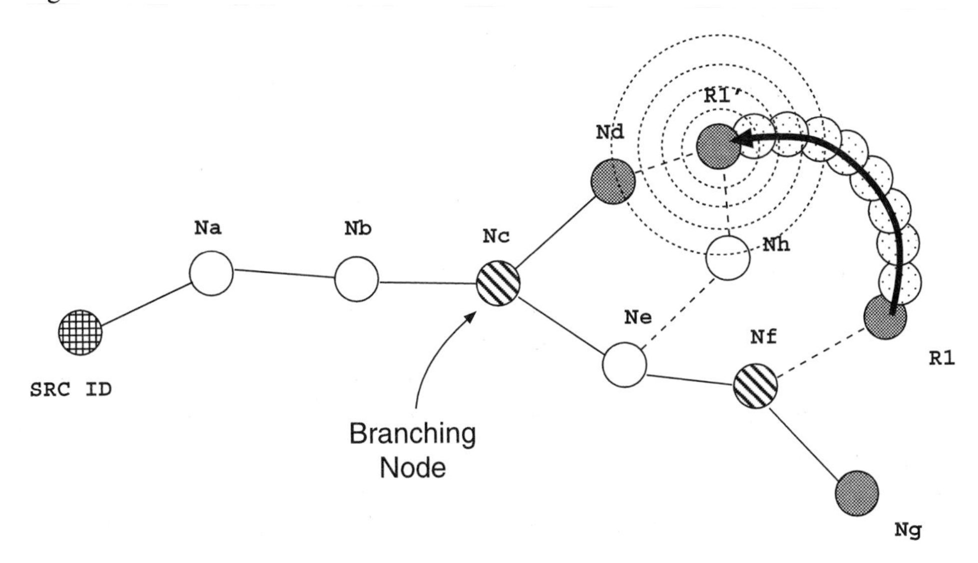

Figure 10.13. ABAM receiver-initiated move-join approach to handle mobility of multicast receivers.

This second approach requires distinctions to be made between a "new-node join" and a "moved-node join." The former is an addition of a new member into the multicast group and an initiation of a new flow to the new member. The latter, however, is concerned with a moved existing member with an interrupted flow. This approach of receiver-initiated move-join is better than the RRC approach because it incurs less signaling traffic.

• Movement by Multicast Source/Sender

When the source node (sender or root) moves, a two-stage process is present. In the first stage, the source node tries to find a partial route to downstream nodes, up to the first branching node. If this is not successful after $\mathcal{P}$ attempts, the second stage is invoked. This requires a BQ process to be initiated, that is, a new multicast tree must be re-established from the source to all intended receivers. The second approach requires re-establishing another stable multicast tree, and hence, may not be the most desirable depending on circumstances.

The source node S may try to perform a "join-like" process to a node on the existing multicast tree, but this can give rise to other problems such as link reversal, where data flowing over parts of the tree has to change direction due to the migrated source joining a different part of the tree. For this approach to work, a route reconfiguration must occur for all other nodes on the multicast tree.

• Movements by Multicast Tree Nodes

If a node belonging to a part of a multicast tree moves, multiple receivers on the tree will be pruned and RRC procedures will have to be changed from the previously mentioned cases. First, the RRC procedures will have to take into account the **furthest leaf node** of the multicast tree. The hop count from the node that invokes the RRC to the furthest leaf node is used to perform the localized broadcast search for partial routes to all affected receivers. In this manner, while searching for possible routes to the furthest node, all other affected nodes will also be considered since they fall within the search radius. Note that this method is superior to that requiring all affected receivers to initiate "join-requests" since it resolves RRC to all affected nodes in a **single** successful search attempt and hence incurs less signaling traffic.

This furthest leaf search method, however, will require that whenever a node joins an existing multicast tree, the distance of this node to all other upstream nodes in the multicast tree must be made known so that RRC procedures are executed correctly. This is accomplished using a route update (RU) message. During RRCs, localized query packets will be sent toward the receivers. Each receiver of the multicast tree will then decide on the best route (based on associativity and other routing metrics) from itself to the node initiating the RRC (NRRC). The NRRC will then consider all the best route replies from all the affected multicast receivers to derive a shared partial tree, that is, the NRRC node will execute a similar tree selection algorithm as in ad hoc multicast tree establishment. However, if none of the selected routes have shared paths, then separate partial branches will be originated from the NRRC to the respective receivers.

Questions arise when the search does not successfully find route paths to all affected receivers. For example, consider the case when there are four receivers, N_a, N_b, N_c, and N_d, and one of the tree nodes moves and an RRC is initiated. Assuming the furthest node is N_d, and N_d is five hops away from NRRC, the search has a diameter of 10 hops. If only N_a, N_b, and N_c are reachable from NRRC, then the RRC process should continue and partial routes should be established to N_a, N_b, and N_c. Because N_d is no longer connected to the multicast tree, it will time out and initiate a receiver join process similar to the leaf move-join process mentioned earlier. In this manner, affected receivers as a result of tree node moves can be resolved in a collaborative manner, combining both RRC and receiver-initiated move-join approaches. This is summarized by the flowchart shown in Figure 10.14.

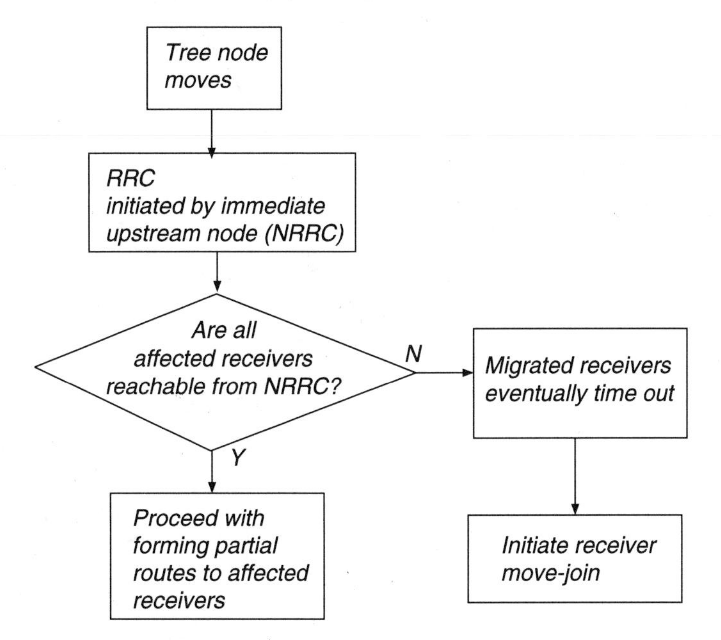

Figure 10.14. Handling multicast tree node migration by using a combination of RRC and receiver-initiated move-join approaches.

• Concurrent Node Movements

In the case of multiple concurrent node movements (tree nodes, source nodes, or receiver nodes), only one RRC will eventually succeed. To avoid the need to frequently reconstruct routes, the degree of associative stability will be used as the primary routing metric during path selection. If the tree selected is associatively stable, then there is no need to perform RRC; hence, no overhead is involved.

10.4.4 Deletion of Multicast Tree

Once a multicast tree is established, it remains active until:

(a) the source node decides to end the multicast session, or

(b) there are no more receivers in the multicast session.

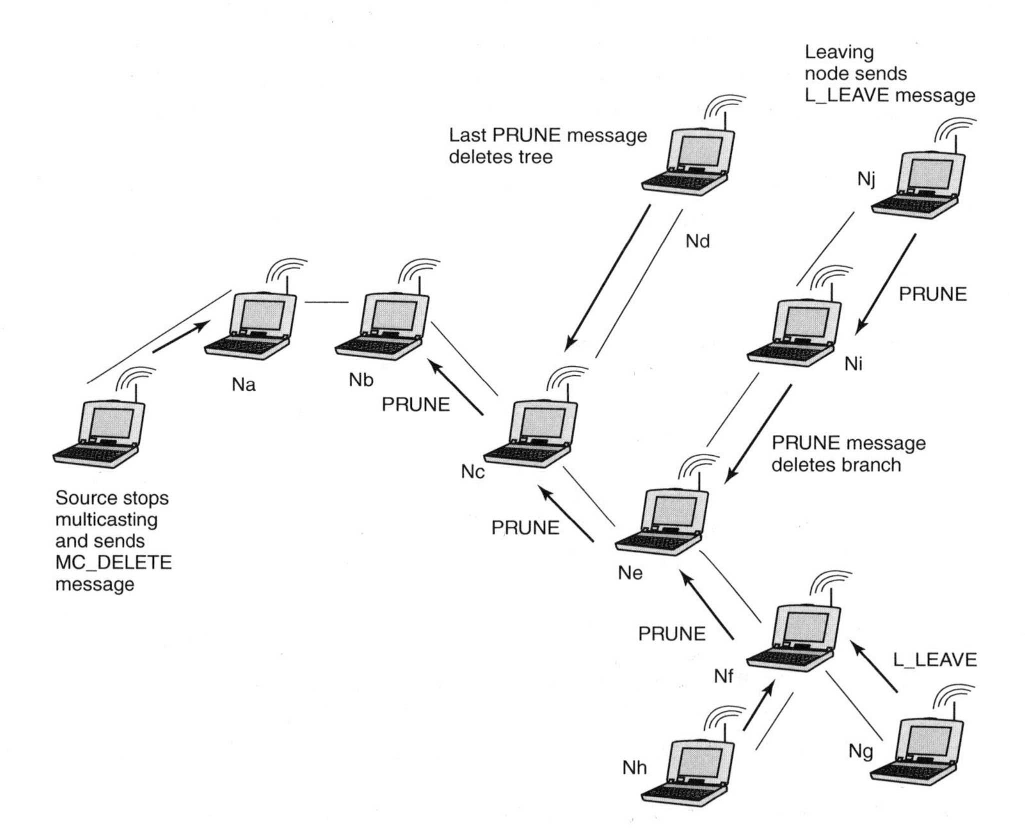

Figure 10.15. Tree and branch deletion in ABAM protocol.

The deletion of a multicast tree can be performed in two ways, namely:

(a) explicit broadcast of a tree deletion message, and/or

(b) route entry expiration upon timeout (soft state).

The former is initiated by the source and all nodes that are part of the multicast tree will delete the appropriate routing entries upon receiving the route delete packet. This is illustrated in 10.15. The latter approach, however, requires nodes of the multicast tree to monitor traffic behavior of on-going multicast packets and when there is an absence of traffic over a threshold period, the route entry for that particular multicast tree is automatically erased. Hence, the latter approach trades off the need for wireless broadcast with higher packet processing load. However, it does not consider the "desires" of the source node, should the source node decide to pause before resuming the transmission of multicast information to receivers. The format of a multicast tree delete message is shown in Figure 10.16.

Figure 10.16. Format of multicast tree delete message.

10.4.5 ABAM Tree Reconfiguration

Tree reconfiguration is required when the associative property is violated. ABAM employs a local repair strategy by having the node upstream of the broken link broadcast an LQ (local query) message. The maximum number of hops an LQ message can propagate is determined by the number of hops to the farthest receiver on the broken branch. The receiver replies with an LQ-REPLY message and the MC-SETUP message is again used to set up the subtree. If the LQ process fails, the immediate upstream node will be notified to start another LQ process. This backtracking process continues until the LQ succeeds or either the branching node or receiver node is reached. The receivers that are left over by the LQ process will finally time out and initiate their own JQ (join query) process. RN (route notification) and RU (route update) messages are used to facilitate the reconfiguration process by notifying nodes about route failures.

10.4.6 Complexity of ABAM

Having discussed the principles of operation of ABAM, the time and communication complexity is examined and shown in Table 10.3. Performance comparions of ABAM vs. ODRMP via simulations can be found in [78].

Let N be the number of nodes in the mobile network.
Let E be the number of links in the network.
Let d be the network diameter.
Let X be the number of receivers affected by a topological change.
Let Z be the diameter of the directed path where the REPLY packet transits.
Let R be the number of receivers in the multicast tree.
Let L be the number of nodes affected by a topological change.
Let Q be the route length of the node furthest away from the source node of the multicast tree.
Let F be the route length of the furthest node on the multicast tree from the node wishing to join the tree.
Let S be the diameter of the directed path between the selected node on the tree to the node wishing to join/leave.
Let P be the route length of the receiver node that is furthest away from the NRRC.
Let K be the number of nodes in existing multicast tree.
Let B be the number of nodes in the newly established subtree.
Let C be the number of nodes in the disconnected subtree.
Let q be the maximum route length from the NRRC to the migrated receiver node.
Let T be the number of on-tree nodes affected by a topological change.

Complexities Evaluation of ABAM	**Communication Complexity (CC)**	**Time Complexity (TC)**
• Ad Hoc Multicast Setup	$O(N + \sum_{i=1}^{R} Z_i + K)$	$O(3Q)$
• Membership Join	$O(N + \sum_{i=1}^{K} Z_i + S)$	$O(2F+S)$
• Membership Leave	$O(S)$	$O(S)$
• RRC for Migration by a Tree Node	$O(L + \sum_{i=1}^{X} Z_i + B)$	$O(3P)$
• RRC for Migration by SRC Node		
- partial RRC to first branching node	$O(L + \sum_{i=1}^{T} Z_i + S)$	$O(2F + S)$
- reestablish the whole tree	$O(N + \sum_{i=1}^{R} Z_i + K)$	$O(3Q)$
• RRC for Migration by a LEAF node		
- using NRRC initiated localized query	$O(L + Z)$	$O(q + Z)$
- using receiver-initiated move-join	$O(N + \sum_{i=1}^{K-1} Z_i + S)$	$O(2F + S)$

Table 10.3. Complexities Associated with ABAM Protocol.

10.5 Comparisons of Multicast Routing Protocols

In this section, the multicast routing protocols discussed earlier will be qualitatively compared based on the following:

(a) basic protocol characteristics,

(b) multicast operations,

(c) communication performance, and

(d) control overhead incurred.

They are presented in Table 10.4, 10.5, and 10.6 respectively.

10.5.1 Protocol Differences

Table 10.4 shows some important characteristics of each multicast routing protocols discussed earlier. In terms of *multicast delivery structure*, DVMRP is the only one using the source-based *tree* structure. AODV can be classified as cored-based tree protocol although the term *core* is not used in AODV terminology. This is because, normally, the first node that requests to join the multicast group performs the function of the core node by representing the starting point for the shared tree establishment, in addition to generating the group sequence number which is the mechanism of AODV to eliminate outdated routing information. In contrast to tree, *multicast mesh* is used in CAMP and a *group* based structure is used in ODMRP. While ODMRP uses control messages broadcast from senders/receivers to create forwarding groups, nodes' geographical positions are used in LBM.

Table 10.4. Characteristics of Various Ad Hoc Mobile Multicast Routing Protocols.

Parameters	DVMRP	AODV	CAMP	ODMRP	LBM
Multicast Delivery Structure	Source-Based Tree	Core-based Tree	Multicast Mesh	Group-based	Location based
Use Centralized Node	No	Yes (Multicast Group Leader)	Yes (Core Nodes)	No	No
Core Node Recovery	N/A	Yes	Yes	N/A	N/A
Routing Scheme	Table-driven	On-demand	Table-driven	On-demand	On-demand
Dependence on Unicast Routing Protocol	No	No	Yes	No	No
Routing Approach	Flat	Flat	Flat	Flat	Flat
Routing Metric	Shortest path	Shortest path to another multicast member along the existing shared tree	Shortest path	Shortest path	Shortest path

The use of core nodes (or any centralized nodes) introduces robustness problem when the core node is not present or is not reachable. AODV and CAMP are two such protocols which utilize centralized nodes (a multicast group leader and core nodes respectively). Both protocols provide recovery mechanism when their core nodes create problem.

Nodes in AODV will randomly nominated themselves as the new multicast group leader when the periodic announcement from the current multicast group leader is not received for a timeout period. Contention mechanism based on node ID is used to resolve the case when several nodes become multicast group leaders of the same multicast group which is possible when the network is partitioned and merged.

In contrary, CAMP allows multiple core nodes to be present in the network and these core nodes are not mandatory in the operation of CAMP. This means that CAMP can still operate even if all the core nodes are not present. However, CAMP provides a mechanism similar to AODV to randomly select new core nodes since core nodes can help to eliminate join request flooding problem in CAMP.

In terms of the *underlying unicast routing protocol*, all protocols except CAMP provide their own embedded mechanisms required for the operations of the protocols. DVMRP uses the distance-vector routing algorithm, which is a table-driven algorithm, for reverse path forwarding calculation. On-demand routing is used in AODV and ODMRP scheme. LBM also operates in an on-demand fashion and does not require any unicast routing information.

All multicast routing protocols discussed so far use the flat routing approach i.e., there is no hierarchy. All of them creates their multicast delivery structures based on the shortest-path metric.

10.5.2 Operation and Performance Differences

Table 10.5 summarizes the basic operations performed by each ad hoc multicast routing protocols. The *session initiation procedure* refers to the procedures executed by the source node so as to enable it to transmit multicast packets for that particular multicast session.

We use the term "session" here to refer to the communication from a specific multicast source node to a specific multicast group. In this case, there can be several multicast sessions for each multicast group depending on the number of multicast source nodes in that multicast group.

The *multicast member join procedure* refers to the procedures executed by the node that wishes to receive multicast packets for a particular multicast group. The *repair procedure* refers to the method used by each protocol to maintain on-going multicast sessions when link breakages are encountered. Finally, the *session termination procedure* are the procedures that will be executed when the multicast session is no longer desired.

Flooding is the process of disseminating a piece of data to all nodes in the network. Protocols in mobile ad hoc networks attempt to eliminate flooding as much as possible due to its high control overhead. As seen in Table 10.5, *periodic flooding* is used in one way or the other in these multicast protocols.

DVMRP times out the pruned-off branches periodically. The net effect can be compared to periodic flooding of multicast data. In AODV, only the multicast group

Table 10.5. Comparisons of Ad Hoc Mobile Multicast Routing Protocols.

Parameters	DVMRP	AODV	CAMP	ODMRP	LBM
Session Initiation Procedure by Multicast Source	Flood data packet	Join the existing multicast tree	Join the existing multicast mesh	Start flooding periodic announcement	Broadcast data packet with location information
Member Join Procedure	Graft to all source trees	Join the existing multicast tree	Join the existing multicast mesh	Start collecting join request; periodically broadcast join table	Nodes have to be within multicast region
Repair Procedure	Periodically flood data packet	Node downstream to broken link rejoins the tree	Use alternate path	Use alternate path	None
Session Termination Procedure	None	Request to prune	Request to prune	Stop periodic announcement; remove information upon timeout	None
Use Periodic Flooding	Yes (data)	Yes (sequence number)	Sometimes (transient cores)	Yes (join request)	No
Multiple Routes	No	No	Yes	Yes	Yes
Session Initiation Latency	Low	High	Moderate	Moderate	Low
Member Join Latency	Moderate	High	Moderate	Moderate	N/A
Repair Latency	Moderate	Moderate	Low	Low	N/A

leaders periodically flood the updated sequence number. In CAMP, two kinds of core nodes are defined: static and transient. The transient core nodes periodically flood announcements about their presence. ODMRP depends heavily on periodic join request flooding by source nodes to create and maintain the forwarding group. LBM, however, does not flood the whole network but instead restrict the effect to its forwarding zone. Nonetheless, positional inaccuracy and lack of GPS service can lower routing throughput significantly and even disable routing functions completely.

In terms of utilizing *multiple routes*, CAMP, ODMRP, and LBM can exploit the presence of multiple routes. The advantage is that there is no disruption to ongoing traffic when the primary route is broken. However, these protocols result in unnecessary packet forwarding unlike DVMRP and AODV, which provide only a single route for each multicast session.

In terms of *latency*, DVMRP and LBM result in low session initiation latency because multicast data are broadcast as soon as the session is started. For other protocols, signaling messages are sent to initialize the session before multicast data can be transmitted. CAMP typically requires the exchange of control messages with a core node to join a multicast group and initiate a session. Similarly, in ODMRP, control messages are exchanged between a source node and receiver nodes before multicast paths are available. These message exchanges incur some latency during session initiation. AODV, however, incurs a slightly longer delay since it uses a 3-way message exchange and a small waiting time is needed to collect multiple reply packets.

A similar explanation can be made regarding membership-join latency. Note that the latency for membership join in DVMRP is moderate. This is because the graft message can be used to connect the pruned-off branch quickly. In the case when the pruned-off branches are changed due to nodes mobility, DVMRP will immediately flood the data to new branches and quickly discover the new member if the path is not interleaved by the valid pruned-off branch. Otherwise, the new member has to wait until the timers expire on those pruned-off branches. The same explanation applies to the moderate repair latency in DVMRP. CAMP and ODMRP exhibits low repair latency since alternate paths are normally available in the multicast mesh and forwarding group. Note that we did not evaluate LBM in terms of member join and repair since it does not require these mechanisms. If a receiver node happens to move out of the multicast region or there is no connectivity within the forwarding zone, the packet will be silently discarded without requiring any action.

10.5.3 Comparing Protocol Overhead

Table 10.6 shows the characteristics concerning the scalability of the multicast protocols discussed. Scalability refers to the ability of such protocols to operate properly in a large scale ad hoc mobile network. This can be analyzed by considering the rate at which network resources are consumed as the size of the network increases. The comparison can be based on various parameters, such as *storage overhead* and *communication overhead* which indicates how much memory is needed and the amount of traffic load generated.

In terms of *storage overhead*, DVMRP is rated poor because each node has to store information about each and every source-based multicast tree in the network. The number of source-based multicast trees grows rapidly as the the number of source nodes and the number of multicast groups increases. In other protocols, only one set of information is kept for each multicast group that a node is participating in, be it as a sender, receiver, or forwarding node. In this manner, the storage requirement is greatly reduced. AODV and CAMP keep track of links that constitute the shared tree and multicast mesh respectively, while only a single forwarding flag is maintained in ODMRP. LBM does not need to store such information since the decision to rebroadcast the packet is based solely on the positional information attached to the packet and its own geographical location.

Comparing *overall communication overhead*, DVMRP is still rated poor since it wastes a lot of bandwidth flooding multicast data during its operation. ODMRP is also rated poor because of the periodic flooding from the multicast source nodes and redundant forwarding of multicast data in the forwarding group. In LBM, the

Table 10.6. Comparisons of Overhead in Ad Hoc Mobile Multicast Routing Protocols.

Parameters	DVMRP	AODV	CAMP	ODMRP	LBM
Information stored at each node	Prune status of all adjacent links *	On-tree nodes record all its upstream link and downstream links	Mesh members record all its upstream links and downstream links; and list of core nodes	forwarding flag	none
Amount of information at each node	One set for each source node in each group	One set per group	One set per group	One set per group	none
Scalability in terms of storage overhead	poor	moderate	moderate	good	very good
Scalability in terms of overall communication overhead	poor	good	moderate	poor	moderate

* If RPF is used as in original DVMRP, each node has to store a routing table to all nodes in the network.

degree of communication overhead depends on the configuration and the distribution of nodes in the forwarding region. Since the determination of forwarding region does not take into account signal and link connectivity, the resulting communication overhead can be high. In contrast, AODV uses flooding only in some operations and multicast data are forwarded based on the predetermined tree structure. Hence, the overall communication overhead is lower than LBM. CAMP also uses flooding in some operations and data are redundantly forwarded along the multicast mesh. This, therefore, incurs a higher communication overhead than AODV, but is still better than LBM.

10.5.4 Time and Communication Complexity

The time and communication complexity associated with the protocols discussed is shown in Table10.7. In this table, *time complexity* is defined as the number of steps required to perform a protocol operation, and *communication complexity* is defined as the number of messages required to perform a protocol operation. These complexities have been analyzed in terms of:

(a) session initiation by the source node,

(b) multicast member join operation, and

(c) multicast route reconfiguration upon link changes.

Note that these complexity values represent *worst-case* behaviors. As shown in Table 10.7, DVMRP has the most communication overhead due to the use of broadcast-and-prune approach and the presence of multiple source-based trees in each multicast group. Other protocols exhibit a similar degree of overhead, which is less than that of DVMRP. In LBM, only the session initiation and forwarding of multicast packets are specified due to the nature of its operation.

Table 10.7. Time and Communication Complexity Comparisons of Ad Hoc Mobile Multicast Routing Protocols.

Complexities	DVMRP	AODV	CAMP	ODMRP	LBM
Time Complexity Session Initiation by Source Node	0	$O(3l_g)$	$O(4l_g)$	$O(2r_f)$	0
Time Complexity Join Multicast Group	$O(s_f)$	$O(3l_g)$	$O(4l_g)$	$O(s_f)$	N/A
Time Complexity Reconfiguration	$O(s_f)$	$O(3l_g)$	$O(2r)$	$O(s_f)$	N/A
Communication Complexity Session Initiation by Source Node	$O(2N - T)$	$O(N + \sum_{i=1}^{T} l_i + l_g)$	$O(N + 3l_g)$	$O(N + \sum_{i=1}^{M} r_i)$	$O(T)$
Communication Complexity Join Multicast Group	$O(S * N)$	$O(N + \sum_{i=1}^{T} l_i + l_g)$	$O(N + 3l_g)$	$O(\sum_{i=1}^{S} s_i)$	N/A
Communication Complexity Reconfiguration	$O(S * N)$	$O(N + \sum_{i=1}^{T} l_i + l_g)$	$O(2r)$	$O(\sum_{i=1}^{S} s_i)$	N/A

N = Total number of nodes in the network
S = Number of multicast source nodes in a multicast group being considered
M = Number of multicast group members in a multicast grou p being considered
T = Number of node on the multicast delivery structure
$l_{i,0<i\leq T}$ = Route length from the ith node on the multicast tree
l_b = Route length from the pruning node to the nearest branching node upstream
l_g = Route length from the joining node to the chosen node to graft to on the multicast tree
r = Route length from the joining node to the source node
$r_{i,0<i\leq N}$ = Route length from the ith mult icast member to the source node
r_f = Route length from the source node to the farthest multicast group member
$s_{i,0<i\leq S}$ = Route length from the joining node to the ith source node
s_f = Route length from the joining node to the farthes t source node

10.6 Conclusions

Ad hoc wireless multicasting is a challenging communication feature that requires new protocols to support it. Most multicast routing protocols designed for the Internet assume that routers in the network are static and they have no provision to deal with tree fragmentation and repair over time when mobility occurs. Some proposals are plain extensions of distance-vector and link-state routing protocols. Such proposals, however, do not perform well in ad hoc mobile networks due to the presence of mobile links at each portion of the route. In addition, multicasting is complicated by the presence of multicast group dynamics. However, multicasting is essential if multiparty communications and multiparty multimedia-based applications are envisioned for wireless ad hoc networks. Hence, significant research effort is needed prior to realizing this dream. Protocol implementation and the building of testbeds are crucial to understanding the performance and practicality issues.

Chapter 11

TCP OVER AD HOC

11.1 Introduction to TCP

The transmission control protocol (TCP) is one of the most popular and widely used end-to-end protocols for the Internet today. Unlike routing, where packets are relayed hop-by-hop toward their destination, TCP actually provides reliable end-to-end transmission of transport-level segments from source to receiver. Transport segments arrive in sequence and lost segments are recovered. Hence, TCP provides *flow* and *congestion* control functions, in addition to reliable transmission (i.e., through error recovery mechanisms).

Reliability in transmission involves the use of some form of handshake between the sender and receiver. Also, sequence numbers can be used to ensure in-sequence delivery of segments and help to identify lost or corrupted segments. Retransmission can be used to resend lost or corrupted segments. Hence, a retransmission timer is needed to determine when to initiate a resend. For TCP, an adaptive retransmission mechanism is employed to accommodate the varying delays encoun-

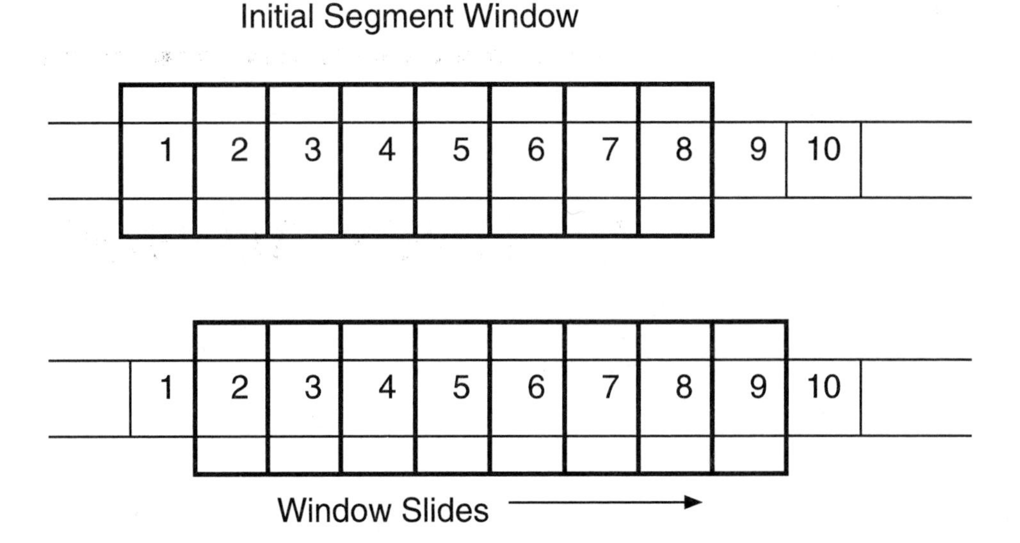

Figure 11.1. The TCP sliding window mechanism.

tered in the Internet environment. The timeout parameter is adjusted accordingly by monitoring the delay experienced on each connection.

Each TCP segment has a segment header, which contains the source node's endpoint address (i.e., the IP address and its corresponding port number), the destination address, a sequence number, and a payload. Application data are fragmented into segments and appended with a TCP segment header. At the TCP receiver, these segments are then reassembled back into messages.

11.1.1 TCP Flow Control

TCP provides reliable connected-oriented service. A virtual circuit connection (VCC) must be established hop-by-hop from the source to the destination prior to data transmission. The source node gradually transmits more and more data if acknowledgements (ACKs) for previously transmitted segments are received successfully. This regulation of traffic transmission in accordance with the *congestion state* and *connection quality* is known as *flow control*. Transmitting segments at a rate faster than what the receiver can handle will result in receive buffer overflow and information loss. This is similar to the scenario where tap water is running through a pipe that has a big funnel on one end (close to the tap) and a smaller funnel at the receiving end. The small-capacity receiver experiences water overflow.

Several flow control methods are used to ensure reliable and fast delivery of information over the Internet. The TCP sliding window mechanism shown in Figure 11.1 allows the sender to send multiple segments before waiting for an ACK. The window size, therefore, defines the number of packets the sender can send before it receives an ACK back. The window gradually slides open when wider ACKs are successfully returned, as shown in Figure 11.1. Hence, by keeping track of which segments sent are ACKed and which are not, flow control is introduced since the sender cannot continue to send unless the receiver has responded with ACKs.

In reality, the window size can be made adaptive and can vary over time. If the receiver's buffer is becoming full, it sends a small window size advertisement to the sender. This results in the sender reducing its window size to avoid receiver buffer overflow. In extreme cases, a receiver can advertise a window size of zero, thus causing the sender to stop transmission. Such mechanisms allow end-to-end flow control in TCP. In short, the TCP receiver's window refers to the range of sequence numbers that the receiver may accept while the TCP sender's window refers to the range of consecutive sequence number of segments that may be sent before an ACK is required.

11.1.2 TCP Congestion Control

Scanning through the recent past, several methods have been used to control congestion. Slow start [79], Tri-S [80], DUAL [81], and TCP Vegas [82] view the network as a black box, and they detect congestion through packet loss and changes in round-trip time or throughput. Random Early Detection [83] and Explicit Congestion Notification [84] schemes depend on gateways to provide indications of congestion. Hence, the former group is categorized as *source-based* schemes while the latter group is known as *gateway-based* methods.

TCP congestion control consists of: (a) *slow start* (SS), (b) *congestion avoidance* (CA), and (c) *fast retransmit/fast recovery*. The endpoint node concludes that congestion exists when an increase in end-to-end delay is observed. Retransmissions can further aggravate congestion since more packets are injected into the network.

Upon starting a connection, or restarting after a packet loss, the congestion window (cwnd) size is set to one packet. As shown in Figure 11.2, the TCP sender gradually increases the cwnd size by one packet upon receipt of an ACK, until the first sign of congestion is detected. Thereafter, *backoff* occurs and the window size is reduced to half the current window size (down to a minimum of one segment). The SS process then begins again gradually. However, the SS threshold is introduced, which changes the increment gradient of segment transmission with respect to time, as shown in the diagram. Each ACK received results in increasing the

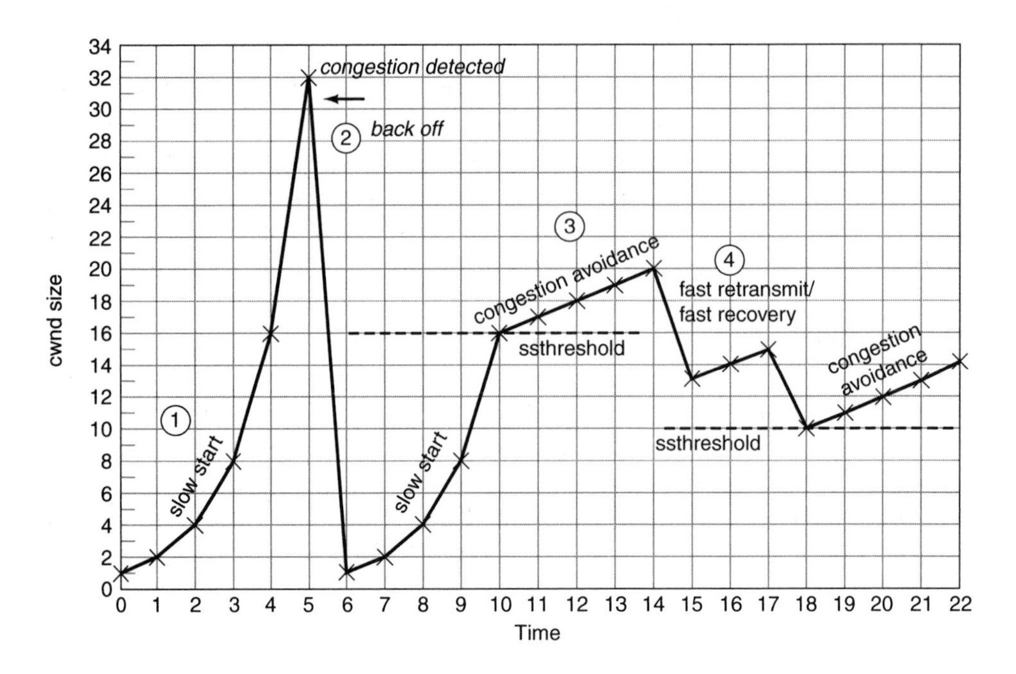

Figure 11.2. TCP congestion control.

window by 1/cwnd-size. In summary, an additive increase (SS)/multiplicative decrease (backoff) policy is used to avoid congestion in TCP. Fast retransmit and fast recovery mechanisms are enhancements which will be discussed later.

11.1.3 Some Issues with TCP

Although TCP is capable of providing flow control, its ability to share a bottleneck fairly and efficiently decreases with the number of flows. Performance of TCP suffers when the number of active TCP flows exceeds the network bandwidth-delay product.

When the TCP sender's congestion window size drops below four, it can no longer be able to recover from a single packet loss since the *fast retransmit* mechanism needs at least three duplicate acknowledgements (ACKs) to get triggered.

There is also this question of fairness. The window-based flow control mechanism of TCP relies on the fact that each TCP sender is responsible for regulating the sending of packets without exchanging information with other TCP connections. Hence, realizing *fair* bandwidth allocation among multiple TCP connections is difficult. Research into these issues is currently in progress.

11.2 Versions of TCP

11.2.1 TCP Reno

TCP Reno [85] employs the SS and CA mechanisms. The sender window size is gradually increased until packet losses are experienced. Thereafter, the window size is halved and a linear, less gradual increase of packet transmission occurs. This proceeds until further packet losses are detected. Consequently, this additive increase and multiplicative decrease lead to periodic oscillations in the cwnd, round-trip delay, and queue length.

The original retransmission mechanism of TCP is based on a timeout where round-trip time (RTT) and variance estimates are computed by sampling the time between when a segment is sent and when an ACK arrives. However, in BSD implementations, a clock with a 500ms interval is used to time the round-trip time. This coarseness implies that the time interval between sending a segment that is lost until there is a timeout and the segment is resent is now much longer than necessary. Hence, *fast retransmission* and *fast recovery mechanisms* are incorporated into the Reno implementation. TCP Reno will not only retransmit when a coarse-grained timeout occurs, but also when it receives three ACKs from the receiver. In Reno, a duplicate ACK is sent whenever a receiver cannot acknowledge incoming new segments due to the failure of arrivals of previous segments.

11.2.2 TCP Tahoe

The early TCP implementations followed a Go-Back-N model using the cumulative positive acknowledgement scheme, which required a retransmit timer to expire to resend lost data. TCP Tahoe [81] utilizes the SS, CA and fast retransmit algorithms (RFC 2001).

In the slow start algorithm, a cwnd is added to the per-connection state. When starting or restarting after a packet loss, the cwnd is set to one packet. Thereafter, on each ACK for new data, the cwnd is increased by one packet.

Congestion avoidance in TCP Tahoe relies on setting the cwnd to half the current window size on timeout. Thereafter, on each ACK for new data, the cwnd is increased by 1/cwnd. In addition, information about the receiver's advertised window and cwnd is also sent. This is, in fact, the congestion avoidance and control method suggested by Van Jacobson and Michael J. Karels.

Finally, the *fast retransmit* algorithm works by monitoring the reception of duplicate ACKs for the same TCP segment. From this, the TCP sender infers that a packet loss has occurred and will retransmit lost packets without having to wait for the retransmission timer to expire.

11.2.3 TCP Vegas

Proposed in 1994, TCP Vegas [82] [86] is different from TCP Reno in the sense that: (a) a new retransmission mechanism is used, (b) an improved congestion avoidance mechanism that controls buffer occupy, and (c) a modified slow start mechanism. In TCP Vegas, all changes are confined to the sending end, and it does not involve any changes to the TCP specification.

Compared to Reno, Vegas uses a more accurate way of determining RRT. It records and reads the source system clock each time a segment is sent. Upon receiving an ACK, it reads the clock again and performs the RRT calculation. When a duplicate ACK arrives at the source, Vegas checks if the difference in the current time and the timestamp recorded for the related segment is greater than the timeout value. If so, it retransmits the segment without having to wait for n duplicate ACKs. This, therefore, speeds up the retransmission process. For cases when a non-duplicated ACK is received, if it is the first or second ACK after a transmission, Vegas will check to see if the time interval since the segment was sent is larger than the timeout value. If so, Vegas then retransmits the segment. Hence, Vegas' new retransmission mechanism does not merely reduce the time needed to detect lost packets from the third duplicated ACK to the first or second duplicated ACK, but it also detects lost packets when there are no second or third duplicated ACKs. Investigations show that this mechanism reduces the number of coarse-grained timeouts in Reno by more than half.

To allow for the detection and avoidance of congestion during slow start, Vegas limits exponential growth only by every other RTT. In between RTTs, the congestion window stays fixed so that a valid comparison can be made between the *expected* and *actual* rates. When the *actual* rate falls below the *expected* rate by a certain threshold γ, Vegas will change from slow start to linear increase/decrease mode.

In Vegas, the expected rate is given by {Expected = Window size / Base RTT}, where Base RTT is the minimum of all measured round-trip times. This commonly corresponds to the RTT of the first segment sent over the TCP connection. "Window size" here corresponds to the current cwnd. The *actual* sending rate is determined by dividing the number of bytes transmitted by the sample RTT. If the *difference* between the actual and expected values is less than α (a threshold), Vegas will increase the cwnd linearly during the next RTT. If the *difference* is greater than β (another threshold), Vegas will decrease the cwnd linearly during the next RTT. No action is taken if the rate *difference* is greater than α but less than β.

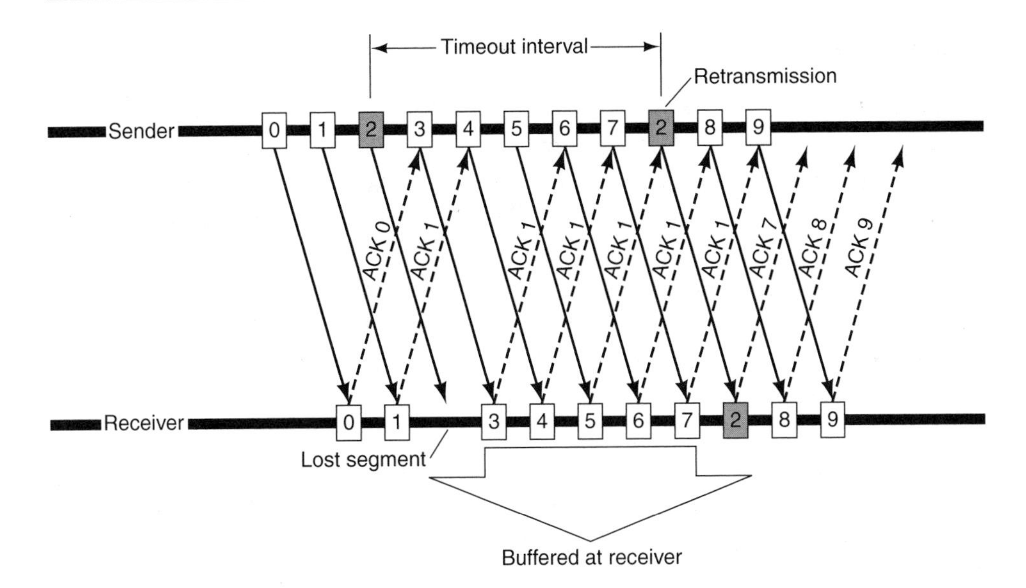

Figure 11.3. TCP window-based selective repeat mechanism.

11.2.4 TCP SACK

TCP with selective acknowledgement (SACK) [87] is an improvement over the postive ACK with retransmission (PAR) scheme.

In PAR, the sender waits for an ACK from the receiver for each packet sent. Upon successful reception of the ACK, the sender transmits the next packet. If an ACK for a packet sent does not arrive within a predetermined timeout period, the packet is retransmitted. While PAR is simple, it is not perfect. Network congestion and delay can cause ACK replies to be delayed. When this happens, the sender will time out and the last transmitted packet will be resent again, resulting in duplicates. Note that PAR uses sequence numbers to correctly associate packets with ACKs.

In SACK (RFC 2018), a selective retransmission scheme is introduced. Despite the introduction of a *window*, TCP can still suffer from poor performance when multiple packets are lost from one window of data. In such situations, the TCP sender will only learn about a single-packet loss after a round-trip time. Although an aggressive sender could choose to retransmit such packets earlier, such retransmitted packets may have already been successfully received. In SACK, the receiver sends back ACK packets to inform the sender that the data has been successfully received so that the sender can then retransmit only the missing packets. This is illustrated in Figure 11.3, where only the missing segment 2 is retransmitted while other correctly received segments are buffered at the receiver.

11.3 Problems Facing TCP in Wireless Last-Hop

When cellular-style networks were introduced, the supported traffic was primarily voice. However, with the presence of data over voice technologies, mobile terminals can now access the Internet and transmit data reliably. However, the use of TCP can cause problems since the TCP connection now comprises fixed and wireless links.

A data connection originating at a mobile terminal or device occurs over a wireless link from the device to the radio base station and from the base station to the static destination host/server via several fixed Internet routers. Hence, delays experienced at the wired and wireless links are different, and this can affect TCP flow and congestion control.

A wireless link over a cellular or wireless LAN is usually shared by multiple devices. Hence, the link delay varies with time. In addition, wireless transmissions are subject to multipath fading and signal jamming, which contribute to packet loss. All these can affect the estimated round-trip time or timely arrival of TCP ACK packets. Hence, some provisions are needed to make TCP wireless-aware so that it can adapt accordingly, without significantly affecting communication performance.

11.3.1 Indirect TCP

In mobile IP [88], mobility is handled at the network layer, where packets are tunneled from the *home agent* to the *foreign agent* when a mobile host moves. While this principle of tunnelling works for datagram flows, TCP flows will be affected by mobility since TCP is an end-to-end protocol. Movements by a host can fragment this end-to-end connection since it is a concatenation of wired and wireless links.

TCP provides virtual circuit service, where a data path is established prior to data transmission. An *endpoint* is defined by a *socket* at the transport layer. A socket contains the source and destination addresses, along with their port numbers. Applications use these sockets to send and receive data. When a TCP connection is established, it remains active until it is disconnected. Hence, during the lifetime of a fixed host-to-mobile host TCP connection, the mobile host could have moved to another location. This breaks the connection and TCP has no way of handling such a change.

Indirect TCP (I-TCP) [59] is a protocol proposed to resolve the disconnection issues in a mobile Internet environment where one of the links in a TCP connection is a wireless link. I-TCP requires the transfer of connection states from one mobile support router (MSR) to the other. I-TCP partitions the mobile TCP connection into two segments, namely:

(a) A regular TCP connection segment between the fixed host and MSR

(b) A dynamic segment between the MSR and mobile host

This approach, therefore, does not follow the end-to-end semantics of original TCP, where TCP functions reside only at the source and destination nodes. In I-TCP, the MSR has to perform some transport-layer functions. The dynamic TCP segment is designed with larger timeouts to suit the delay associated with intermittent wireless links. To hand off a mobile TCP connection, I-TCP uses a *socket migration* technique, where two new sockets with the same endpoint parameters as those in the old MSR are created at the new MSR. Since the wireless and wired endpoints do not change after a handoff, there is no need to re-establish the connection at the new MSR.

11.3.2 TCP Snoop

TCP Snoop is a link-aware transport protocol proposed by UC Berkeley for wireless last-hop networks. TCP Snoop addresses packet loss issues due to the presence of wireless links. Such losses cause TCP to back off and time out, resulting in poor end-to-end communication performance. With the help of a snoop agent present at the radio base station, lost segments are detected and retransmitted locally, without intervention by the sender. Also, last-hop round trip times are estimated. The suppression of duplicate ACKs corresponding to wireless losses from the TCP sender avoids unnecessary invocations of congestion control procedures by the sender. For data flow from the mobile to a fixed host in the backbone wired network, a mechanism known as Explicit Loss Notification (ELN) is used. ELN allows the decoupling of retransmissions from congestion control. At the base station, packets that were lost in a single transmission window are detected and negative acknowledgments (NACKs) are sent back to the mobile host. The NACK implementation is similar to TCP SACK, which was discussed earlier.

11.4 Problems Facing TCP in Wireless Ad Hoc

In ad hoc wireless networks, when a route is broken due to the mobility of nodes in the route, a route reconstruction or reconfiguration procedure is invoked. A delay is incurred during this time when the route is repaired. However, the TCP sender is unaware of this incident. Hence, it mistakes this delay of ACK arrival, or the increase in RTT, as signs of network congestion. Accepting this belief implies that the source node begins to reduce its transmission window size and initiates SS, which significantly reduces communication throughput performance unncessarily.

11.5 Approaches to TCP over Ad Hoc

11.5.1 TCP Feedback (TCP-F)

Introduced in 1998, TCP-F [89] allows the source to be informed of a route disconnection as a result of node mobility (see Figure 11.4). When a link in a route is broken, the upstream node that detects the disconnection will send a Route Failure Notification (RFN) message back to the source. Upon receiving this message, the source enters SNOOZE state. This is a new state introduced into the TCP state machine, as shown in Figure 11.4.

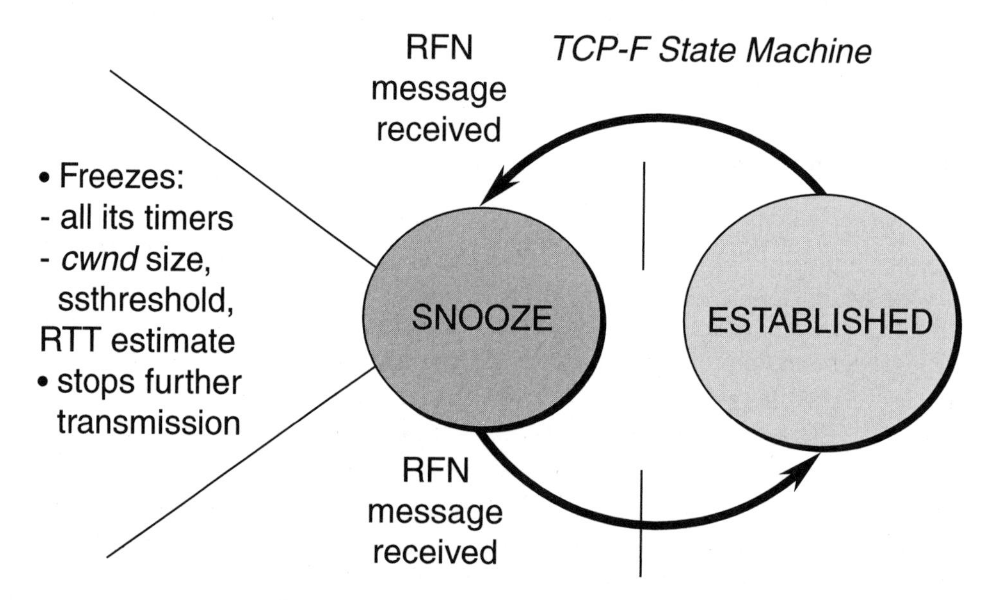

Figure 11.4. TCP-F protocol state machine.

When the TCP source enters SNOOZE state, it performs the following:

(a) The source stops transmitting all data packets (i.e., be it new or retransmitted data).

(b) The source freezes all its timers, the current cwnd size, and values of other state variables, such as the retransmission timer value. The source then initiates a route failure timer, whose value will depend on the worst-case route repair time.

(c) When the route repair complete message is received, data transmission will be resumed and all timers and state variables will be restored.

11.5.2 TCP-BuS

The TCP principle deals with end-to-end connections. However, an ad hoc wireless connection comprises multiple wireless links. Hence, trying to provide flow and congestion control at the source and destination nodes is neither sufficient nor situable.

I-TCP breaks the TCP semantics by allowing the base station to perform transport-layer functions, that is, a TCP connection is now further broken into two segments. This is necessary since the link between the base station and mobile terminal is wireless and special treatment is needed. If this "segment" model is further extended to an ad hoc wireless connection, then flow and congestion control can be performed in the same vein, but in a distributed fashion.

The five enhancements introduced in TCP-BuS [90] include:

- Explicit notifications
- Extension of timeout values
- Selective retransmission
- Avoidance of unnecessary requests for fast retransmission
- Reliable transmission of control messages

Explicit notifications are used to differentiate between *network congestion* and *route failure* as a result of mobility. The node that detects a route disconnection sends an Explicit Route Disconnection Notification (ERDN) message back to the source. The source then stops transmission. When the route reconfiguration or repair process is completed, an Explicit Route Successful Notification (ERSN) message is sent back to the source via the pivoting node. On receiving this message, the source resumes data transmission. This is further illustrated in Figure 11.5.

Further extension of the timeout value is necessary to account for the time needed for route reconfiguration or repair. In TCP-BuS, timeout values for buffered packets at the source and nodes along the path to the pivoting node are doubled. The optimal timeout value will depend on a variety of factors and this demands further research.

Currently in TCP, retransmission of lost packets on the path due to congestion relies on a timeout mechanism (i.e., by considering the RRT). However, since each node in an ad hoc route now has a TCP layer, the timeout values for buffered packets at these nodes should be doubled. In this manner, retransmission for timeout packets is localized and the source is spared from excessive retransmission.

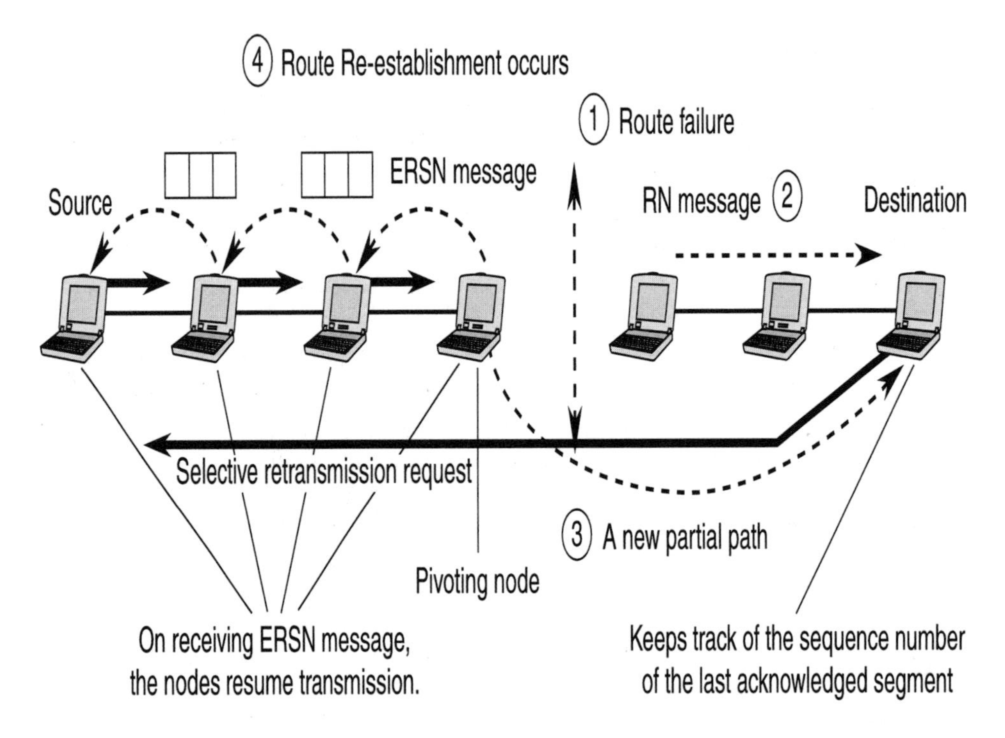

Figure 11.5. Sequence of events by TCP-BuS after a successful route reconfiguration.

In ad hoc routing, prior to a route disconnection, packets are in transit from the source toward to the destination. These buffered packets will be flushed when a route disconnection occurs. However, this causes extensive retransmission to be performed by the source node. Instead, the destination node should continue to send ACK packets containing expected sequence numbers until the expected in-sequence packets arrive at the destination. Hence, unnecessary requests for fast retransmissions are avoided.

When a node in an ad hoc route detects a route disconnection, it notifies the source node by sending a control message. However, the source node can only take action if the control message is sent reliably hop-by-hop. Also, intermediate nodes receiving this message will stop transmitting their buffered packets. Reliable transmission can be achieved via sensing if the control message has been relayed by the receiving node. If not, a timer will expire and per-hop retransmission will occur. Similarly, when the route repair process is completed, another control message will be sent by the pivoting node back to the source so that it can resume data transmission.

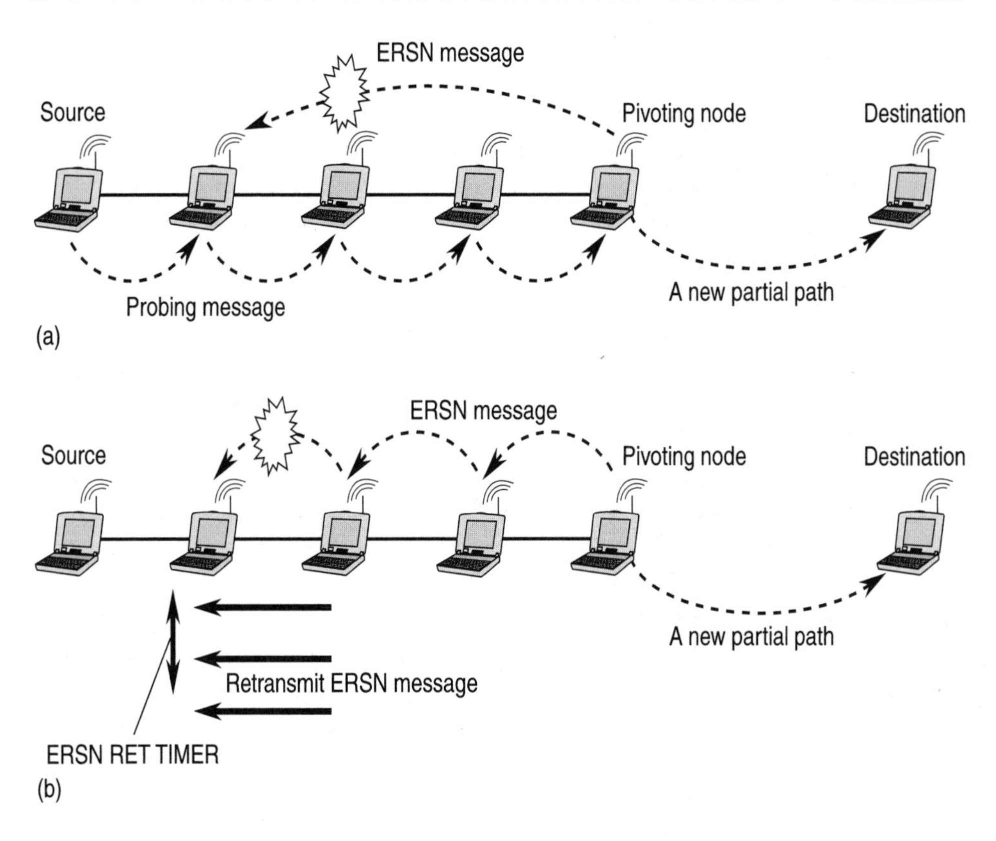

Figure 11.6. Assurance of reliable transmission of TCP-BuS control messages to allow proper flow control using: (a) probing, and (b) per-hop data integrity control.

As shown in Figure 11.6, TCP-BuS implements reliable transmission of control messages through two possible approaches:

- The source periodically sends PROBE messages to check if a privoting node has successfully acquired a new partial path to the destination.
- Each intermediate node is responsible for sending an ERSN message reliably to to its upstream node until it receives data packets.

A detailed report on the performance of TCP-BuS and comparisons with other transport protocols can be foud in [90].

11.6 Conclusion

Building a reliable transport mechanism on top of the unreliable datagram delivery service provided by IP is appropriate. TCP has fulfilled this function over the Internet with the exception of wireless networks. Delays and packet losses can significantly impact the proper operation of TCP, in particular flow and congestion control.

This chapter highlighted the basic working principles of TCP, followed by a description of later versions, including TCP SACK, TCP Tahoe, TCP Reno, and TCP Vegas. The discussion then proceeded to TCP over wireless last-hop networks, where again, packet loss and delays cause problems. Two solutions to resolve these problem include I-TCP and TCP Snoop.

Finally, the reader was exposed to TCP problems in ad hoc, which are more severe. TCP-F is a solution where the TCP source has its timeout values extended and its state preserved when a route is broken. Transmission is subsequently resumed when the route is repaired. TCP-BuS extends this concept further by introducing mechanisms for reliable transmission of feedback control messages, further extending timeout values at each node in the route by avoiding unnecessarily fast retransmissions.

Chapter 12

INTERNET & AD HOC SERVICE DISCOVERY

12.1 Resource Discovery in the Internet

Resource discovery over the Internet has been a popular topic in the past. With the explosion in the amount of information available on the Internet, it is important to have tools to assist us to locate or search for the information we want (to an acceptable degree of accuracy and detail) quickly. Several search engines have evolved (such as *www.altavista.com*, *www.yahoo.com*, and *www.excite.com*) to provide such facilities. There is no true standard existing for service discovery designed specifically for ad hoc wireless networks. Instead, we will present here what has been done by the IETF. The IETF formed a working group to look into service location back in 1988. In 1997, Version 1 of the Service Location Protocol (SLP) was documented as RFC 2165. The SLP architecture shall be presented below. Service discovery for Bluetooth devices will be discussed in a later chapter. In addition, service discovery for ad hoc networks will be highlighted toward the end of this chapter.

12.2 Service Location Protocol (SLP) Architecture

SLP [91][92] is a protocol that was developed to allow for the discovery of resources present in a network. The SLP architecture has *user*, *directory*, and *service agents*. User agents (UAs) interact with directory agents (DAs) that contain lists of service agents (SAs). A DA returns information about who the user should contact regarding services requested or needed. Services can have a lifetime, and hence, DAs should remove expired services and keep available services up-to-date.

Service advertisements are sometimes used to inform other users of available services. Service advertisements include information like the host address and port number of the service provider. They are also commonly referred to as service access points. In addition to service access information, information about the *service type* and corresponding *attributes* are enclosed.

12.2.1 User Agent

A UA can serve multiple end-users. It is a software entity and it acts as an agent to search for requested services. It can even cache service information so that subsequent requests can be fulfilled immediately. The UA is responsible for interrogating service availability. It does not inform the user what access methods to be used, but merely points the user toward the service provider to contact. As shown in Figure 12.2, the UA can contact an SA or DA for service information. For implementation considerations, one UA can exist per subnetwork.

12.2.2 Directory Agent

To allow a service location architecture to scale, having just a UA and an SA is insufficient. With a large network, a UA may receive thousands of replies about the service it is requesting. This can give rise to discovery implosion and is particularly bad for a wireless environment. The introduction of a DA helps to alleviate this problem. The DA consolidates all the service replies and caches them into a directory.

As shown in Figure 12.2, acting as a proxy, the DA can reply back to the UA's directly, hence avoiding the discovery acknowledgment implosion problem. Another way to address scalability is the use of administrative domains. For example,

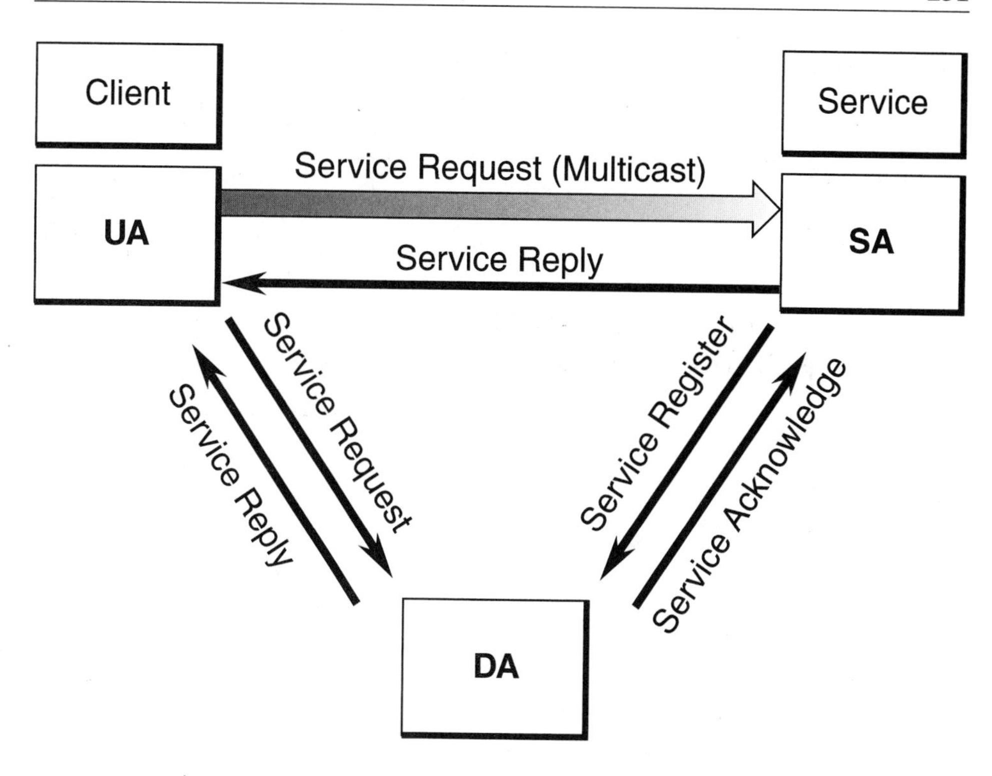

Figure 12.1. Interactions among DAs, UAs, and SAs.

printers belonging to a classified department would only serve people/users within that department. Hence, UAs searching for services within a permitted administrative domain only see those devices available in that domain. Administrative domains, therefore, limit the visibility of services.

12.2.3 Service Agent

The SA acts as an agent for the service provider. It advertizes available services either directly to the UAs or to DAs. If affiliated with certain DAs, an SA can periodically register available services with the DAs. An SLPv1 network could consist of merely SAs and UAs. In such a scenario, the UAs would send out service request queries via multicast. An alternative SLPv1 architecture consists of UAs, SAs, and DAs. SAs can register their advertisements with DAs, and UAs can query DAs via unicast communications.

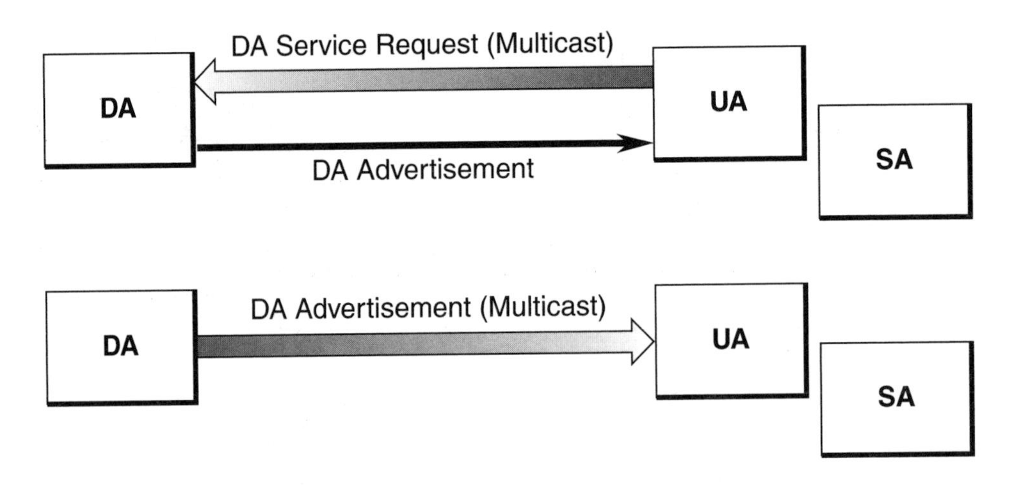

Figure 12.2. The interactions between a DA and a UA/SA in SLP.

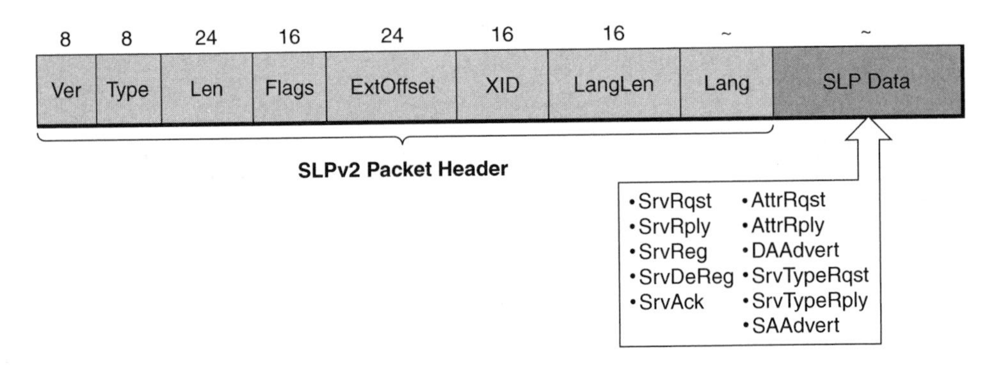

Figure 12.3. The SLPv2 packet structure.

12.3 SLPv2 Packet Format

Several problems associated with SLPv1 were discovered later. The two major ones are *scalability* and *security*. The former is attributed to unscoped advertisements and DAs, which result in considerable overload in the network and at the DAs.

Scoped DAs need to deal with advertisements within their scopes. The lack of security lies in DA advertisements. An attacker could act as a DA, intercept SA registrations, and send faulty DA replies to UAs. Hence, some form of signature

Table 12.1. Message Types and Function Codes Used in SLPv2

Message Type	Abbreviation	Function Code
Service Request	SrvRqst	1
Service Reply	SrvRply	2
Service Registration	SrvReg	3
Service Deregistration	SrvDereg	4
Acknowledgment	SrvAck	5
Attribute Request	AttrRqst	6
Attribute Reply	AttrRply	7
DA Advertisement	DAAdvert	8
Service Type Request	SrvTypeRqst	9
Service Type Reply	SrvTypeRply	10
SA Advertisement	SAAdvert	11

check is necessary so that UAs and SAs can validate DA advertisements. These problems are addressed in SLPv2 [93].

As shown in Figure 12.3, each SLPv2 packet is prefixed by a common header. The header identifies the version of the protocol and type of control message. It also specifies the length of the entire SLPv2 message. The "XID" field refers to the transaction ID. As in SLPv1, the transaction ID is unique for each service request and service replies are made with reference to the transaction IDs concerned. The next field is the language tag. A language tag must always be specified in an SLP message. In addition, the language tag contained in a reply packet must correspond to the tag in the request message. This ensures proper interpretation of such messages.

12.4 Jini

Jini [94][95] provides a platform for clients to locate services. A client can join a Jini lookup service and request information about a particular device, such as a printer. Jini was designed to operate over distributed systems in wired networks; hence, low latency is assumed. It is unclear what would be the consequences of applying Jini directly to wireless distributed ad hoc systems. The increase in link delay would certainly impact client/server interaction time. In addition, Jini is bulky and may not be suitable for low computation and memory devices, such as the palm

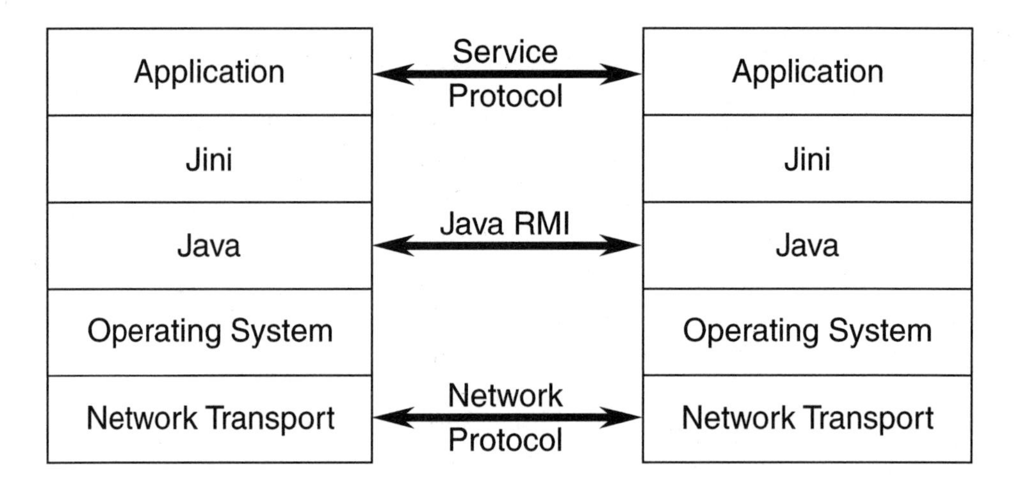

Figure 12.4. The Jini protocol stack.

pilot.

As shown in 12.4, Jini 'sits' on top of Java. Jini introduces the concept of "federation" where a group of devices register with each other to share services. Each Jini subsystem contains a set of lookup services that maintain dynamic information about services in the network. The location of these services can be known in advance or discovered using multicast. Clients can use Jini lookup service as a simple discovery service or they can download Java code which allows them to access the service. In the former, clients can query the lookup service and use a proprietary protocol to access services. In the latter, the server must have uploaded a Java proxy that the clients can download. To manage service access, Jini services are leased for a period of time when a client needs them. This lease, however, has to be renewed periodically in order to have continous access to the desired service.

12.5 Salutation Protocol

The Salutation Protocol [96] predates SLPv1. In fact, it is a specification made by the Salutation Consortium. Part 1 of the Salutation Protocol version 2.0 specification [97] was released in 1996, and it defines the architecture, general framework, and information about the Salutation Manager Protocol (SMP). Part 2 of the specification, however, defines the Salutation Personality Protocol and attributes.

The goals of salutation are to address problems associated with service discov-

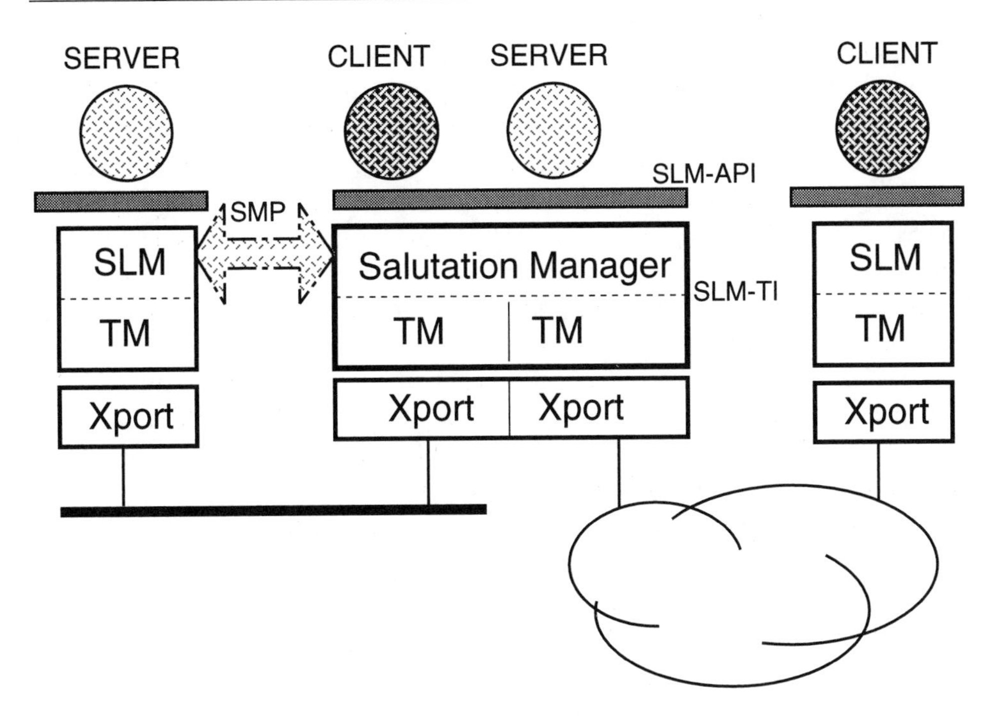

Figure 12.5. The saluation manager protocol architecture.

ery and utilization among the appliances/devices in an environment of widespread connectivity and mobility. At the heart of the salutation architecture is the *salutation manager* (SLM), which is an entity that acts as a service broker for other devices. A client discovers services and sends requests through an SLM. The service provider is known as the *server* while the service user is called the *client*. SLMs use SMP to communicate. SMP uses the Remote Procedure Call (RPC) version 2 protocol. The SLM and transport manager (TM) both collectively perform the service broker role.

The SLM provides: (a) a service registry, (b) service discovery, (c) service availability, and (d) service session management functions as a service broker. The service registry contains information about services discovered. These services may reside in the location or on a remote device. An SLM can discover other SLMs and their associated services through the service discovery mechanism, which occurs over RPC. SLMs perform periodic checks via RPC message exchange to ensure service availability is updated.

In salutation, service access must occur over a service session established between the client and server. Different "personality" protocols can be used during message exchange in a service session. In the *native personality* mode, the SLM is only responsbile for establishing the session pipe. The management of messages and data formats is left to the client and server. In the *emulated personality* mode, the SLM establishes the pipe and manages the message stream, but is not involved with data formats, which is left to the client and server. Finally, in the *salutation personality* mode, the SLM establishes the pipe, manages the message stream, and provides the data format definition for client/server interaction.

As shown in Figure 12.5, the SMP architecture defines an SLM-API, which is the application programming interface provided by the SLM to applications. This API provides transport layer independence and facilitates application development and portability. The semantics defined for this API can be found in [97].

The salutation specification also defines service records, namely: (a) service description record, (b) functional unit description record, and (c) attribute record. A service description record can include several functional units. Attributes further specify the capability of functional units.

12.6 Simple Service Discovery Protocol (SSDP)

SSDP [98] was proposed by researchers from Microsoft. As part of UPnP (Universal plug and play) [99], SSDP is used for the discovery of resources in a wired network. A service provider announces its presence upon service creation and clients request services when desired. Multicast communications are used for both service announcements and requests. Unicast communications are used by service providers to inform clients that they can provide a requested service.

SSDP aims to function without needing any configuration, management, or administration. Using multicast allows the presence of a multiparty channel. Clients can use this channel to express what they need and await responses. SSDP also allows HTTP clients and HTTP resources to discover each other in a LAN. SSDP clients discover SSDP services using a reserved local adimistrative multicast address (239.255.255.250) over an SSDP multicast channel/port.

Service discovery occurs when an SSDP client multicasts an HTTP UDP discovery request to the SSDP multicast channel/port. SSDP services listen to the SSDP multicast channel/port to hear such discovery requests. If an SSDP service provider hears an HTTP UDP discovery request that matches the service it is offering, it then responds with a unicast HTTP UDP message.

SSDP is an event-driven protocol. This means that a new service provider that comes on-line should declare its presence. Thereafter, no more service announcements are needed until the service provider goes off-line. Service providers do not periodically send announcements. Instead, a *service lifetime* is specified in the form of seconds, minutes, days, weeks, months, and years. Service discovery messages are sent based on-demand, when a client desires a service.

SSDP cannot use port 80 since one cannot grab a particular port on a particular multicast address without owning the same port on the local unicast address. A lookup on /etc/services should reveal that port 80 (TCP and UDP) is reserved for HTTP. Designers should note this.

12.7 Service Discovery for Ad Hoc

12.7.1 Limitations of Existing Schemes

Our discussions in the earlier sections are concerned with service location for the Internet, primarily for wired networks or single-hop-based Bluetooth piconets.

Such protocols, however, have not taken into account the latency and packet loss issues associated with the presence of several wireless links in a route and the presence of mobility by nodes/routers in the route. Jini, UPnP, and Saluation are higher layer protocols that do not specifically address wireless and mobility issues.

In addition, not all devices are the same. Some devices are small and have limited computation, memory, and storage capability. They can only act as clients, not servers or service providers. The presence of power constraints would mean that service location protocols should not incur excessive messaging over the wireless interface. In addition, service location architectures have to be properly designed, taking into account the feasibility of using centralized directory agents.

12.8 Ad Hoc Service Location Architectures

12.8.1 Service Co-Ordinator Based Architecture

Assuming an ad hoc wireless environment where several heterogeneous mobile devices exist, certain mobile devices will be better equipped than others. Such devices can be selected to act as *service co-ordinators* (SCs). SCs act as brokers or service collection points. They advertise their presence to other nodes in the network so that servers or service providers in the neighborhood are aware of the SCs. Servers then register their available services and access information with the SCs. How far a SC should advertise itself (in terms of radio hops) is, therefore, an important design issue.

Advertisements that are too narrowly scoped will result in less servers registered per SC. However, this implies less burden on the SC. A too widely-scoped advertisement will result in many servers replying to the SCs, which would incur more control traffic.

The frequency of advertisement by the SC will reflect the validity of the SC over time and space. Should an SC power down or isolate itself from the rest of the network, its advertisement will stop and expected advertisement would not arrive at neighboring nodes. Similarly, the frequency of registration by servers with the affilated SC/s will reflect how updated the list of available services are in the SCs. Outdated service entries in the SCs will result in a service access failure. Deriving optimal advertisement and registration frequencies to yield reasonable performance is a challenge that demands future research.

Figure 12.6 illustrates how servers and clients interact with SCs in an ad hoc mobile network environment. In Figure 12.6a, SCs advertise their presence (i.e., sending a broadcast message via $\mathcal{X}$ hops, for example, and enclosing their addresses) to their neighbors. Servers that receive the advertisement may then reply to register with one or more neighboring SCs with information about available services. Clients that are interested in locating available services will send query messages to their closest and affiliated SC/s. If after a certain timeout period, a client receives no reply from its affiliated SC/s, it concludes that its desired service may not be currently available and may attempt to retry later. Should it receive a reply from an affiliated SC, it can then proceed with accessing the service provider directly, as shown in Figure 12.6d.

12.8.2 Distributed Query-Based Architecture

In this architecture, there are no service coordinators (SCs). Clients locate available services directly from servers. As shown in Figure 12.7, clients interested in locating services broadcast messages to neighboring nodes. Servers that receive the query message will send a reply via a directed broadcast. It is possible that the same service may be available from two neighboring servers and hence it is up to the client to select the appropriate server.

The use of SCs allow service management functions to be distributed and managed by different nodes acting as SCs. This is attractive in terms of scalability. Providing a scalable service location architecture implies the need for distributed and localized control. However, a remote server may not be registered with an SC near to the service-desiring client and hence denying the client possible available service. In addition, mobility of SCs can also cause problems since service-server mappings would have to be updated.

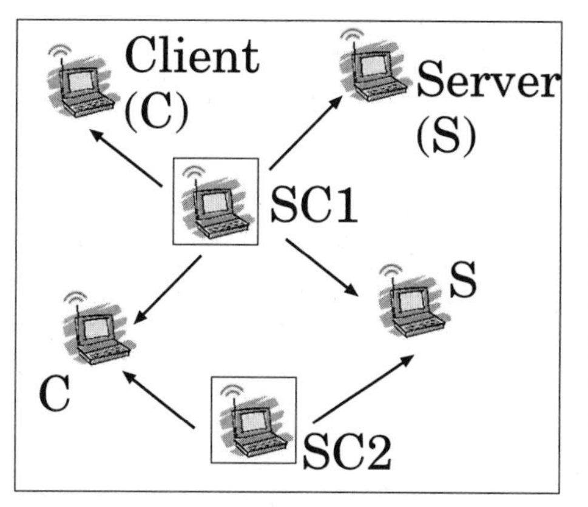

(a) SCs advertise themselves.

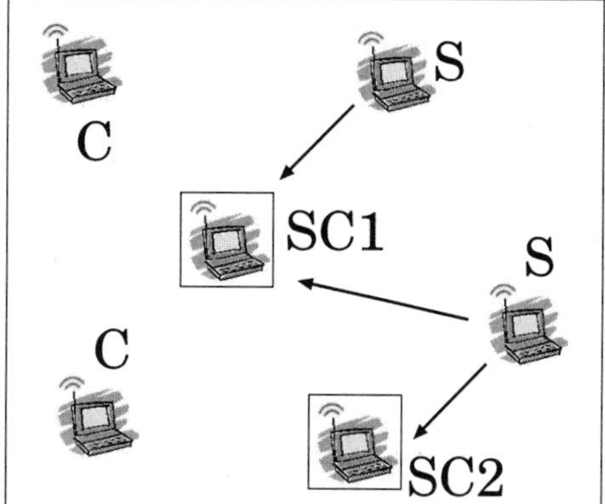

(b) Servers register with SCs.

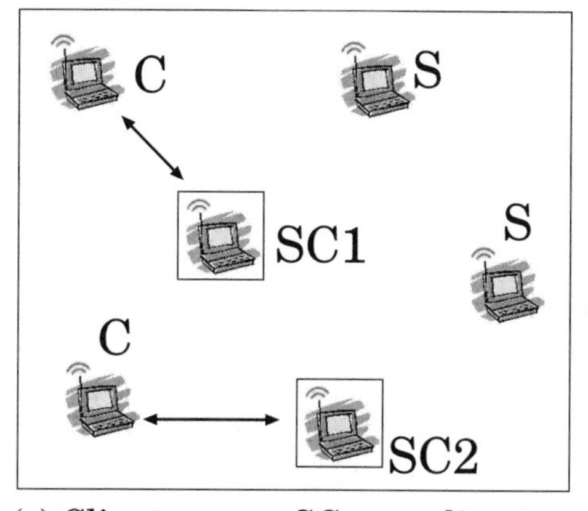

(c) Clients query SC according to their chosen SC.

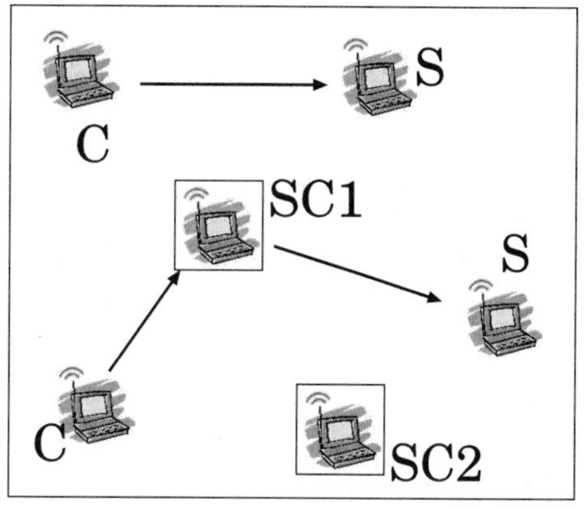

(d) Clients access specific server for services.

Figure 12.6. Architecture based on service co-ordinators.

The distributed query approach, however, does not rely on the existence of SCs. Without the presence of SCs, clients now have to perform broadcast query in search of servers that can provide desired services. The success of service search will also depend on the scope of the broadcast and also the percentage and location distribution of servers in the network. Hence, this approach can only scale to a certain extent. Compared to the SC-based approach, the distributed query approach is more robust in dealing with mobility of nodes (be it clients or servers).

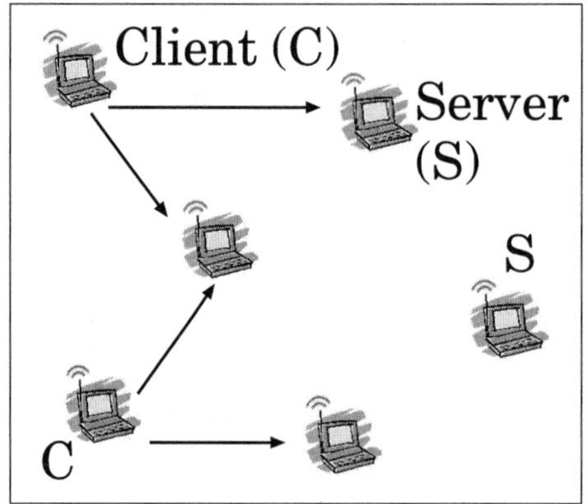

(a) Clients broadcast their service queries to neighbors.

(b) An available server responds to the corresponding client.

Figure 12.7. Architecture based on distributed query.

12.8.3 Hybrid Service Location Architecture

The hybrid approach combines the principles of SC-based and distributed query-based service location techniques. It further improves *service availability* beyond that attainable by the SC-based scheme alone. At the same time, it supports scalability through the use of SCs in the network.

In an ad hoc mobile network, *service availability* is a useful metric for evaluating performance of service location operations since it defines how many service query requests made by clients can be fulfilled by SCs or servers in the network. This is particularly important from both network operators' and end users' point of view.

As shown in Figure 12.8, SCs advertise their presence to neighboring nodes. Servers that are acquainted with nearby SCs registered with them. Servers that are not acquainted with any SCs will merely await for either SC advertisements or client-service queries. A client in the neighborhood of an SC can send its service query to the SC. However, a client that does not have any neighboring SCs will rely on the distributed query approach to contact servers in the neighborhood. Figure 12.8c shows that a client's service request message may be forwarded by other clients to reach a neighborhood SC. If the desired service is indeed handled by that SC, it can reply back to the client. Hence, both SCs and servers that can provide the desired service or that have information on how to access the desired service can reply back to the client. This technique, therefore, enhances service availability.

An evaluation of the performance of SC-based, distributed query, and hybrid

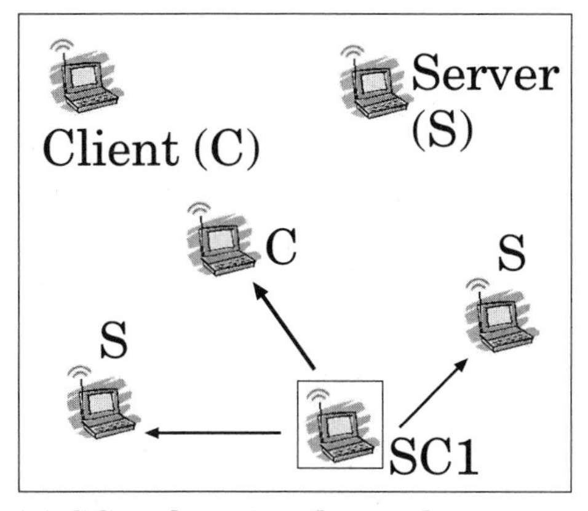

(a) SCs advertise themselves.

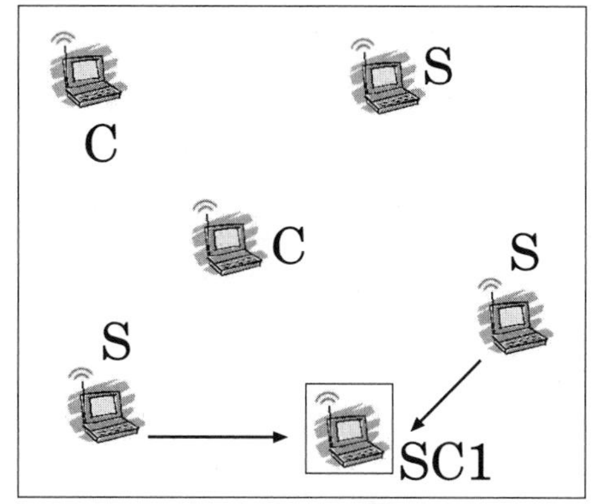

(b) Servers register with SC.

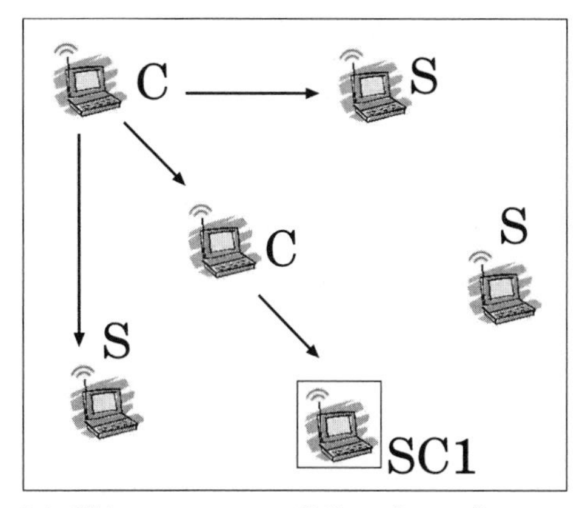

(c) Clients query SC or broadcast service query to the network.

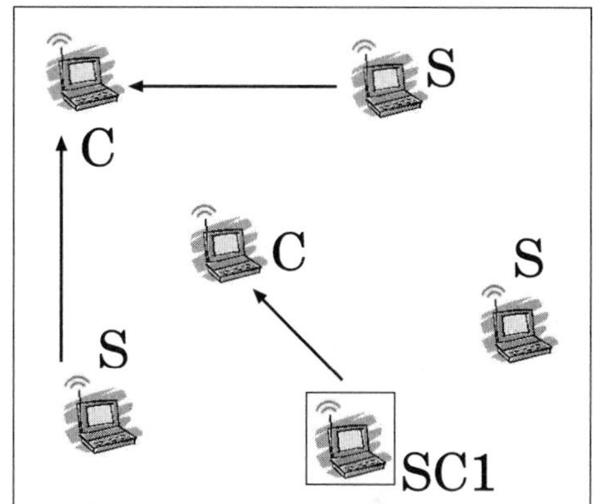

(d) SCs or servers respond with service information.

Figure 12.8. Architecture based on hybrid service location.

service location approaches is reported in [100]. Factors that affect service availability include: (a) percentage of servers present, (b) service query search radius, (c) percentage of service co-ordinators present, (d) rate of mobility, (e) frequency of SC advertisment, and (f) frequency of servers registration. Simulation results reported in [100] showed that the hybrid approach yields better average service availability with a lower overhead compared to the SC- and distributed query-based protocols.

12.9 Conclusions

Service discovery mechanisms involve basic operations like *service request*, *service advertisement*, and *service reply*. Some of these operations require the use of multicasting or broadcasting. However, current efforts are limited to the Internet environment and are not addressing ad hoc wireless networks. Bluetooth service discovery, which we will discuss later, allows a device to discover services present in other devices in a wireless fashion. Issues related to security and scalability are important and need to be addressed. Ad hoc wireless networks are particularly vulnerable to power and bandwidth limitations; hence, service discovery mechanisms designed for such networks must be efficient, simple, and effective. Further research is necessary to derive suitable service discovery and access schemes for ad hoc wireless networks.

Chapter 13

BLUETOOTH TECHNOLOGY

Currently, the wireless voice industry supports mainly mobile phone services. However, the computer industry is currently very engrossed with ad hoc wireless networking. Bluetooth aims to merge the two industries and their services into one common device. Wireless devices can now be interfaced to other voice and data-capable mobile devices. Bluetooth is not a new technology; it is a new use of existing technologies.

Today, there are many choices for wireless networks:

- Big LEOs (Low Earth Orbit)
- Little LEOs
- SkyTel 2-Way
- CSMA/AMPS Cellular
- GSM-PCS

- Wireless LANs
- Metricom
- TDMA-PCS

Bluetooth operates on a transparent mode that is based on proximity RF to allow communications among mobile devices with a wireless modem.

13.1 Bluetooth Specifications

Bluetooth [101] specifies a 10m radio range and supports up to 7 devices per piconet. There is a controller within each piconet. The radio frequency is centered at 2.45GHz (RF channels: 2420+k MHz, k=0,...78). The transmitter's initial center frequency must be +-75khz from the carrier frequency, $\mathcal{F}_c$. Frequency hopping is used to combat the presence of interference and fading in a wireless environment. The TDD (time division duplex) scheme is used for full-duplex communications.

The sleeping power specified is 30 micro amps; transmission power is 800 micro amps; standby mode is 300 micro amps. It can support three full-duplex voice channels simultaneously per piconet with a data rate of 721 kbps. The price for a Bluetooth chip is about $5 or less. It can operate on both *circuit* and *packet* switching modes, providing both *synchronous* and *asynchronous* data services. Within a piconet, Bluetooth provides point-to-point communications. A device is picked as the MASTER and the rest serve as SLAVES. Multi-hop communications in Bluetooth are achieved through the *scatternet* concept, where multiple masters must establish links to each other. The MASTER is, therefore, the bottleneck!

If we examine throughput performance, sometimes Bluetooth is faster than cellular data (9.6 to 14.4 kbps). However, compared to wired or wireless LAN (10 Mbps), Bluetooth is slower.

Bluetooth must be able to:

- Recognize any other Bluetooth device in radio range
- Permit easy connection of these devices
- Be aware of device *types*
- Support service discovery
- Support connectivity-aware applications

13.2 Bluetooth Architectures

13.2.1 Bluetooth Piconet

The Bluetooth architecture is based on a *cluster* or *grouping* of ad hoc wireless devices. The clustering allows coordination at the media access layer and also routing. As shown in Figure 13.1, the piconet has a cluster head, which acts as the central controller and MASTER. All other devices within the cluster become the SLAVES. The specification limits the number of SLAVES within a piconet to seven; hence, there are no more than eight active devices in a piconet. If there are other devices within the piconet, then they should not be active or they will be considered "parked." Figure 13.1 shows the presence of a "parked" member.

All devices on a piconet adhere to the same frequency hopping sequence and timing of the MASTER. Note that SLAVES within a piconet only have links to the MASTER and not between SLAVES in the piconet. This sort of setup implies that communication between SLAVES within the same piconet has to occur via the MASTER. The MASTER sets the frequency hopping sequence and SLAVES synchronize to the MASTER in *time* and *frequency* by following the MASTER's hopping sequence.

Every Bluetooth device has a unique device/hardware address and a Bluetooth clock. An algorithm is used to derive the hopping frequency based on the MASTER device address and clock. The MASTER performs channel access management by scheduling time slots to SLAVES, which then influence how often a SLAVE can communicate. The number of slots assigned will depend on the QoS requirement. For voice traffic, accessible slots are reserved and available on a periodic basis. Hence, channel access is coordinated through the MASTER and a form of dynamic TDMA scheme is used.

13.2.2 Bluetooth Scatternet

If a Bluetooth device is within the locality of two piconets, then time-sharing has to occur, that is, it will spend a few slots on one piconet and a few slots on the other. The cascading of one piconet to another results in *scatternets*.

As shown in Figure 13.2, a MASTER of a piconet can act as a SLAVE of another MASTER in another piconet. It is not possible for a device to serve as a MASTER for two piconets. For a SLAVE in a piconet to talk to another SLAVE in another piconet, both MASTERs have to be involved in relaying the packets across piconets.

A device, however, can be a SLAVE for two MASTERs on two different piconets. In such a configuration, that SLAVE acts as the communication bridge/router

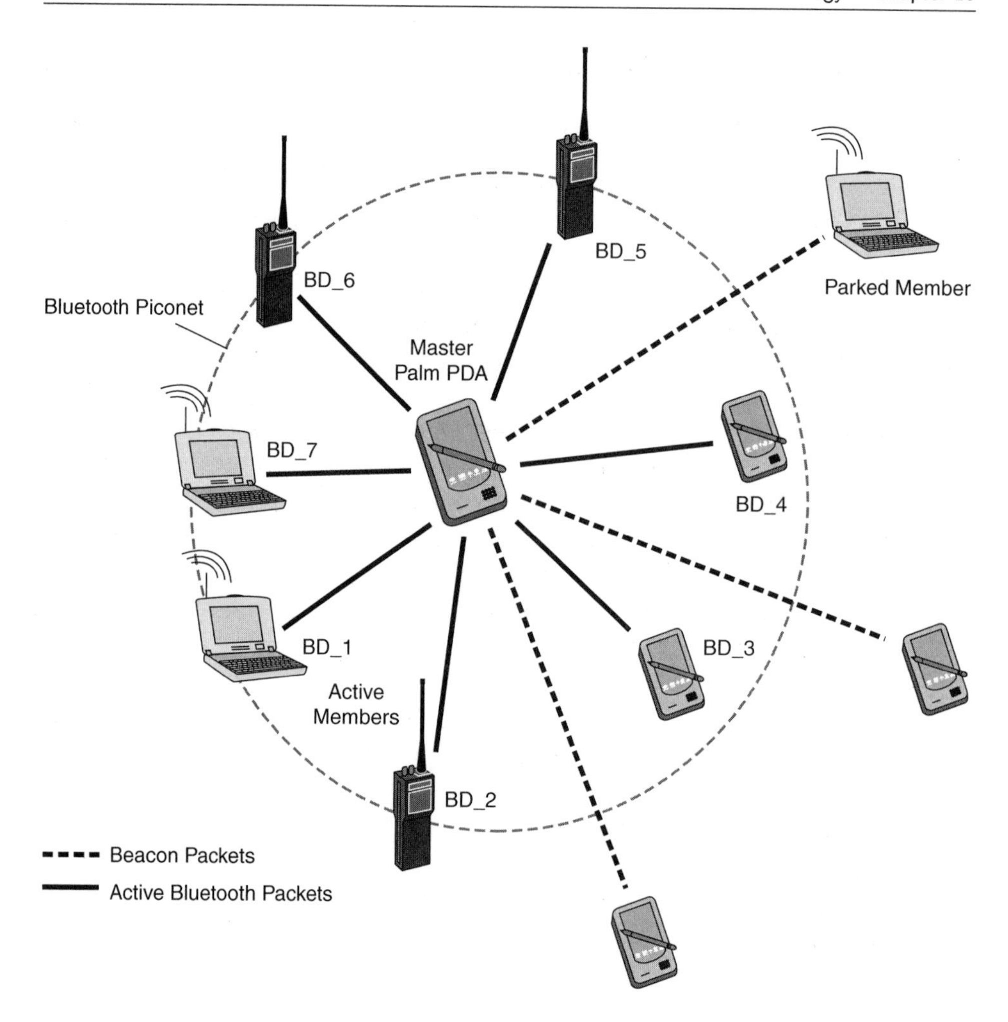

Figure 13.1. The Bluetooth piconet architecture.

if communications span across piconets. Hence, it is clear that there is a limitation as to the communication performance of multi-hop routes over Bluetooth scatternets due to the presence of time switching among piconets. Currently, there is no deeper clustering other than piconets and scatternets in Bluetooth.

13.3 Bluetooth Protocols

Bluetooth defines not only the radio system, but also communication protocols. The radio baseband handles bits of information and presents them in a suitable form for

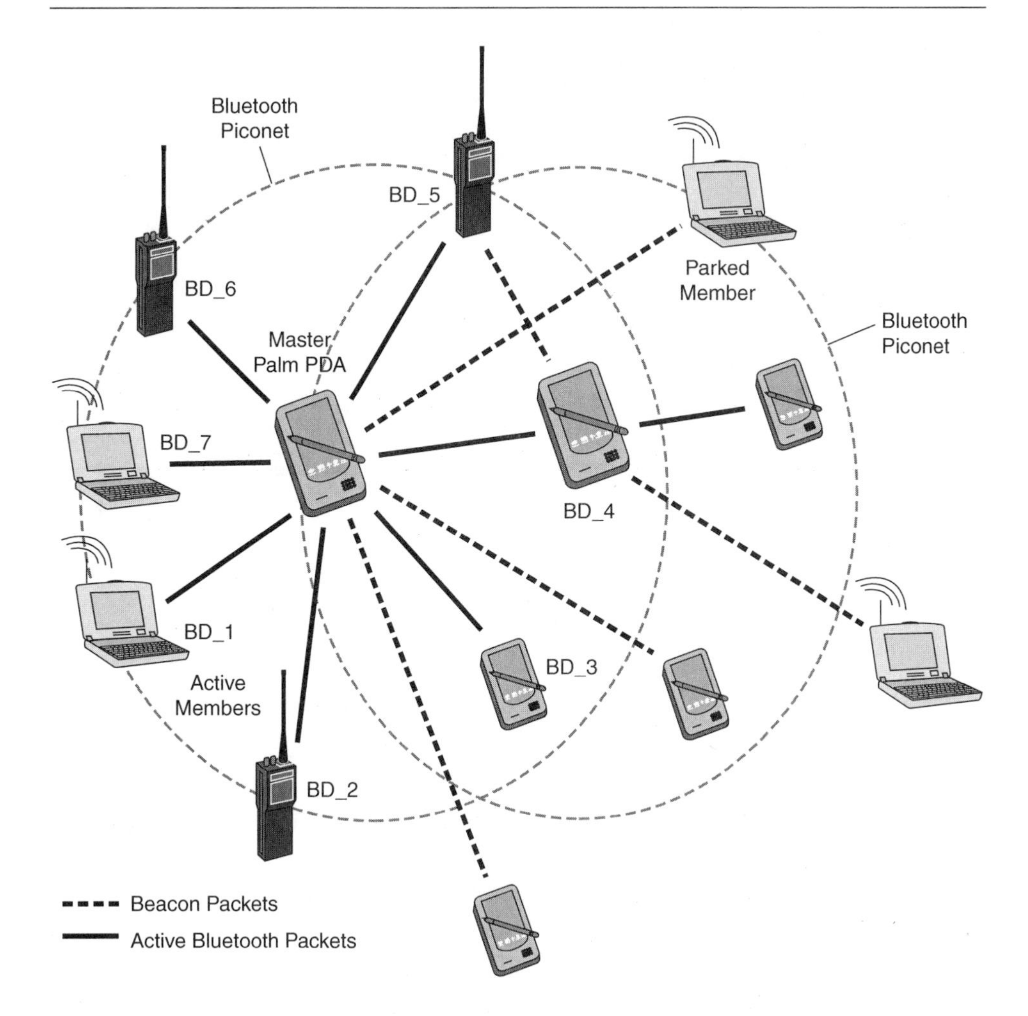

Figure 13.2. The Bluetooth scatternet architecture.

radio transmission. This involves coding/decoding and modulation/demodulation.

The link control layer supports link establishment and provides link control. The link management (LM) layer controls and configures links to other Bluetooth devices. The LM is responsible for attaching SLAVES to a piconet, and allocating their active member addresses. It establishes ACL (asynchronous connectionless) data and SCO (synchronous connection oriented) voice links and is capable of putting connections into low-power modes. The host controller interface (HCI) allows the implementation of lower Bluetooth functions on the Bluetooth device and higher protocol layer functions on a host machine. Interactions between these

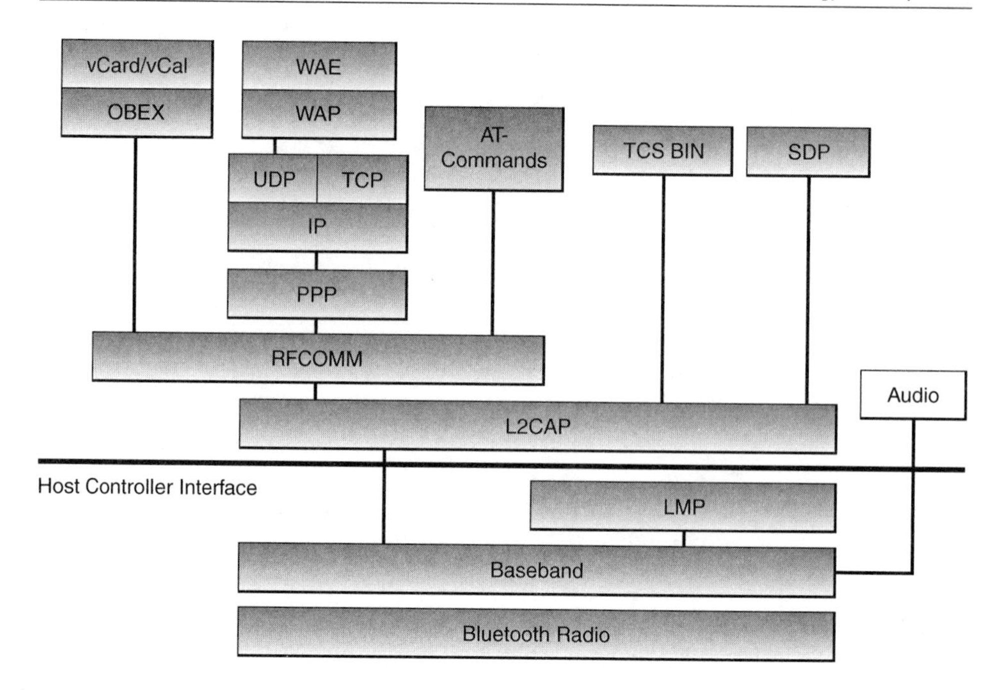

Figure 13.3. The Bluetooth protocol architecture.

two entities occur over the HCI. Hence, the HCI is useful for testing, debugging, and also separating functions into different hardware modules.

Multiplexing and assembly/disassembly functions are performed by the logical link control and adaptation layer (L2CAP). RFCOMM is a reliable transport protocol providing framing, multiplexing, and flow control functions. It can emulate serial connections familiar to RS-232 serial ports since Bluetooth involves mostly point-to-point links. OBEX is the object exchange session protocol that enables the exchange of data objects and supports dialogues between two devices. vCard, vCalender, vMessage, and vNotes state the formats for the electronic business card, electronic calendar, scheduling, messaging, and mail.

As shown in Figure 13.3, WAP-related protocols can also be supported over a Bluetooth platform, although this may not be the most optimal thing to do. If we were to compare the OSI ISO seven-layer protocol model with the Bluetooth protocol stack, the baseband layer would correspond to the data link layer, while link control and management layers would correspond to the network and transport layers.

At the physical layer, the license-free ISM band at 2.4GHz is being used for Bluetooth. The operating band of frequencies is divided into 1MHz-spaced sub-

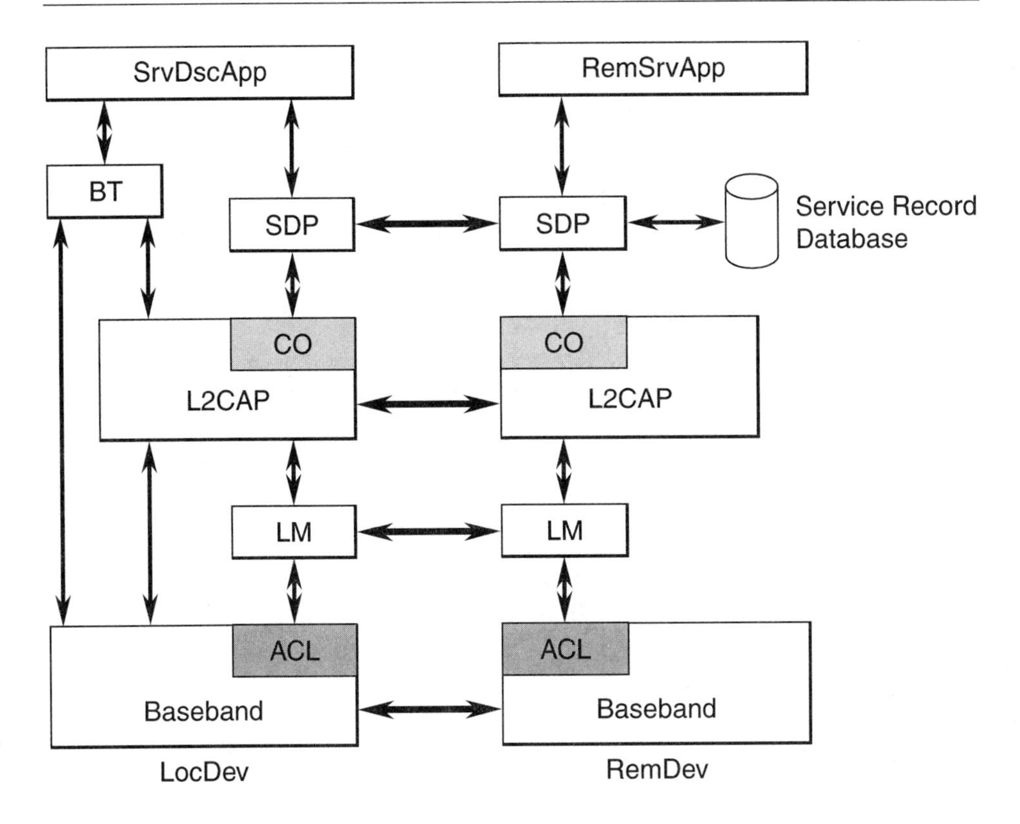

Figure 13.4. Bluetooth service discovery.

bands, and using GFSK with a signaling data rate of 1M symbol per second, 1 Mbps is available.

13.4 Bluetooth Service Discovery

Bluetooth employs *paging* and *inquiry* mechanisms to detect/sense the presence of other Bluetooth devices so that communication links can be established. The page or inquiry scan rate occurs once every 1.28 seconds.

Each device periodically transmits a series of inquiry packets via broadcast. Nodes in the neighborhood reply with a frequency hop synchronization (FHS) packet. This packet informs the inquiry source how to establish communication with it. It also contains specific information about the device, such as the device's class. Eventually, each device knows which other devices are present and their FHSs.

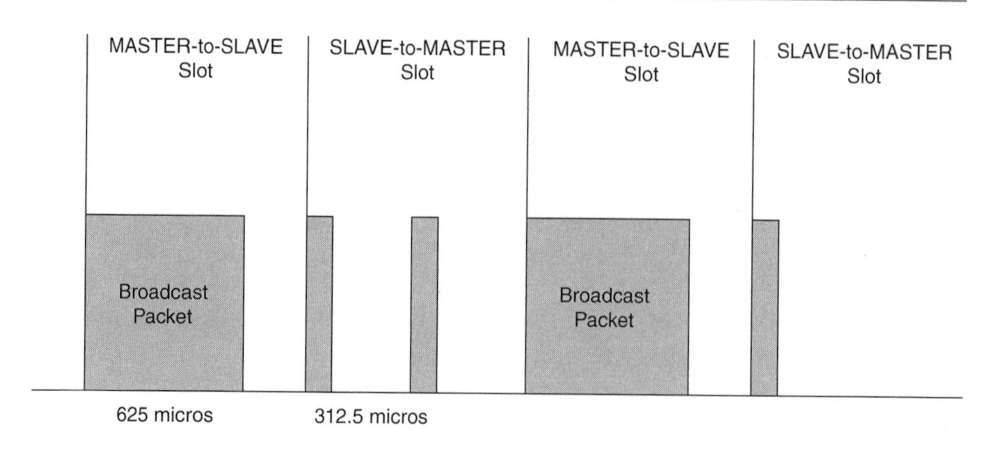

Figure 13.5. Bluetooth time slots.

Detecting the presence of other Bluetooth devices is just half the story. To discover services present in those devices, a service discovery protocol (SDP) is needed. In this protocol, a device first pages the correspondent device. If the correspondent device is scanning for pages, it will respond. The L2CAP channel is used to establish a connection to the service discovery server on the other device, thereby obtaining information about available services on that device. If it decides to access a particular service, it then has to establish an L2CAP connection to the device where the service resides. The interactions of the client/server and the respective protocol layers involved in SDP are shown in Figure 13.4.

Essentially, an SDP client has to search for a specific desired class of service to retrieve attributes associated with that service so as to allow the establishment of a connection to access that service. Devices can implement policies so that they do not respond to inquiries and hence can remain invisible to other devices. They may do so for a variety of reasons, such as power conservation and privacy.

13.5 Bluetooth MAC

The Bluetooth link control and management layers are responsible for channel access control and link-level medium access control. TDM is used, where each time slot is 625 μsecs. The slots toggle between the MASTER and SLAVEs, that is, when the MASTER is using the slot for transmission, the SLAVE is receiving and cannot transmit. A packet can be multiple slots in length. For example, a three-slot packet is 1875μsec long. Since this is a time-based system, time synchronization is important. Bluetooth has a 28-bit counter, which is incremented every 312.5μsecs

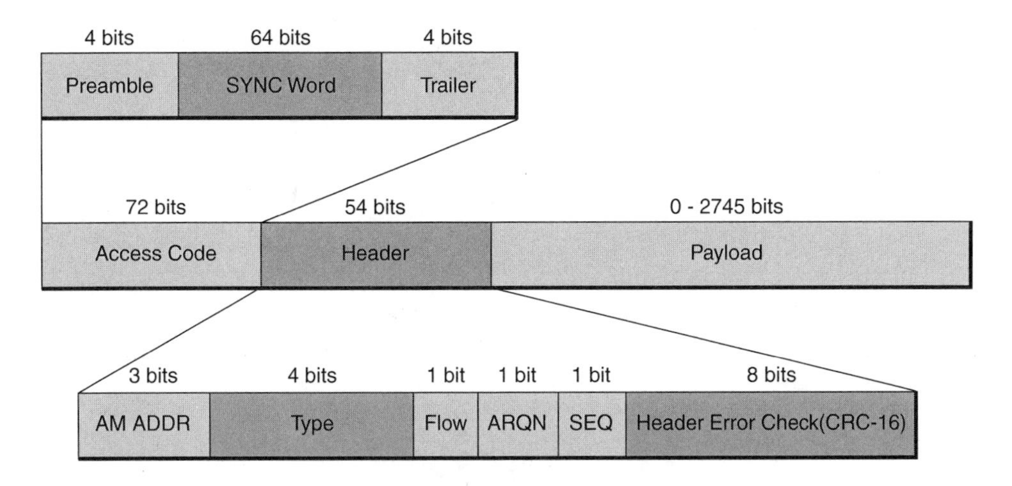

Figure 13.6. Bluetooth packet format.

and the counter value wraps once per day. The channel is divided into time slots of 625 μsecs each and the slots are numbered according to the Bluetooth clock, from 0 to 2^{27} - 1, which is cyclic. The MASTER transmits in even time slots, while SLAVEs transmit in odd time slots. Data are transmitted at a symbol rate of 1 M symbols/sec.

13.6 Bluetooth Packet Structure

As shown in Figure 13.6, the Bluetooth packet structure consists of an access code (68 or 72 bits), a 54-bit header, and a payload (from 0–2745 bits). The access code is used to detect presence and distinguish packets. The header contains control information (for addressing, link and error control, etc.), while the payload contains data. The packet type field in the header defines the traffic type carried by the current packet and the type of error correction mechanism used for the payload. Two basic types of data packets are: (a) ACL (asynchronous connectionless), and (b) SCO (synchronous connection-oriented). The flow flag is used to indicate if a receive buffer is full. Both ARQN and SEQN implement flow control via checking packet acknowledgments. The HEC field is used for CRC operation.

13.7 Bluetooth Audio

Bluetooth has the TCS (Telephony Control Protocol Specification), which defines how telephone calls can be sent across a Bluetooth link [101]. It provides call

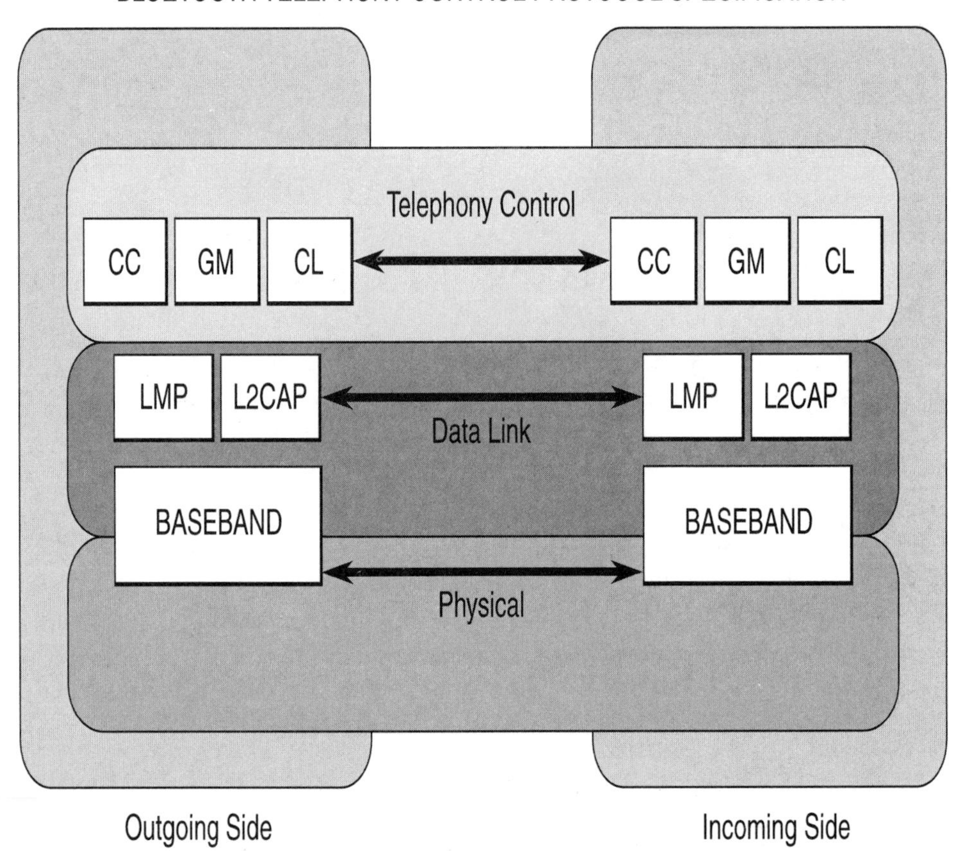

Figure 13.7. Bluetooth audio/voice via TCS.

control (CC), group management (GM), and connectionless signalling (CTCS) to support audio and voice. TCS is based on the existing ITU-T Q.931 [102] protocol. TCS CC is responsible for call setup and release, including both point-to-point and point-to-multipoint calls. GM handles groups of devices and manages inquiry, paging, and scanning. A wireless user group (WUG) is a collection of devices that supports TCS and at the same time has a working relationship. A device can join and leave a WUG. In Bluetooth, voice communications occur over the SCO channels.

Audio is an important element of Bluetooth functions and three full-duplex audio channels can occur simultaneously. To compensate against transmission corruption due to the presence of interference, the CVSD (Continuous Variable

Slope Delta Modulation) technique is used. The use of 64kbps CVSD provides toll-quality speech. Bluetooth, however, also uses log PCM coding (A-law and μ-law). Higher quality audio, such as hi-fi music, is best compressed and sent over ACL channels.

13.8 Bluetooth Addressing

Each Bluetooth device has a 48-bit IEEE MAC address, known as the device address. As shown in Figure 13.8, this address is further split into the non-significant address part (16-bit NAP), upper address part (8-bit UAP), and lower address part (24-bit LAP). The NAP is used to initialize the encrytpion algorithm. The UAP is used for HEC, CRC, and frequency hopping calculations. Finally, the LAP is used for sync word generation and calculation of frequency hopping.

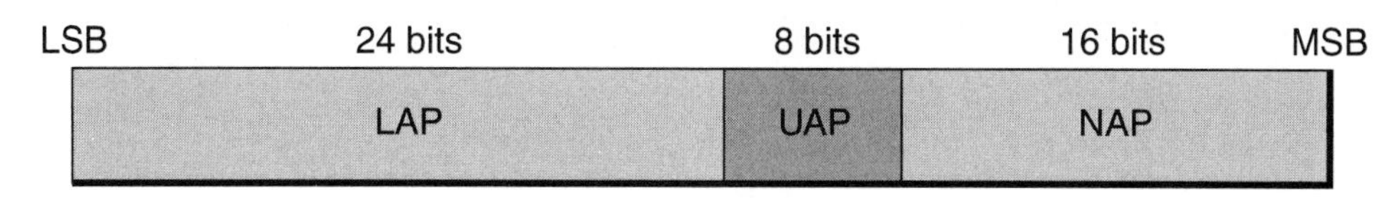

ULAP: Lower Address Part
UAP: Upper Address Part
NAP: Non-significant Address Part

Figure 13.8. Bluetooth addressing.

13.9 Bluetooth Limitations

Bluetooth does not address routing. It does not specify which unicast or multicast ad hoc routing schemes to use. In fact, most network functions are pushed ridiculously into the link layer. It does not support multi-hop multicasting. Bluetooth's MAC protocol is based on frequency hopping, which requires some form of hop synchronization. Bluetooth's MAC protocol does not address how to cope with mobility. A response to mobility is unknown in Bluetooth. In Bluetooth, the MASTER node is the bottleneck! The Bluetooth architecture also limits the number of nodes in a piconet! Bluetooth does not address interoperability issues. It is not clear how Bluetooth devices can interoperate with WAP (Wireless Application Protocol) devices. Bluetooth also ignores power-saving methods done at upper layers, above the link layer. Bluetooth cannot operate in a multiple-air interface environment.

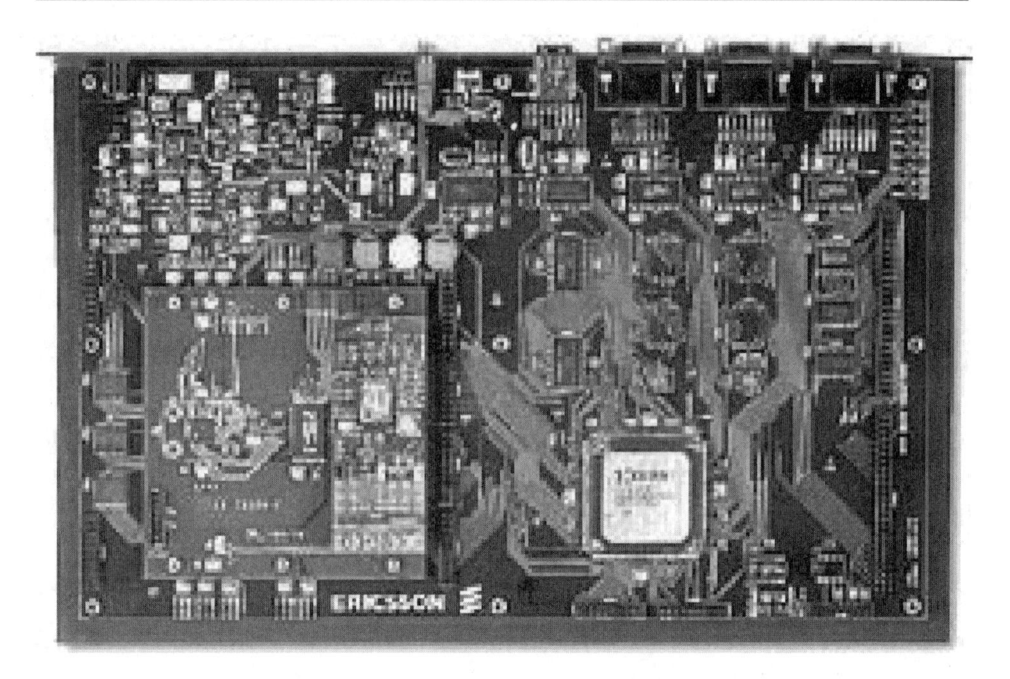

Figure 13.9. Bluetooth development kit (source Ericsson).

13.10 Bluetooth Implementation

Many companies are making Bluetooth-related products, such as Bluetooth baseband chips. Oki, for example, is implementing a CMOS Bluetooth chipset solution. The chipset consists of a CMOS RF transceiver and a baseband controller. The RF transceiver is housed in a 48-pin BGA, while the baseband chip is a 144-pin BGA. The latter has a 32-bit ARM RISC processor and a PM-CVSD (Continuous Variable Slope Delta Modulation) transcoder for voice communications. A UART is included, as well as a USB interface.

Ericsson, on the other hand, is also producing Bluetooth-related products. In particular, their Bluetooth development kits can assist other companies to build their own Bluetooth applications and products (see Figure 13.9). An upgrade kit to allow the handling of point-to-multipoint communications and scatternet support is also available. The Bluetooth application toolkit aids software development, while the Bluetooth application and training toolkit has been developed for educational bodies to provide hands-on training on Bluetooth technologies. Other tools include the Bluetooth HCI toolbox, the Bluetooth script engine, and the Bluetooth log analyzer.

13.11 Conclusions

Bluetooth is a major industrial effort to address wireless device-to-device technology. It is the next major effort after WAP. Devices, although heterogeneous in construction, will be able to "talk" to one another through the help of Bluetooth technology. The concept of implanting Bluetooth technology onto a single chip will allow many devices to be Bluetooth-enabled. Bluetooth specifications address radio, link control, link management, transport, interfaces, service discovery, and application issues. We must wait to see if indeed Bluetooth will impact the way we work, live, and interact with devices. Many applications have yet to be developed and limitations of Bluetooth may have to be addressed in the future.

Chapter 14

WIRELESS APPLICATION PROTOCOL (WAP)

While the evolution of cellular networks has resulted in many mobile services, such services are primarily for voice. Mobile phone users do have the desire to access the Internet. Hence, efforts were made to enhance the capability of mobile phones and devices. In this chapter, we will discuss another commerical effort known as WAP, the Wireless Application Protocol.

WAP is an open protocol for wireless multimedia messaging. WAP allows the design of advanced, interactive, and real-time mobile services, such as mobile banking or Internet-based news and travel services. WAP [103] products include the Ericsson R320 phone with a WAP browser.

Internet protocols are not designed to operate efficiently over mobile networks. Standard HTML web content cannot be displayed fully on the small-size screens of wireless devices, pagers, and mobile phones. WAP addresses these issues nicely.

WAP is a license-free wireless protocol standard that brings data information and telephony services to wireless devices.

The evolution of WAP is probably of interest to readers of this book. In the mid 1990s, Ericsson made advances in value-added services on mobile networks through the creation of the Intelligent Terminal Transfer Protocol (ITTP). Nokia and others, however, made advances in device user interfaces, such as the Handheld Device Markup Language (HDML) and HDTP (Handheld Device Transport Protocol). HDTP can be viewed as a new, lightweight protocol optimized for client/server transactions over wireless links. Further, Nokia again made another advancement through the introduction of the smart short message services (SMS) concept, which allows GSM users to access services present in the Internet. With such fragmentation of effort by different companies, a joint effort for a widely acceptable standard became a necessity. Hence, WAP was born.

14.1 The WAP Forum

In 1997, Ericsson, Motorola, Nokia, and Unwired Planet formed the WAP Forum. More than 90 companies in the wireless telecommunications business are members of the WAP Forum. WAP is the standard developed by the WAP Forum, a consortium formed by *device* manufacturers, *service* providers, *content* providers, and *application* developers. WAP specifies an application framework and protocols for wireless devices. WAP is a fusion of mobile networking techologies and Internet technologies.

The WAP Forum's objectives include:

- To bring Internet content and advanced data services to digital cellular phones and other wireless terminals
- To create an interoperable wireless protocol specification that will work across differing wireless network technologies
- To enable the creation of content and applications that could scale across a wide range of wireless bearer networks and device types
- To embrace and extend existing standards and technologies

The key features provided by WAP include:

- A programming model similar to the Internet
- Wireless Markup Language (WML)

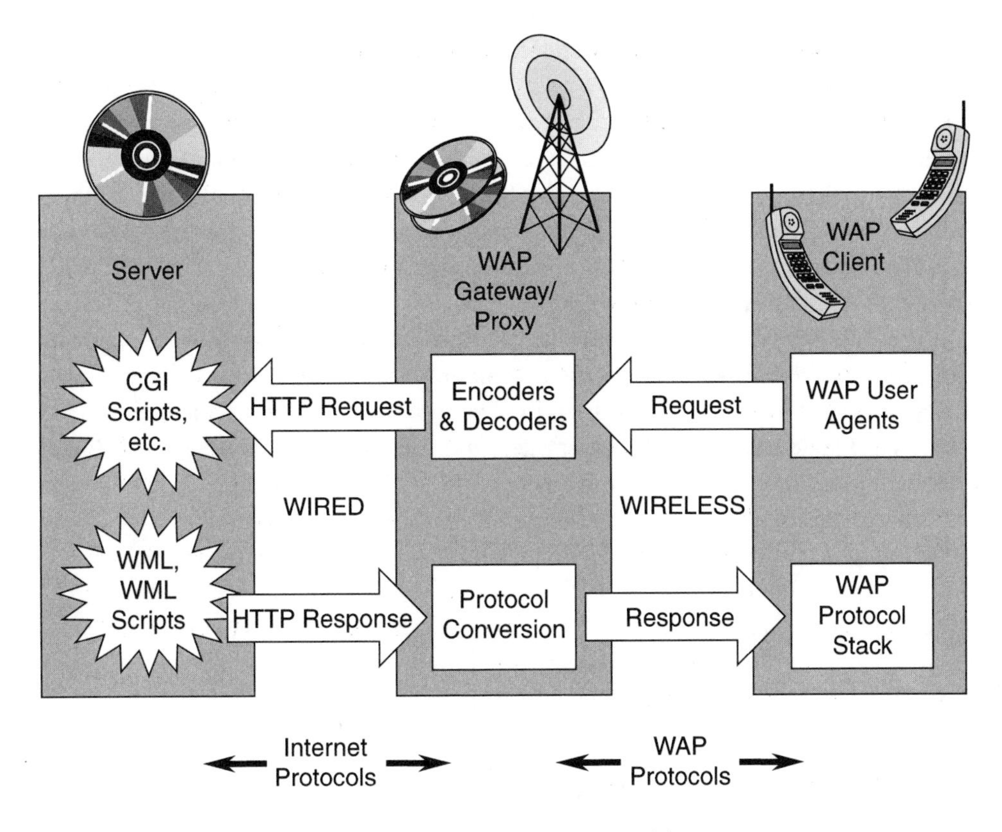

Figure 14.1. The WAP system architecture.

- WMLScript
- Wireless Telephony Application (WTA)[104]
- Optimized protocol stack

WAP attempts to look at bandwidth and power constraints, unlike existing Internet protocols.

14.2 The WAP Service Model

In the current Internet model, the client runs a copy of the Web browser, which uses the underlying Internet protocols to access useful content residing in a server in the network. Such interactions occur through using HTTP request and reply messages.

WAP is based on an Internet technology that has been optimized to address the constraints of wireless links and wireless devices. Services created by HTML do not usually fit well on small handheld wireless devices due to their display limitations. In addition, such devices do have limited storage and computing capability, and this implies that excessive or redundant information is not welcome. Hence, WML is used instead of HTML. WML pages can also be encoded in binary format to reduce the amount of data to be transmitted over the wireless interface. The WAP service model is shown in Figure 14.1.

The WAP service model reveals the presence of a WAP proxy, which is responsible for protocol conversion and data formatting. It acts as the interface between the wired and wireless worlds. These two environments have extreme differences, such as available bandwidth, bit error rates, and storage and processing capabilities. When a mobile device requests information via the WAP protocol, it is intercepted and interpreted by the WAP proxy, which then forwards the request via HTTP on behalf of the mobile device to the appropriate server in the network. When the proxy receives the information in response to its earlier request, the information is stored and converted (formatting) to a suitable form for processing and display by the mobile device. After the conversion, the proxy sends the processed information to the mobile device using the WAP protocol.

14.3 The WAP Protocol Architecture

The WAP architecture provides a scalable and extensible environment for application development on mobile communication devices. It achieves this through a layered protocol design, covering protocols at Layer 4 and above. The WAP protocol stack is independent of the underlying network, which could take the form of GSM, CDMA, CDPD, iDEN, etc. Hence, WAP is essentially an *application* stack specification; it is not network-centric. Figure 14.2 shows the WAP protocol architecture and how it differs from the Internet protocol stack. The WAP layers are elaborated below.

- **Wireless Application Environment (WAE)**
  Generally, WAE [105] enables a spectrum of applications to be supported over WAP. WAE has two main elements, namely: (a) user agents, and (b) services and formats. The former includes the WML and WTA user agents. The latter consists of WML scripts, image formats, etc. A user agent can take the form of a Web browser. The WML user agent is responsible for the interpretation of WML and WMLScript. WAP employs the same addressing model as in the Internet, that is, it uses Uniformed Resource Locators

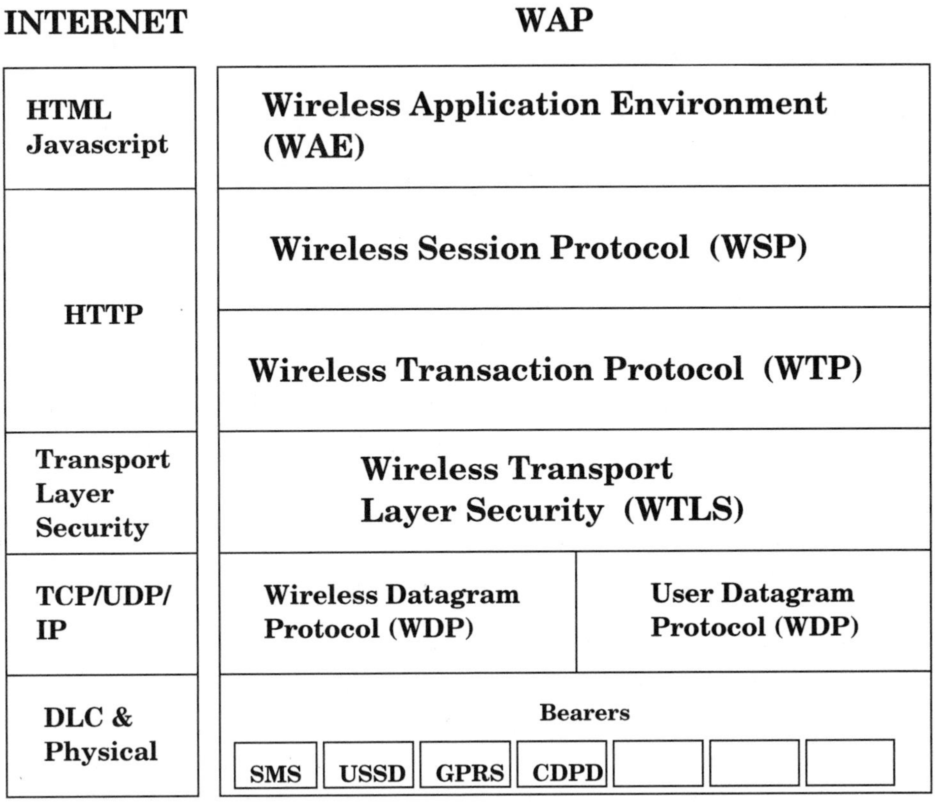

INTERNET	WAP	
HTML Javascript	Wireless Application Environment (WAE)	
HTTP	Wireless Session Protocol (WSP)	
	Wireless Transaction Protocol (WTP)	
Transport Layer Security	Wireless Transport Layer Security (WTLS)	
TCP/UDP/ IP	Wireless Datagram Protocol (WDP)	User Datagram Protocol (WDP)
DLC & Physical	Bearers: SMS, USSD, GPRS, CDPD	

Figure 14.2. WAP and Internet protocol architectures.

(URLs). A URL uniquely identifies an available resource. WAP also uses Uniform Resource Identifiers (URIs) to address resources that are not accessed via well-known protocols.

- **Wireless Session Protocol (WSP)**
  The WSP [106] provides both connection-oriented and connectionless services. It is optimized for low-bandwidth networks with relatively long latency. WSP is a binary version of HTTP version 1.1, but with the additions of: (a) session migrations, (b) header caching, etc. WAP connection mode allows the establishment of sessions between a client and the WAP gateway or proxy. It can handle session interruptions as a result of mobility and re-establish session states at a later point in time. Header caching allows better bearer utilization since in HTTP, most of the requests contain static headers

that need to be re-sent again.

- **Wireless Transaction Protocol (WTP)**
  WTP [107] is designed for transaction-style communications on wireless devices. In a *transaction*, users express their intentions and financial commitments to service providers for processing. Very often, such transactions demand reliable, fast, and secure communications. WTP is a lightweight protocol suitable for implementation in *thin* clients. WTP implements selective retransmission of lost segments.

- **Wireless Transport Layer Security (WTLS)**
  WTLS [108] is needed for WAP to ensure data integrity, privacy, authentication, and protection from denial-of-service. It is based on Transport Layer Security (TLS) 1.0, but optimized for wireless channels. It provides transport layer security between a WAP client and the WAP gateway/proxy. Digital certificates are used for authentication and nonrepudiation of server and client. Encryption is also used to enhance the degree of confidentiality.

- **Wireless Datagram Protocol (WDP)**
  WDP is the transport layer protocol in WAP. It has the same functionality provided by the Internet User Datagram Protocol (UDP). Whether WAP uses UDP or WDP, datagram delivery services are provided by port number functionality and the characteristics of different bearer services are hidden from the upper layers. WDP can be extended to provide segmentation and reassembly functions.

14.4 The WWW Programming Model

Having presented WAP architectures and protocols, let's re-examine the fundamental difference between the WWW and WAP models. The WAP programming model is very much tailored toward the web-based programming model. This is because Web-based applications are foreseen to be the main platform for both wired and wireless users.

The original WWW model comprises a Web client and server. It is based on an RPC paradigm, where clients convey their intentions and requests to servers for processing and execution. In the WWW, applications and content are presented in standard data formats, and are browsed by clients known as Web browsers. In this model, the Web browser sends requests for named data objects to a network server, which responds with the data encoded using the standard format.

The WWW standard has a *naming model* based on the URL (refer to RFC 1738 [109] and RFC 1808 [110]). WWW content has a specific type, thereby allowing Web browsers to correctly process the content based on its type (refer to RFC 2045 and RFC 2048). Currently, most Web browsers adhere to a set of standard content formats, such as HTML and the JavaScript [111] language. The WWW has a transport protocol that enables reliable data transmission over the Internet, known as the Hypertext Transport Protocol or HTTP (refer to RFC 2068 [112]).

The WWW specifications define three classes of servers:

- **Origin Server**—The server where the resource or content resides.
- **Proxy**—This *middleman*, or *broker*, is a piece of intermediary software that acts as both a server and a client. Requests sent to the proxy are either serviced by the proxy itself or passed on to the actual servers. The proxy typically resides between the clients and servers that have no direct means of communication.
- **Gateway**—This is different from the proxy since the gateway treats requests it receives as if it is the original intended server.

14.5 The WAP Programming Model

As shown earlier, the WAP programming model is similar to the WWW programming model. WAP uses *proxy-based* technologies to connect between the wireless domain and the WWW. The WAP proxy acts as a protocol gateway, which is responsible for translating requests from the WAP protocol stack (WSP, WTP, WTLS, and WDP) to the WWW protocols (HTTP and TCP/IP). It also performs encoding and decoding, which make Web access over the wireless interface efficient and compact.

WAP devices use a micro-browser that is more compact and lightweight, but is analogous to the WWW browser. The micro-browser can be viewed as a reduced version of JavaScript called WMLScript. WAP also supports a standard naming model, such as the WWW-standard URLs. WAP supports content typing and includes display markup, images, and a scripting language. WAP uses the underlying network protocols to enable communication of browser requests from wireless terminals to network Web servers. Figure 14.3 shows the Nokia 9290 communicator, which is a fully integrated mobile terminal combining phone, fax, e-mail, calendar, and imaging. It also offers Internet access via WAP and HTML-based WWW browser.

Figure 14.3. A commerical WAP-enabled phone (source Nokia).

14.6 Conclusions

In this chapter, we examined WAP, a protocol that has been standardized and deployed. WAP allows the introduction of mobile Internet services into mobile wireless devices via mobile cellular networks. WAP interfaces with different entities through the use of a gateway/proxy and a set of lightweight data presentation/formatting scripts. Such scripts allow information to be formatted in such a manner that is suitable for transmission over wireless and for presentation on a small wireless device with limited display capability. News, stock rates, shopping, and advanced calling services can all be done via WAP-enabled mobile devices.

Chapter 15

AD HOC NOMADIC MOBILE APPLICATIONS

15.1 In the Office

Mobile ad hoc devices can automatically recognize the presence of other wireless devices through sensing the presence of neighboring beacons. They can initiate synchronization of a PDA (Personal Digital Assistant) with a desktop application over a wireless interface. This allows the transfer of files, emails, and even personal schedules seamlessly from the PDA to the desktop. The user can then proceed to work on the desktop computer if she or he chooses to do so. With machines capable of "talking" to each other, the distinction of which machine/device to work on has become blurred. We are one step closer to a machine-independent working environment.

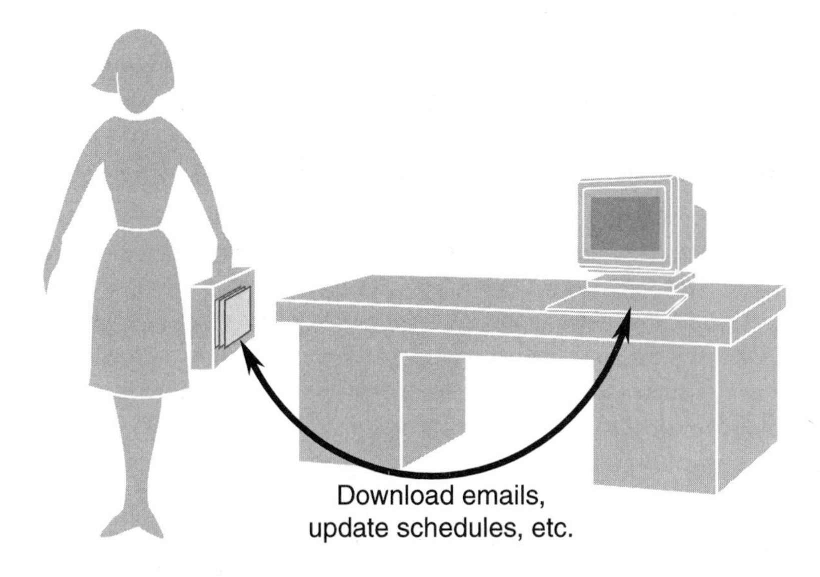

Figure 15.1. In the office.

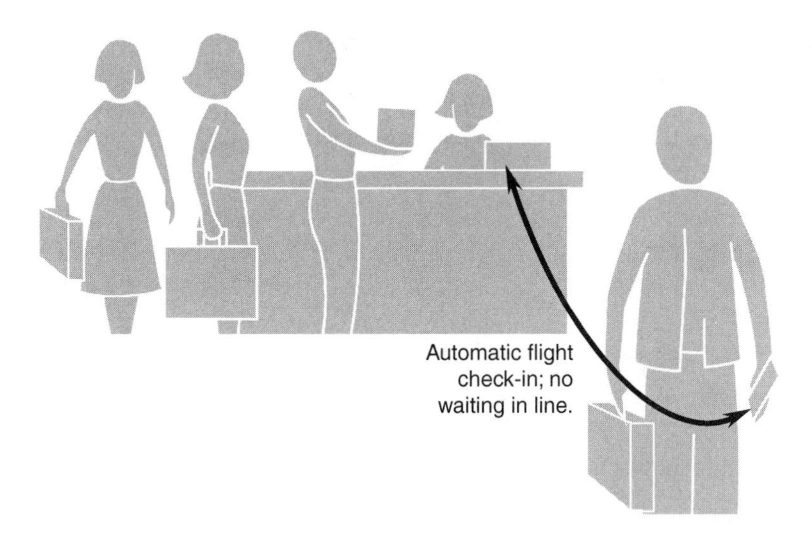

Figure 15.2. While travelling.

15.2 While Traveling

Lately, we have seen computerization in the airline industry. Most flight reservations can be done on-line via the Web, and bookings of flights can be made via credit card transactions performed on-line. However, when you arrive at the airport, you still have to wait at the gate for a boarding pass. If you have luggage to check in, you still have to wait at the main lobby check-in area. In the future, the airline reservation system should be able to respond intelligently to personal wireless ad hoc devices worn or carried by passengers. When you step into the airline terminal, your wireless device will automatically communicate with the airline system via an ad hoc wireless access point, which is also connected to the airline wired network. Since you have a reservation with the airline, a confirmed seat number can then be assigned and conveyed to you. In such situations, you will no longer need to wait for a boarding pass.

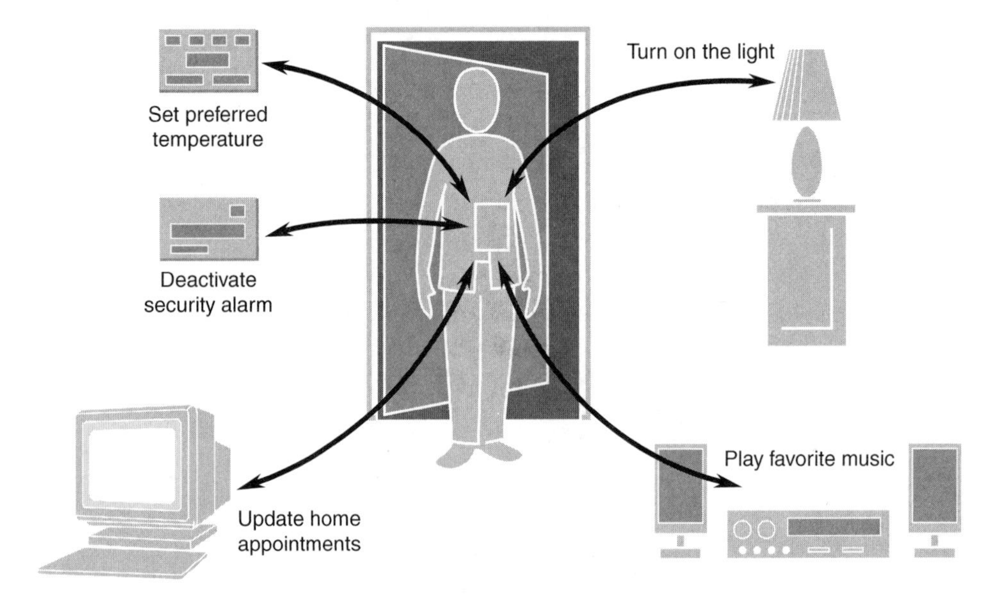

Figure 15.3. Arriving home.

15.3 Arriving Home

Home devices could also be "ad hoc wirelessly enabled." A user's ad hoc device could communicate with home wireless devices to unlock doors, activate lights and home audio and video entertainment units, adjust heating/cooling settings to preset preferences, and also deactivate security alarms. Ad hoc devices worn by different members of the family could be programmed to have different levels of control over household electronic devices and the system. The head of household, for example, could override the activation of the television and radio systems set by his/her children.

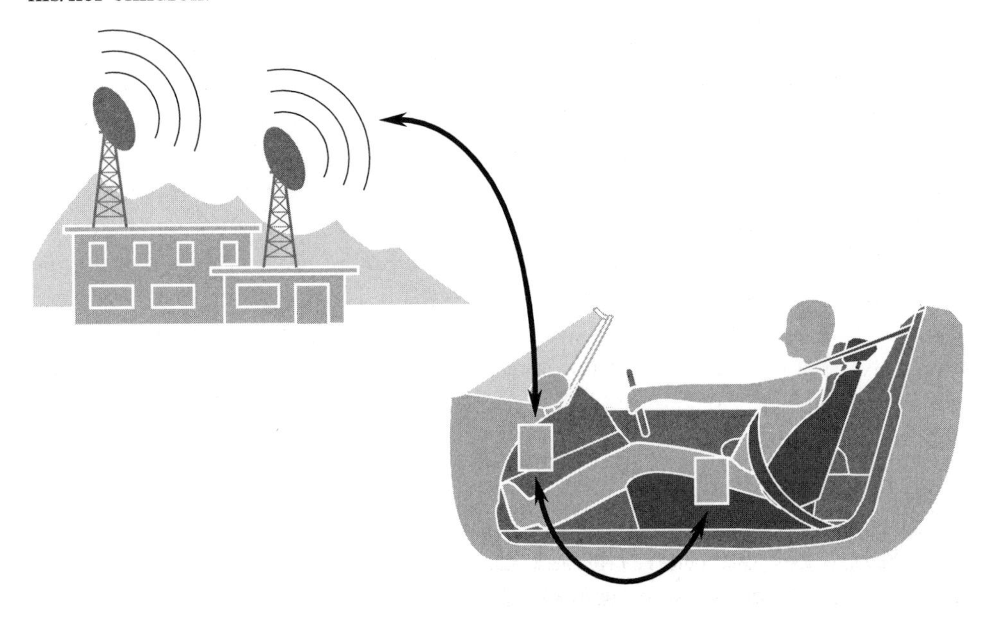

Figure 15.4. Ad hoc mobile communications and service transactions while in the car.

15.4 In the Car

A user driving a car in a foreign state can receive email messages via the wide area wireless data system installed in the car. Most wide area wireless data systems have base stations distributed throughout many areas to provide radio coverage and connectivity back to the Internet. An ad hoc mobile device can, therefore, interoperate with a wireless last-hop device to enable delivered data communications. Accessing email and Web-based information can, therefore, be made even while on the move. Another possible application is the access of roadside shopping/food/store information while driving down the road. Imagine roads of the future with have lamp posts enhanced with ad hoc communications ability. Each store will advertise sale and service information to be broadcast out to the nearest lamp post. Drivers can enter keywords into their ad hoc mobile devices to search for the nearest post office or supermarket. Such requests will be relayed in an ad hoc fashion via several lamp posts. Also, one could even make dining, hotel, or movie reservations prior to arrival!

15.5 Shopping Malls

Wireless RF tags are small, easy to manufacture, and relatively cheap. As shown in Figure 15.5, they can be used to provide pricing, sale, and stock information. RF tags are programmable and can be updated via a wireless interface. Very often, RF tags use an LCD-type display due to their thin and energy-efficient construction. Certain tags can act as master tags to relay information to neighboring tags. This approach has, in fact, been realized by NCR Corporation. Such tags are very simple to fabricate and they are cheap and long-lasting. They have a small form factor and a very low power consumption. Customers carrying handheld wireless devices can locate and obtain price information of specific products they are interested in, even if such products are a distant away. This will greatly enhance our shopping lifestyle and increase the productivity of inventory staff.

15.6 The Modern Battlefield

The modern digital wireless battlefield demands *robust* and *reliable* communications in many forms. Most communication devices are installed in mobile vehicles, tanks, trucks, etc. Also, soldiers can carry telecomm devices that could talk to a wireless base station or directly to other telecomm devices if they are within radio range. However, these forms of communication are considered primitive. At times when the wireless base station is destroyed by an enemy, a soldier will be prohibited from communicating with other soldiers if the called party is not within radio range. This is the scenario where ad hoc wireless communications come into play. Ad hoc networks are commonly known as self-organizing networks since they are robust when nodes disappear due to destruction or mobility. Through multi-hop communications, soldiers can communicate to remote soldiers via *data hopping* or *data forwarding* from one radio device to another, as shown in Figure 15.6.

Another scenario is the management of minute wireless sensors in a battlefield. Very often, prior to launching an attack, critical surveillance information of the enemy site is needed. Wired sensors are not practical since they can be easily detected by enemies and long wires for connectivity are just not reliable and scalable. Hence, sensors are becoming wirelessly enabled, allowing them to be scattered in targeted zones. Such sensors are able to quickly gather information about the location and

Figure 15.5. Wireless RF price tags (source NCR).

environment and periodically relay this information back to the command, control, and communication center (CCC). This information is then used to influence attack strategy and determine attack decisions.

15.7 Car-to-Car Mobile Communications

Car mobile communicators enabled with ad hoc wireless communications technology will allow the formation and de-formation of ad hoc wireless networks with other cars. As shown in Figure 15.7, information can be relayed among cars. *Group mobility* is often observed here since cars on freeways are constrainted to a particular direction. If a communication session needs to go beyond that of the ad hoc wireless network, data can still be relayed back to the Internet via the nearest lamp post, which acts as a wireless base station. When a car accident had occurred, many cars would be waiting in line for the traffic to be cleared. Hence, an alert message can be sent by a car to other cars via ad hoc mobile communications such

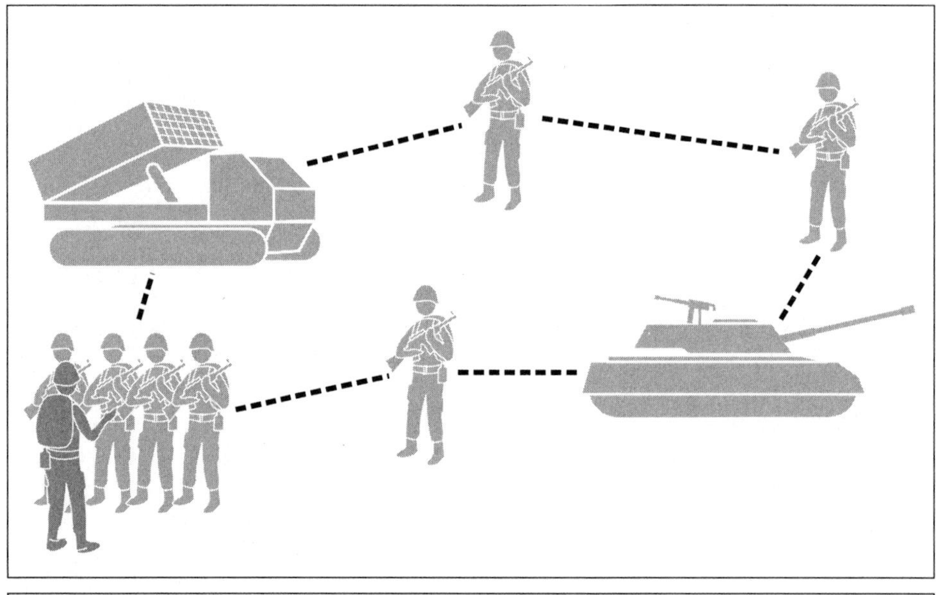

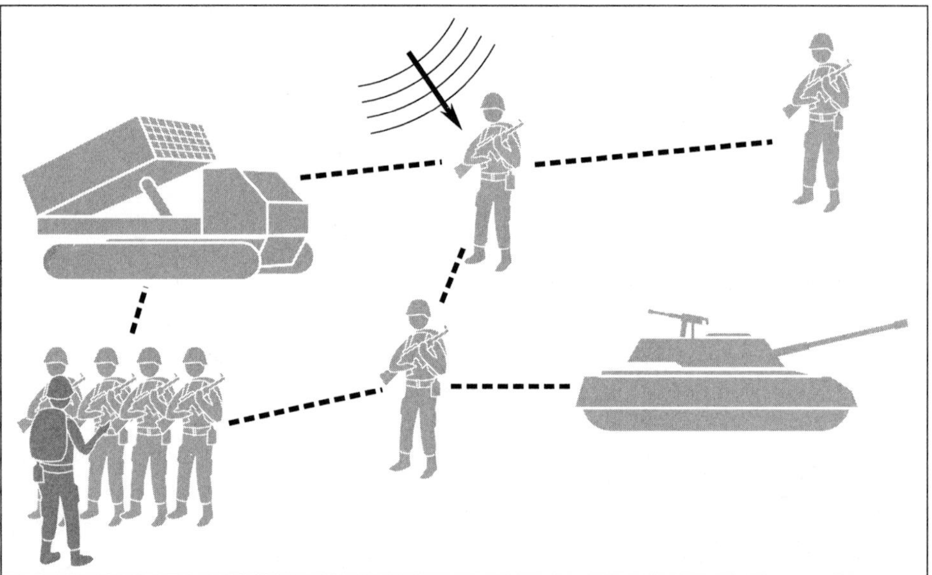

Figure 15.6. The modern battlefield equipped with self-organizing communication systems.

that drivers for on-coming cars may choose to go on an alternate route to avoid the congestion.

Figure 15.7. Car-to-car ad hoc mobile communications and networking.

15.8 Mobile Collaborative Applications

Computer-Supported Collaborative Work (CSCW) has so far been happenning over wired platforms with workstations or computers equipped with multimedia devices to support audio and video. CSCW is particularly important in improving work flow and enhancing the productivity of product design work. It also enhances problem-solving through group effort. As shown in Figure 15.8, mobile users can detect the presence of their work partners and establish ad hoc mobile multicast sessions to enable multi-party discussions, collaborative design work, etc. Synchronization of updates are done transparently for the user. Given the presence of heterogeneity in the type of mobile devices users may carry, the mobile CSCW application must be able to adapt and provide an appropriate GUI (graphical user interface) to users. Further research and investigation in this area is, therefore, necessary.

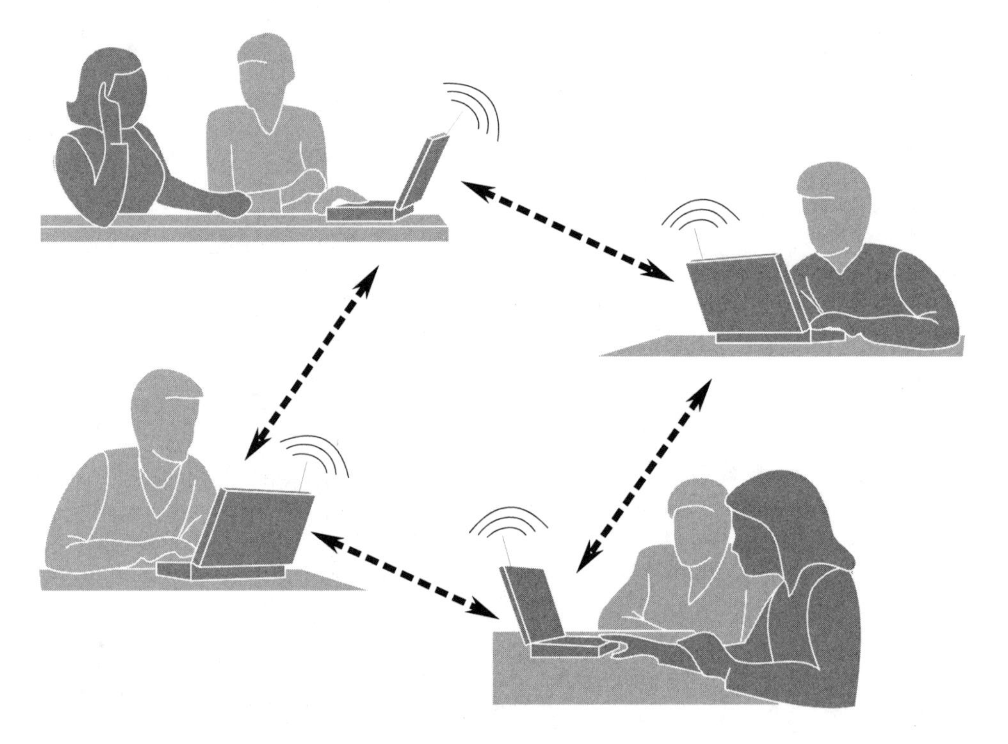

Figure 15.8. Ad hoc mobile collaborative applications.

15.9 Location/Context Based Mobile Services

Ad hoc wireless technologies can also be deployed to support intelligent, location-based service discovery and transactions anytime and anywhere. Mobile users (including mobile phone users) can be equipped with the right information at the right place and the right time. For example, if a user is in a shopping mall, information like the lowest price associated with a product can be made known, enabling him or her to make a decision. Being at the shopping mall allows a user to *see* and *feel* the real product that he/she plans to buy, but he/she can make the buying decision on-line.

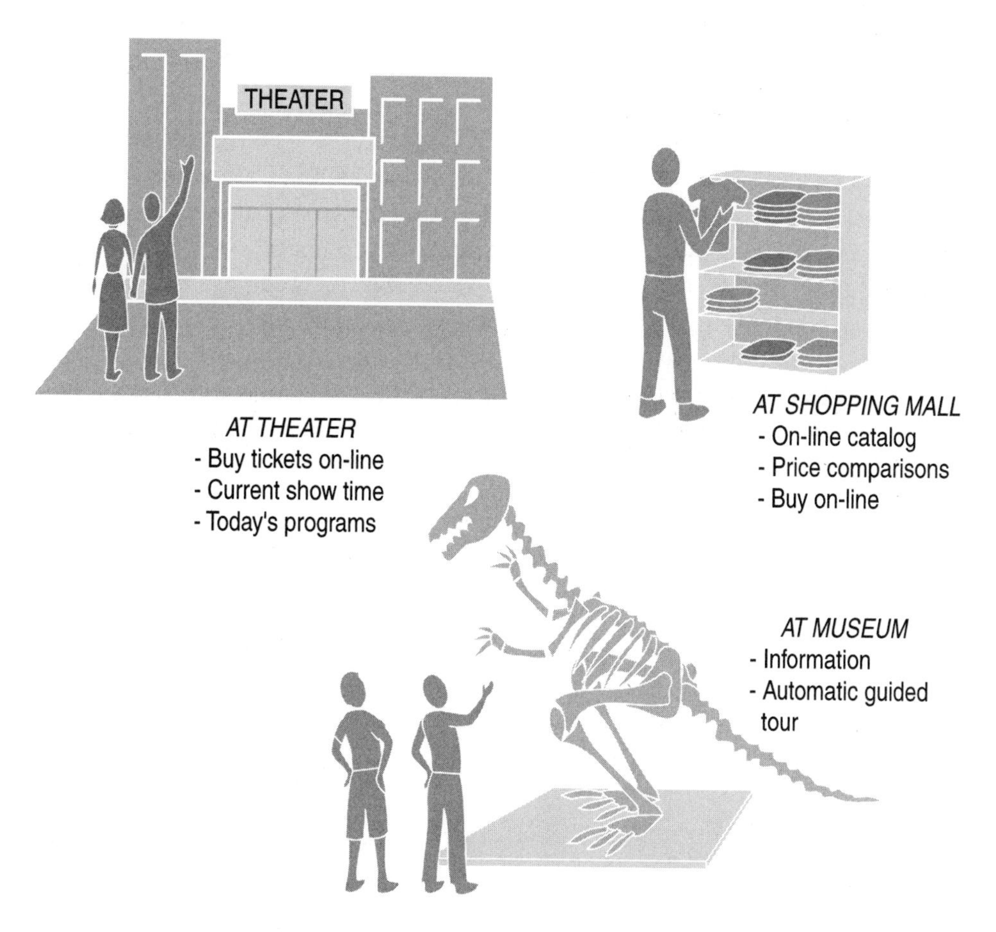

Figure 15.9. Location-based mobile services.

Another scenario would be when the mobile user is at a museum, looking at some remains of dinosaurs. His or her mobile device would be detected by the museum's network system and its location would be tracked down to the section on dinosaurs. Hence, the museum network could start sending information from the database about various dinosaur species to the mobile user. This is a way where only the desired and appropriate information is delivered to the user.

Lastly, as shown in Figure 15.9, mobile users can access theater databases via ad hoc devices while shopping in the mall. They can make movie reservations and even obtain electronic tickets. As they enter the theater, their electronic tickets could be confirmed by the usher, who also has an ad hoc wireless device. Note that messages sent by mobile users could be relayed multiple times via several ad hoc mobile nodes before they reach the theatre transaction/database system.

15.10 Conclusions

Many people have been wondering about the potential applications for mobile ad hoc networks. Many people are fascinated by the term *mobile computing*, but few really experience its impact. Many people think that computing on a laptop is mobile computing, but this is a very narrow view. Also, computing via a wearable computer is not mobile computing. Ad hoc mobile applications combine the characteristics of *neighbor-aware*, *location-aware*, *connectivity-aware*, and *context-aware* applications in varying degrees and forms. The potential applications mentioned in this chapter are by no means exhaustive and it is left to the readers to further imagine application scenarios.

Chapter 16

CONCLUSIONS AND THE FUTURE

16.1 Pervasive Computing

Pervasive computing [113] is a popular term used by many technologists and executives. It refers to the goal of achieving a transparent computing infrastructure that crosses every aspect of an individual's daily life to an extent with few boundaries. The boundaries here refer to both the physical world and connectivity.

To be truly *pervasive* would mean that the *degree of penetration* is fairly high. Pervasiveness can also be measured by how useful the applications are and how interactive the environment can be. Thus, a pervasive computing environment would be virtually everywhere.

Our physical world is pervasive because it surrounds us all the time. Our physical world is a collection of nature and things. Things can be defined as solid objects with or without any computing logic. If I were to go to a park and feel a leaf on a tree, I do not perform any logical computing, although I am interacting with the physical world.

In the author's point of view, an interaction with an object in the physical world could mean more than just human touch. If we can embed "digital life" into these objects, then they too can have a unique presence in our computing, network, and digital worlds.

Some researchers in the past viewed pervasive computing as taking place in the form of *smart spaces*. They advocated that such spaces could only be realized through hardware and software integration. In particular, energy was focused on smart meeting rooms and human-computer interfaces (HCIs). Such rooms were well-equipped with the necessary audio/video/computing facilities so that any meeting could be made intelligent (for example, notes were automatically taken and cameras detected who was talking and then sent a closeup of the speaker). The information gathered as a result of the interactions was then distributed via the Internet or some form of communications infrastructure, be it infrared, wired, or wireless. However, if pervasive computing is confined to only smart work spaces, then its pervasiveness is limited.

Giving objects "digital life" is equivalent to what Bluetooth is trying to do. Implant an object with a Bluetooth chip and it becomes an intelligent networked device. In the past, applications occupied the computing space. Useful applications justified the existence of sophisticated computers. However, in the author's view, the digital world will soon experience another form of revolution. Currently, computers rule the world and computers are getting more and more powerful and complex (microprocessors are composed of very highly integrated circuits). However, in the pervasive computing world soon to evolve, devices or tagged objects are going to "rule" the world.

Humans will soon be able to break the physical barrier and venture out more into the physical world to perform their communications and computing activities. We will see various aspects of *human-centered*, *device-centered*, *application-centered*, and *environmental-centered* development.

16.2 Motorola *PIANO* Project

In January 1999, Motorola announced its intention to merge Sun's Jini technology with a newly developed Motorola networking technology known as *PIANO*. In *PIANO*, Motorola envisioned a future where devices could detect the presence of other devices, initiate information exchange, and hence form temporary wireless

Figure 16.1. The size of a dust mote. (Picture reproduced with permission from Kris Pister, UC Berkeley.)

networks, much like intranets, on-the-fly. Jini technology expands the power of Java technology, enabling spontaneous networking over a wide variety of hardware and software.

Sun and Motorola envisioned that small Jini-equipped pods or gateways would be present, allowing the services of a Jini-enabled world to become accessible. The *PIANO* software technology can work with virtually any radio or infrared technology. This project subsequently diverts its attention to Bluetooth.

16.3 UC Berkeley Sensor Networks: Smart Dust

Sensors have been taken to great heights through an immense reduction in form factor and also an increase in the number of sensors used per square meter. Smart dust [114][115] is a UC Berkeley project involving wireless sensors that are not only minute (millimeter scale, as shown in Figure 16.1), but capable of communicating at a minimum data rate of 1 kbps. Each dust mote consists of a power supply, sensors, and communications circuitry (as shown in Figure 16.2). These motes can be powered by solar cells or thick film batteries.

In the military era, these dust motes are easily concealed from enemies and can contain acoustic, vibration, and magnetic field sensors. Chemical and biological sensors are not excluded. In the commercial era, sensors are already being used in automobiles to detect impact as a result of collision and hence activating air bags and cutting off the engine to save lives.

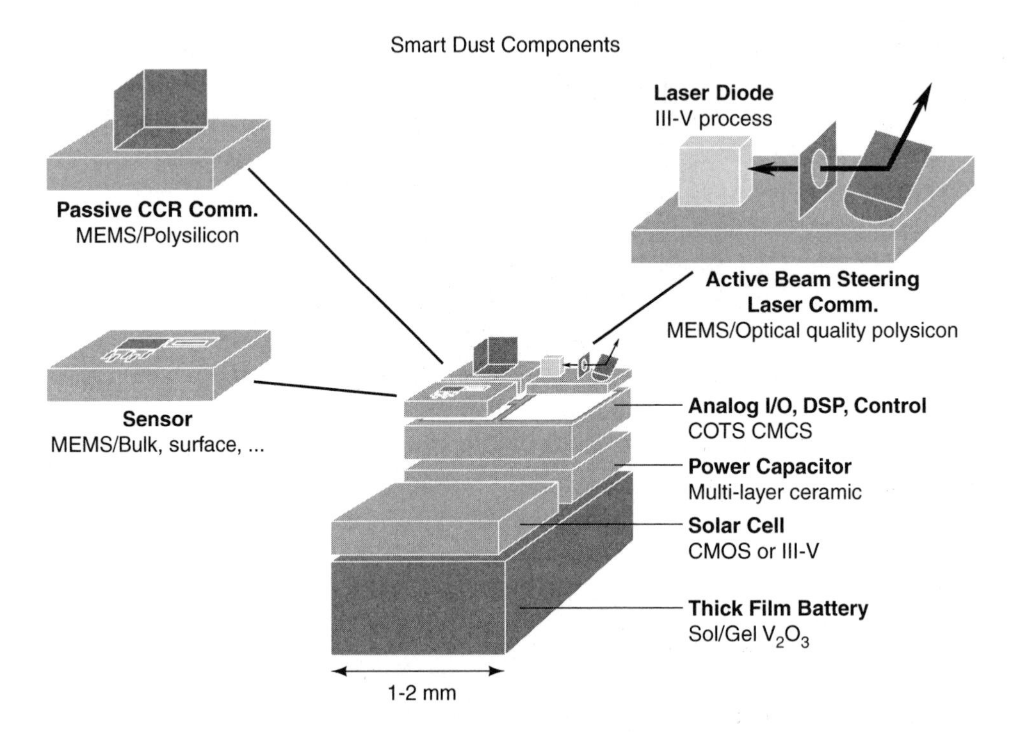

Figure 16.2. The internals of a dust mote. (Picture reproduced with permission from Kris Pister, UC Berkeley.)

Smart dust has both power and size constraints. The use of optical links with active and passive transmitters and RF links are suggested. A dust mote will be acting as the centralized node and will be communicating with up to 1000 dust motes. The uplink (dust motes to the central transceiver) is expected to have an aggregate throughput of 1 Mbps. Multiplexing over the uplinks occurs using the SDMA (Space Division Multiple Access) technique. The downlink can occur via broadcast.

The average power consumption for each dust mote is not expected to exceed 1 μWatts. Using active laser diode-based transmitters could result in high power consumption. Hence, passive optical transmitters consisting of corner-cube retroreflectors (CCRs) are used instead.

In sensor networks, the mobility of sensors is not viewed to be a major issue since once these sensors are deployed, they seldom move. The error control technique used is based on a plain CRC since retransmissions are not considered power-efficient. The channel access protocol is also relatively simple. Motes wait for the signal from the central transceiver to transmit. The transmissions can occur simultaneously via SDMA and the implanted imaging receiver is capable of detecting the signals with negligible mutual interference.

Other issues remaining for smart dust include means of distributing smart dust at destinations and also inhalation of smart dust by humans. Due to the minute size of smart dust, reusability does not seem to be attractive.

16.4 EPFL Terminodes/Large-Scale Networks

The author was a visiting professor at EPFL in Lausanne, Switzerland. Addressing routing in a large scale ad hoc wireless network is the main research focus at the communication systems group at EPFL. The term *terminode* [116] is used to describe ad hoc mobile hosts which act as both a network node and end terminal. Scalability, mobility, and simplicity in routing are key design factors for their routing protocols. Each terminode has a permanent end-system unique identifier (EID), and a temporary location dependent address (LDA).

Mobility management is present in EPFL's ad hoc network architecture. Location management allows the tracking of the locations of communicating nodes. This enables a terminode to also predict the location of corresponding terminodes. With distributed LDA management, a terminode can obtain another terminode's LDA. To communicate, a source terminode sends packets that are enclosed with the destination node's LDA and EUI.

Terminode routing [117] is not achieved via a single routing protocol. In actual fact, several routing protocols are used, depending on situations and circumstances. They are: (a) terminode local routing (TLR), and (b) terminode remote routing (TRR). The former is used when the destination node is within the vicinity of the source node and it does not exploit location information for routing. The latter approach, however, is to enable the transmission of data to remote destinations with the aid of geographical information.

TLR relies on table-driven routing where terminodes periodically exchange EUI and LDA information. The locality is defined by a terminode's local radius. Hence, when a terminode receives a packet destined for another terminode which is within its local radius, it will know how to route the packet towards the destination.

TRR is supported through several procedures, namely: (a) anchored geodesic packet forwarding, (b) friend-assisted path discovery, (c) path maintenance, and (d) multipath routing. A packet may be routed via TLR, TRR or both. The mechanisms are complex. A route vector has to be first derived by the source using a path discovery method. This vector is a chain of *anchored nodes*. Packets are forwarded from one anchor to another via geodesic packet forwarding. If anchors are selected correctly, then packets will have a high chance of arriving at the destination node.

It remains yet to be seen the performance of routing over large scale wide area ad hoc mobile networks. The presence of delays and mobility can affect communication throughput and the lack of uniform access to GPS service may render routing complex.

16.5 802.15 PANs and 802.16 Wireless MANs

The IEEE 802.15 working group has been established to address issues related to personal area networking (PAN), also known as short distance wireless networks. It includes the wireless networking of portable and mobile computing devices such as PCs, Personal Digital Assistants (PDAs), peripherals, cell phones, pagers, and consumer electronics. The IEEE 802.15 Working Group is part of the 802 Local and Metropolitan Area Network Standards Committee of the IEEE Computer Society. The IEEE 802.15 Working Group provides a clear distinction between WLANs and PANs. There exists significant synergy between Bluetooth SIG and the IEEE 802.15. The latter body ties the relationship of PAN protocol stacks to the IEEE 802 Layer 1 and 2 protocol models.

On the other hand, the mission of the IEEE 802.16 working group on broadband wireless standards is to develop standards and recommended practices to support the development and deployment of fixed broadband wireless access systems. There are several task groups within the 802.16 working group. Group 1 is focused on the air interface at 10-66 GHz. Group 2 is addressing the issue of coexistence of broadband wireless access systems. Group 3 is examining the air interface for licensed frequencies from 2-11 GHz. And, Group 4 is concerned with license-exempt bands, particularly at 5-6 GHz. For example, Wireless High-Speed Unlicensed Metroplitan Area Networks (WirelessHUMANs) are part of this group.

An unlicensed spectrum band has the advantage of availability, but it is still subject to power limit constraints. Unlicensed bands include the ISM (Industry, Science, and Medicine), UPCS (Unlicensed PCS), UNII (Unlicensed National Information Infrastructure), and millimeter wave. WirelessHUMAN will include metropolitian area networks (MANs) capable of supporting data, voice, and video traffic.

16.6 Ad Hoc Everywhere?

Our life in the future should be very carefree with little to no hassle. There will be less *searching* and more *fast* and *accurate access* to information when needed. Time and location boundaries will eventually be eliminated, resulting in a true information age style of civilization.

Information will be available when it is needed. Location will no longer be able to disconnect and influence our frequency of communication. Future devices will become more and more intelligent in the sense that they will start to talk among themselves to serve us better, fufilling our communication and even daily needs. They will improve our work efficiency and enhance our living comfort and luxury.

The communication infrastructure of the future will be heterogeneous and ad hoc wireless networks will be one such platform. Our presence will be felt, our desires will be met, and our frustrations will be conquered by an intelligent, robust, and user-sensitive network.

GLOSSARY OF TERMS

ABR	Associativity Based Routing
ANSA	Advanced Networked System Architecture
AMPS	Advanced Mobile Phone System
ATM	Asynchronous Transfer Mode
B-ISDN	Broadband Integrated Service Digital Networks
BGA	Ball Grid Array
CAMP	Core Assisted Multicast Protocol
CBT	Core Based Trees
CCITT	Consultative Committee of the International Telegraph & Telephone
CDMA	Code Division Multiple Access
CDPD	Cellular Digital Packet Data
CORBA	Common Object Request Broker Architecture
CSCW	Computer-Supported Cooperative Work
CSMA	Carrier Sense Multiple Access
DCE	Distributed Computing Environment

DCS Digital Cordless Technology 1800 MHz

DECT Digital European Cordless Telephony or Digitally Enhanced Cordless Telephony

DVMRP Distance Vector Multicast Routing Protocol

EDGE Enhanced Data Rates for GSM Evolution

EDO Extended Data Out Memory

ETSI European Telecommunications Standard Institute

FCC Federal Communications Commission

FDMA Frequency Division Multiple Access

FEC Forward Error Correction

FGMP Forwarding Group Multicast Protocol

FPLMTS Future Public Land Mobile Telecommunication Systems

GEO Geosynchronous (or geostationary) Earth Orbit

GFSK Gaussian Frequency Shift Keying

GPRS General Packet Radio Service

GPS Global Positioning System

GSA Global System For Mobile Communication

HCI Human Computer Interface

HIPERLAN High Performance Radio LANs

HTML HyberText Markup Language

HTTP HyberText Transfer Protocol

IETF Internet Engineering Task Force

INTELSAT International Telecommunications Satellite Consortium

IRDA InfraRed Data Association

ISM Industrial Scientific and Medical

ISP Internet Service Provider

ITU International Telecommunications Union

LEO Low Earth Orbit

NBBS IBM's Networking Broadband Services Architecture

MBONE Multicast Backbone

MBS Mobile Broadband System

MOA Mobitex Operators Association

MOSPF Multicast Open Shortest Path First

MSC Mobile Switching Center

MU Mobile Unit

OEMs Original Equipment Manufacturers

OSF Open Software Foundation

PACS Personal Access Communications System

PAN Personal Area Network

PBX Private Branch Exchange

PCCA Portable Computer and Communications Association

PCMCIA Personal Computer Memory Card International Association

PCN Personal Communications Network (UK)

PCS Personal Communication Systems (USA)

PDN Piblic Data Networks

PDA Personal Digital Assistant

PGA Pin Grid Array

PIM-DM Protocol Independent Multicast Dense Mode

PIM-SM Protocol Independent Multicast Sparse Mode

POTS Plain Old Telephone Service

RACE Research & Technology Development in Advanced Communications Technologies in Europe

RDRN Rapidly Deployable Radio Networks

RRC Route Reconstruction Procedures used in ABR

SMS Short Messaging Service

SUPERNET Shared Unlicensed Personal Radio Network

TACS Total Access Communications System

TAXI Transparent Asynchronous XCVR Interface

TCP Transmission Control Protocol

TETRA Trans European Trunked Radio

TGMS Third Generation Mobile Systems

TIA Telecommunications Industry Association

TACS Telecommunications Analogue Cellular System

TDMA Time Division Multiple Access

UART Universal Asynchronous Receiver Transmitter

USB Universal Serial Bus

UPT Universal Personal Telecommunication

UMTS Universal Mobile Telecommunication System

VSATs Very Small Aperture Terminals

WML Wireless Markup Language

WTA Wireless Telephony Application

REFERENCES

[1] W. R. Young, "Advanced Mobile Phone Service: Introduction, Background, and Objectives," in *Bell Systems Technical Journal*, vol. 58, January 1979.

[2] Theodore S. Rappaport, *Wireless Communications*. Prentice Hall, 1996.

[3] Vijay K. Garg and Joseph E. Wilkes, *Principles & Applications of GSM*. Prentice Hall, 1999.

[4] DECT Forum, "DECT Forum Web Page," 2000.

[5] U. Black, *Mobile and Wireless Networks*. Prentice Hall, 1996.

[6] F. J. Ricci, *Personal Communications Systems Applications*. Prentice Hall, 1997.

[7] V. Li, "Personal Communication Systems (PCS)," in *Proceedings of the IEEE*, vol. 83, pp. 1210–1243, September 1995.

[8] Jayant Kadambi, Ian Crayford, and Mohan Kalkunte, *Gigabit Ethernet - Migrating to High-Bandwidth LANs*. Prentice Hall, 1998.

[9] D. Goodman, *Wireless Personal Communications Systems*. Addison-Wesley Wireless Communications Series, 1997.

[10] A. J. Viterbi, *CDMA: Principles of Spread Spectrum Communication*. Addison-Wesley Wireless Communications Series, 1995.

[11] 3GPP, "Third Generation Partnership Project," 2001.

[12] J. M. McQuillan and D. C. Walden, "The ARPA Network Design Decisions," in *Computer Networks*, vol. 1, pp. 243–289, August 1977.

[13] Defense Advanced Research Project Agency, "Darpa home page."

[14] R. Want and A. Hopper, "The Active Badge Location System," in *ACM Transactions on Information Systems*, vol. 10, pp. 91–102, January 1992.

[15] Deborah Estrin, Ramesh Govindan, and John Heidemann, "Scalable Coordination in Sensor Networks." USC/ISI Technical Report, 1999.

[16] Ya Xu, John Heidemann, and Deborah Estrin, "Adaptive Energy-Conserving Routing for Multihop Ad Hoc Networks." USC/ISI Technical Report, 2000.

[17] Suresh Singh and C. S. Raghavendra, "PAMAS - Power Aware Multi-Access Protocol With Signalling for Ad hoc Networks," in *ACM Computer Communications Review*, July 1998.

[18] F.A. Tobagi and L. Kleinrock, "Packet Switching in Radio Channels: Part II - the Hidden Terminal Problem in Carrier Sense Multiple Access Modes and the Busy-Tone Solution," in *IEEE Transactions on Communications*, vol. 23, pp. 1417–1433, 1975.

[19] Z.J. Haas and Jing Deng, "Dual Busy Tone Multiple Access (DBTMA) - Performance Results," in *Proceedings of IEEE Wireless Communications and Networking Conference (WCNC)*, September 1999.

[20] Z.J. Haas and Jing Deng, "Dual Busy Tone Multiple Access (DBTMA) - Performance Evaluation," in *Proceedings of IEEE Vehicular Technology Conference*, May 1999.

[21] C.-K. Toh, Vasos Vassiliou, Guillermo Guichal, and C.-H. Shih, "MARCH: A Medium Access Control Protocol for Multihop Wireless Ad Hoc Networks," in *Proceedings of IEEE Military Communications*, October 2000.

[22] C. P. . P. Bhagwat, "Highly Dynamic Destination-Sequenced Distance Vector Routing (DSDV) for Mobile Computers," in *Proceedings of ACM SIGCOMM' 94*, pp. 234–244, September 1994.

[23] S. Murthy and J.J. Garcia-Luna-Aceves, "'A Routing Protocol for Packet Radio Networks'," in *Proceedings of ACM First International Conference on Mobile Computing & Networking (MOBICOM'95)*, November 1995.

[24] Jyoti Raju and J.J. Garcia-Luna-Aceves, "'A Comparison of On-demand and Table-driven Routing for Ad Hoc Wireless Networks'," in *Proceedings of IEEE ICC*, June 2000.

[25] C.-C. Chiang, H.-K. Wu, W. Liu, and M. Gerla, "Routing in Clustered Multihop Mobile Wireless Networks with Fading Channel," in *Proceedings of IEEE Singapore International Conference on Networks*, 1997.

[26] C. Perkins and E. Royer, "Ad-Hoc On-Demand Distance Vector Routing," in *Proceedings of 2nd IEEE Workshop on Mobile Computing Systems and Applications*, Feburary 1999.

[27] David B. Johnson and David A. Maltz, *Mobile Computing*. Kluwer Academic Publishers, 1996.

[28] D-K. Kim, C-K. Toh, and Y-H Choi, "RODA: A New Dynamic Routing Protocol using Dual Paths to support Asymmetric Links in Mobile Ad Hoc Networks," in *Proceedings of IEEE International Conference on Computer Communications and Networks (IC3N)*, October 2000.

[29] V. Park and M. Scott Corson, "A Highly Adaptive Dsitributed Routing Algorithm for Mobile Wireless Networks," in *Proceedings of IEEE INFOCOM'97*, March 1996.

[30] M. Scott Corson and Anthony Ephremides, "A Distributed Routing Algorithm for Mobile Radio Networks," *ACM Wireless Networks Journal*, vol. 1, pp. 61–81, January 1995.

[31] R. Dube, et al, "Signal Stability based Adaptive Routing (SSA) for Ad Hoc Mobile Networks," in *IEEE Personal Communication Magazine*, February 1997.

[32] S. Singh, M. Woo, and C. S. Raghavendra, "Power-Aware Routing in Mobile Ad Hoc Networks," in *Proceedings of ACM/IEEE MobiCom'98 Conference*, October 1998.

[33] Z. Haas and M. Pearlman, "The Zone Routing Protocol (ZRP) for Ad Hoc Networks," in *IETF MANET Draft*, June 1999.

[34] J. Moy, "Open Shortest Path First (OSPF)," in *IETF RFC 1247*, July 1991.

[35] C. Hedrick, "Routing Information Protocol (RIP)," in *IETF RFC 1058*, June 1988.

[36] Z. Haas and M. Pearlman, "The Performance of Query Control Schemes for the Zone Routing Protocol," in *Proceedings of ACM SIGCOMM'98*, September 1998.

[37] J.J. Garcia-Luna-Aceves, M. Spohn, and D. Beyer, "Source Tree Adaptive Routing (STAR) Protocol," in *IETF Internet Draft*, October 1999.

[38] J.J. Garcia-Luna-Aceves and M. Spohn, "Source Tree Adaptive Routing in Wireless Networks," in *Proceedings of IEEE ICNP'99*, September 1999.

[39] University of Santa Cruz and DARPA/ITO, "'SPARROW - Secure Protocols for Adaptive, Robust, Reliable, and Opportunistic WINGs'," *Available from: http://www.cse.ucsc.edu/research/ccrg/projects/sparrow.html*, 2000.

[40] C.-K. Toh, "Associativity-Based Routing For Ad-Hoc Mobile Networks," in *Journal on Wireless Personal Communications*, vol. 4, First Quarter, 1997.

[41] C.-K. Toh, *Wireless ATM and Ad Hoc Networks: Protocols and Architectures*. Kluwer Academic Press, 1997.

[42] C.-K. Toh, *Protocol Aspects of Mobile Radio Networks*. PhD thesis, Cambridge University Computer Laboratory, August 1996.

[43] David C. Plummer, "Address Resolution Protocol," in *IETF RFC 826*, November 1982.

[44] "Advanced Power Management (APM), BIOS Interface Specification Rev.1.2," February 1996.

[45] Intel, Microsoft, and Toshiba, "Advanced Configuration and Power Interface (ACPI)," February 1999.

[46] Smart Battery Systems Implementers Forum Webpage, *'http://www.sbs-forum.org'*, June 1999.

[47] "Smart Battery Manager Specification, Rev 1.1," December 1998.

[48] "Smart Battery Selector Specification, Rev 1.1," December 1998.

[49] "Smart Battery Charger Specification, Rev 1.1," December 1998.

[50] "Smart Battery Data Specification, Rev 1.1," December 1998.

[51] INTEL Architecture Labs, *'http://www.intel.com/ial/ipm/'*, 2001.

[52] K. M. Sivalingam, J-C Chen, and P. Agrawal, "Design and Analysis of Low Power Access Protocols for Wireless and Mobile ATM Networks," *ACM/Baltzer MONET Journal*, 1999.

[53] J-C Chen, K. M. Sivalingam, P. Agrawal, and S. Kishore, "A Comparison of MAC Protocols for Wireless Local Networks Based on Battery Power Consumption," in *Proceedings of IEEE INFOCOM*, April 1998.

[54] Suresh Singh, "Power Efficient MAC Protocol for Multihop Radio Networks," in *Proceedings of IEEE PIMRC*, September 1998.

[55] Suresh Singh, Mike Woo, and C. S. Raghavendra, "Power-Aware Routing in Mobile Ad Hoc Networks," in *Proceedings of ACM/IEEE MOBICOM*, October 1998.

[56] Prathima Agrawal, "Energy Efficient Protocols for Wireless Systems," in *Proceedings of IEEE PIMRC*, September 1998.

[57] C.-K. Toh, "A Novel Distributed Routing Protocol To Support Ad-Hoc Mobile Computing," in *Internatonal Phoenix Conference on Computers & Communications (IPCCC'96)*, pp. 480–486, March 1996.

[58] P. Lettieri, C. Fragouli, and M. B. Srivastava, "Low Power Error Control for Wireless Links," in *Proceedings of ACM/IEEE MOBICOM*, September 1997.

[59] Ajay Bakre & B. R. Badrinath, "I-TCP: Indirect TCP for Mobile Hosts," in *Proceeding of 15th International Conference on Distributed Computing Systems (ICDCS)*, pp. 136–143, May 1995.

[60] H. Eriksson, "MBone: The Multicast Backbone," in *Communications of the ACM*, August 1994.

[61] S. Deering, D. Farinacci, V. Jacobson, C. Lui, and L. Wei, "An Architecture For Wide Area Multicast Routing," in *Proceedings of ACM SIGCOMM'94*, September 1994.

[62] D. Waitzman, C. Pratidge, and S. Deering, "Distance Vector Multicast Routing Protocol (DVMRP)," in *Internet Work Group RFC 1584*, March 1994.

[63] S. Deering, D. Estrin, D. Farinacci, V. Jacobson, C. Liu, and L. Wei, "Protocol Independent Multicast (PIM) Version 2 Dense Mode Specification," June 1999.

[64] J. Moy, "Multicast Extension To Open Shortest Path First (MOSPF)," in *Internet Work Group RFC 1584*, March 1994.

[65] Steve Deering, "Multicast Routing in Internetworks and Extended LANs," in *SIGCOMM Summer 1998 Proceedings*, August 1998.

[66] Y. K. Dalal and R. M. Metcalfe, "Reverse Path Forwarding of Broadcast Packets," in *Communications of the ACM*, vol. 21, pp. 1040–1048, December 1988.

[67] A. Ballardie, "Core Based Trees (CBT) - An Architecture for Scalable Inter-Domain Multicast Routing," in *Proceedings of ACM SIGCOMM'93*, pp. 85–95, September 1993.

[68] S. Deering, D. Estrin, D. Farinacci, V. Jacobson, C. Liu, and L. Wei, "Protocol Independent Multicast (PIM) Version 2 Sparse Mode Specification," in *IETF Internet Draft RFC 2362*, June 1999.

[69] Elizabeth Royer and C.-K. Toh, "A Review of Current Routing Protocols for Ad Hoc Mobile Wireless Networks," in *IEEE Personal Communications Magazine*, First Quarter 1999.

[70] S-Y. Ni, Y-C Tseng, Y-S Chen, and J-P Sheu, "The Broadcast Storm Problem in a Mobile Ad Hoc Network," in *Proceedings of ACM/IEEE MOBICOM*, August 1999.

[71] Mingyan Liu, Rajesh R. Talpade, and Anthony McAuley, "'AMRoute: Adhoc Multicast Routing Protocol'," in *University of Maryland Technical Research Report (CSHCN T.R. 99-1)*, 1999.

[72] C. W. Wu, and Y. C. Tay, "AMRIS: A Multicast Protocol for Ad Hoc Wireless Networks," in *Proceedings of IEEE MILCOM*, November 1999.

[73] J. J. Garcia-Luna-Aceves and E. L. Madruga, "The Core-Assisted Mesh Protocol," in *IEEE Journal on Selected Areas in Communications*, August 1999.

[74] Mario Gerla, Guangyu Pei, et al, "On-Demand Multicast Routing Protocol (ODMRP) for Ad Hoc Networks," in *IETF Internet Draft*, November 1998.

[75] Y. B. Ko and N. Vaidya, "Location-based Multicast in Mobile Ad Hoc Networks," in *Technical Report, Texas A&M*, October 1998.

[76] Mario Gerla, Ching-Chuan Chiang, Lixia Zhang, "Tree Multicast Strategies in Mobile, Multihop Wireless Networks," *ACM/Balzter Mobile Networks and Applications Journal*, 1998.

[77] C.-C. Chiang, Mario Gerla, and Lixia Zhang, "Forwarding Group Multicast Protocol (FGMP) for Multihop Wireless Networks," in *ACM/Baltzer Journal of Cluster Computing*, vol. 1, 1998.

[78] C-K. Toh, and S. Bunchua, "Ad Hoc Mobile Multicast Routing using the Concept of Long-lived Routes," *Journal of Wireless Communications and Mobile Computing*, vol. 1, no. 3, 2001.

[79] Van Jacobson, "Congestion Avoidance and Control," in *Proceedings of ACM SIGCOMM'98*, August 1998.

[80] Zheng Wang and Jon Crowcroft, "A New Congestion Control Scheme: Slow Start and Search (Tri-S)," in *ACM Computer Communication Review*, vol. 21, January 1991.

[81] Zheng Wang and Jon Crowcroft, "Eliminating Periodic Packet Losses in the 4.3-Tahoe BSD TCP Congestion Control Algorithm," in *ACM Computer Communication Review*, vol. 22, April 1992.

[82] Lawrence S. Brakmo, and Sean W. O'Malley, "TCP Vegas: New Techniques for Congestion Detection and Avoidance," in *Proceedings of ACM SIGCOMM'94*, October 1994.

[83] Sally Floyd and Van Jacobson, "Random Early Detection Gateways for Congestion Avoidance," in *IEEE/ACM Transactions on Networking*, vol. 1, August 1993.

[84] Sally Floyd, "TCP and Explicit Congestion Notification," in *ACM Computer Communication Review*, vol. 24, October 1995.

[85] Jitendra Padhye, Victor Firoiu, Donald F. Towsley and James F. Kurose, "Modeling TCP Reno performance: a simple model and its empirical validation," *IEEE/ACM Transactions on Networking*, vol. 8, no. 2, 2000.

[86] Lawrence S. Brakmo, and Larry Peterson, "TCP Vegas: End-to-End Congestion Avoidance on a Global Internet," in *ACM Computer Communication Review*, vol. 24, October 1995.

[87] M. Mathis, J. Mahdavi, et. al., "TCP Selective Acknowledgment Options," in *Internet RFC Draft*, October 1996. Available from http://sunsite.dk/RFC/rfc/rfc2018.html.

[88] James D. Solomon, *'Mobile IP - The Internet Unplugged'*. Prentice Hall, ISBN 013-856-246-6, 1998.

[89] K. Chandran, et. al., "A Feedback Based Scheme for improving TCP Performance in Ad Hoc Wireles Networks," in *Proceedings of ICDS*, 1998.

[90] D. Kim, C-K. Toh, and Y. Choi, "TCP BuS: Improving TCP Performance in Wireless Ad Hoc Networks," *Journal of Communications and Networks*, vol. 3, no. 2, 2001.

[91] J. Verizades, E. Guttman, C. Perkins, and S. Kaplan, "Service Location Protocol," in *IETF RFC 2165*, 1997.

[92] James Kempf and Pete St. Pierre, *Service Location Protocol for Enterprise Networks.* John Wiley & Sons, 1999.

[93] E. Guttman, C. Perkins, J. Veizades, and M. Days, "Service Location Protocol version 2," in *IETF Request for Comments 2608*, June 1999.

[94] SUN Microsystems, "Jini Architecture Specification Version 1.1," October 2000.

[95] Steve Morgan, "Jini to the Rescue," *IEEE Spectrum*, April 2000.

[96] Salutation Consortium, "Salutation Architecture Specification Version 2.1," 1999.

[97] S. Consortium, "Salutation Architecture Specification Version 2.0," in *Salutation Consortium*, June 1999.

[98] Y. Goland, T. Cai, P. Leach, Y. Gu, and S. Albright, "Simple Service Discovery Protocol," in *IETF Internet Draft*, October 1997.

[99] M. Corporation, "Universal Plug and Play Device Architecture Version 0.92," May 2000.

[100] Guillermo E. Guichal, "Service Location Architectures for Mobile Ad Hoc Networks," Master's thesis, Georgia Institute of Technology, June 2001.

[101] Jennifer Bray and Charles F. Sturman, *Bluetooth: Connect Without Cables.* Prentice Hall, 2001.

[102] Uyless Black, *'ATM, Volume II - Signaling in Broadband Networks'*. Prentice Hall, 1998.

[103] S. Education, *Mobile Networking with WAP*. Friedr. Vieweg & Sohn Verlagsgesellschaft mbH, 2000.

[104] W. Forum, "WAP Wireless Telephony Application," in *WAP Forum Specification*, November 1999.

[105] W. Forum, "WAP Wireless Application Environment," in *WAP Forum Specification*, 1999.

[106] W. Forum, "WAP Wireless Session Protocol," in *WAP Forum Specification*, November 1999.

[107] W. Forum, "WAP Wireless Transaction Protocol," in *WAP Forum Specification*, November 1999.

[108] W. Forum, "WAP Wireless Transport Layer Security," in *WAP Forum Specification*, November 1999.

[109] T. Berners-Lee, et al, "Uniform Resource Locators (URL)," in *Internet RFC Draft*, December 1994. Available from ftp://ftp/isi.edu/in-notes/rfc1738.txt.

[110] R. Fielding, "Relative Uniform Resource Locators," in *Internet RFC Draft*, June 1995. Available from ftp://ftp/isi.edu/in-notes/rfc1808.txt.

[111] David Flanagan, *JavaScript: The Definitive Guide*. O'Reilly and Associates, Inc., 1997.

[112] R. Fielding, "Hypertext Transfer Protocol - HTTP/1.1," in *Internet RFC Draft*, June 1997. Available from ftp://ftp/isi.edu/in-notes/rfc2068.txt.

[113] IBM Corporation, "IBM Pervasive Computing Web Page," 2000. http://www-3.ibm.com/pvc/.

[114] V. S. Hsu, J. M. Kahn, and K. S. J. Pister, "Wireless Communications for Smart Dust," in *UC Berkely Electronics Research Laboratory Memorandum M98/2*, January 1998.

[115] J. M. Kahn, R. H. Katz, and K. S. J. Pister, "Next Century Challenges: Mobile Networking for Smart Dust," in *Proceedings of the ACM MOBICOM*, November 2000.

[116] L. Blazevic, L. Buttyan, S. Capkun, S. Giordano, J. P. Hubaux, J. Y. Le Boudec, "'Self-Organization in Mobile Ad-Hoc Networks: the Approach of Terminodes'," *IEEE Communications Magazine*, vol. 39, June 2001.

[117] L. Blazevic, S. Giordano, J. Y. Le Boudec, "Self-Organizing Wide-Area Routing," in *Proceedings of SCI 2000/ISAS 2000*, July 2000.

INDEX